# AN ILLUSTRATED THEORY OF NUMBERS

# AN ILLUSTRATED THEORY OF NUMBERS

MARTIN H. WEISSMAN

AMERICAN MATHEMATICAL SOCIETY
Providence, Rhode Island

2010 *Mathematics Subject Classification.* Primary 11A05, 11A07, 11A15, 11A41, 11A51, 11D04, 11D09, 11E16, 11E41.

For additional information and updates on this book, visit
**www.ams.org/bookpages/mbk-105**

**Library of Congress Cataloging-in-Publication Data**
Names: Weissman, Martin H., 1976-
Title: An illustrated theory of numbers / Martin H. Weissman.
Description: Providence, Rhode Island : American Mathematical Society, [2017] | Includes bibliographical references and index.
Identifiers: LCCN 2017003379 | ISBN 9781470434939 (alk. paper)
Subjects: LCSH: Number theory. | AMS: Number theory – Elementary number theory – Multiplicative structure; Euclidean algorithm; greatest common divisors. msc | Number theory – Elementary number theory – Congruences; primitive roots; residue systems. msc | Number theory – Elementary number theory – Power residues, reciprocity. msc | Number theory – Elementary number theory – Primes. msc | Number theory – Elementary number theory – Factorization; primality. msc | Number theory – Diophantine equations – Linear equations. msc | Number theory – Diophantine equations – Quadratic and bilinear equations. msc | Number theory – Forms and linear algebraic groups – General binary quadratic forms. msc | Number theory – Forms and linear algebraic groups – Class numbers of quadratic and Hermitian forms. msc
Classification: LCC QA241 .W354 2017 | DDC 512.7–dc23 LC record available at https://lccn.loc.gov/2017003379

**Copying and reprinting.** Individual readers of this publication, and nonprofit libraries acting for them, are permitted to make fair use of the material, such as to copy select pages for use in teaching or research. Permission is granted to quote brief passages from this publication in reviews, provided the customary acknowledgment of the source is given.

Republication, systematic copying, or multiple reproduction of any material in this publication is permitted only under license from the American Mathematical Society. Permissions to reuse portions of AMS publication content are handled by Copyright Clearance Center's RightsLink® service. For more information, please visit: `http://www.ams.org/rightslink`.

Send requests for translation rights and licensed reprints to `reprint-permission@ams.org`.

Excluded from these provisions is material for which the author holds copyright. In such cases, requests for permission to reuse or reprint material should be addressed directly to the author(s). Copyright ownership is indicated on the copyright page, or on the lower right-hand corner of the first page of each article within proceedings volumes.

©2017 Martin Hillel Weissman. All rights reserved
Printed in the United States of America.

∞ The paper used in this book is acid-free and falls within the guidelines
established to ensure permanence and durability.
Visit the AMS home page at `http://www.ams.org/`

10 9 8 7 6 5 4 3 2 1    22 21 20 19 18 17

I PROPOUND THIS EASY PROCESS OF COMPUTATION, DELIGHTFUL BY ITS ELEGANCE, PERSPICUOUS WITH WORDS CONCISE, SOFT AND CORRECT, AND PLEASING TO THE LEARNED.

BHASKARA II, FROM *Lilavati*, 1150

IT IS CHARACTERISTIC OF HIGHER ARITHMETIC THAT MANY OF ITS MOST BEAUTIFUL THEOREMS CAN BE DISCOVERED BY INDUCTION WITH THE GREATEST OF EASE BUT HAVE PROOFS THAT LIE ANYWHERE BUT NEAR AT HAND AND ARE OFTEN FOUND ONLY AFTER MANY FRUITLESS INVESTIGATIONS WITH THE AID OF DEEP ANALYSIS AND LUCKY COMBINATIONS.

CARL FRIEDRICH GAUSS, 1817

I HOPE THAT POSTERITY WILL JUDGE ME KINDLY, NOT ONLY AS TO THE THINGS WHICH I HAVE EXPLAINED, BUT ALSO AS TO THOSE WHICH I HAVE INTENTIONALLY OMITTED SO AS TO LEAVE TO OTHERS THE PLEASURE OF DISCOVERY.

RENÉ DESCARTES, FROM *La Géométrie*, 1637.

THUS, EVEN IF YOUR PROBLEM IS NOT A PROBLEM OF GEOMETRY, YOU MAY TRY TO *draw a figure*. TO FIND A LUCID GEOMETRIC REPRESENTATION FOR YOUR NONGEOMETRICAL PROBLEM COULD BE AN IMPORTANT STEP TOWARD THE SOLUTION.

GEORGE PÓLYA, FROM *How to Solve It*, 1946.

THAT THE GEOMETER'S MIND IS NOT LIKE THE PHYSICIST'S OR THE NATURALIST'S, ALL THE WORLD WOULD AGREE; BUT MATHEMATICIANS THEMSELVES DO NOT RESEMBLE EACH OTHER; SOME RECOGNIZE ONLY IMPLACABLE LOGIC, OTHERS APPEAL TO INTUITION AND SEE IN IT THE ONLY SOURCE OF DISCOVERY.

HENRI POINCARÉ, FROM *The Value of Science*, 1905

# Contents

0   Seeing Arithmetic       1

I   Foundations        23

1   The Euclidean Algorithm      25

2   Prime Factorization      47

3   Rational and Constructible Numbers      75

4   Gaussian and Eisenstein Integers      99

II   Modular Arithmetic       125

5   The Modular Worlds      127

6   Modular Dynamics      153

7   Assembling the Modular Worlds      173

8   Quadratic Residues      193

## III  Quadratic Forms  223

9  *The Topograph*  225

10  *Definite Forms*  259

11  *Indefinite Forms*  281

*Index of Theorems*  305

*Index of Terms*  309

*Index of Names*  313

*Bibliography*  317

# *Preface*

THIS BOOK IS about numbers, mostly whole numbers, equality and order, addition and multiplication. Our explorations will lead to a study of prime numbers, linear and quadratic equations, and to number systems which are larger or just stranger than the whole numbers. After finishing this book, the reader will be able to answer questions and problems like the following:

- How many prime numbers are there? And about how many prime numbers are there between 1 and 1 000 000 000?
- Find a square number $x$ such that $x + 1$ is three times a square number. Can you find another such number?
- Is there a square number which is 30 more than a multiple of 37?

The problems in this book are about numbers and their relations to each other. These sorts of questions were interesting for mystical reasons in the ancient Greek world, for astronomical reasons to Indian mathematicians, for reasons of agriculture and government to ancient Chinese mathematicians. Today they are important to anyone who wants to understand data security.

But beyond utility or mysticism, we find these questions most interesting because of the beautiful variety of techniques used to answer them. You might not find these questions interesting now, but their answers and explanations could change your point of view. To this end, the book you're reading *illustrates* the techniques of number theory.

THIS BOOK IS NOT a particularly useful book. It will not help you balance your checkbook, nor will it help you understand political polls or medical tests. It's just not that kind of book. But it might change the way you think about numbers, and you might see something beautiful that very few people have seen before.

This book is not about numerology. Some readers might be interested in the Gospel of John's description of 153 fish caught by the resurrected Jesus. We find it more interesting that 153 is the sum of the first 17 counting numbers: $153 = 1 + 2 + 3 + \cdots + 15 + 16 + 17$. The first statement belongs to numerology, and the second to number theory. Perhaps the author of the Gospel of John was aware of the second statement, but such conjectures are not the focus of this book.

This book is not a philosophy of numbers. We won't touch any questions like "What is a number?" or provide guidance on "What numbers could not be."[1] We assume the reader uses numbers to count things, and sometimes to measure things, and we leave it at that. Our presentation might exhibit a brand of Platonism, but this is for reasons of pedagogy rather than philosophy.

[1] See P. Benacerraf, "What numbers could not be" in *The Philosophical Review*, Vol. 74, No. 1 (1965), pp. 47–73.

This book is not a semiology of numbers. We won't discuss how numbers have been written by different peoples in different times. We won't discuss numerals,[2] nor sexagesimal, nor the "discovery" of zero. Instead we focus on questions about numbers that are independent of semiotics – questions that a hypothetical number theorist from another planet might ask regardless of how numbers are written in its part of the universe.

[2] "There is a big difference between numbers, which are concepts, and numerals, which are written symbols for numbers." From *Where Mathematics Comes From* by George Lakoff and Rafael E. Núñez, Basic Books (2000), p.83.

PLEASE ENJOY this book, and spend ample time with the illustrations. The best math books are meant to be read slowly, with a pen and notebook, with ample time for staring out into space. A window is advisable. Be comfortable in an occasional state of confusion, and confident that clarity will follow someday.

## Illustrations

There are different kinds of illustrations in this book. Some are data visualizations, others are visual explanations – aids to logical reasoning – and others might be called visual mnemonics. All were created with the PGF/TikZ language developed by Till Tantau – this allows for tight integration of graphics and text. For complicated graphics, I wrote Python scripts to create and analyze data and output PGF/TikZ code. I became interested in design and visualization from the famous one-day course of Edward Tufte; the entire book has been typeset using Tufte-LaTeX,[3] a package to imitate the layout of Tufte's books.

[3] Tufte-LaTeX was developed by Kevin Godby, Bill Kleb, and Bill Wood.

DATA VISUALIZATIONS are in fashion, along with hot phrases like "big data" and "data science." I would argue that the prime numbers form the most interesting data set in mathematics. In a book at this level, one can only *prove* a few coarse statistical properties of primes (e.g., their infinitude, and infinitude within a few progressions). But one can *see* more. Data visualizations provide the reader – at an elementary level – with a window to see the most important recent research in number theory.

The nature of number-theoretic data introduces some challenges for visualization. For example, there are large data sets with low dimensionality but one is interested in subtle features. The prime numbers are the simplest *type* of data, a one-dimensional distribution, but one is interested in subtle statistics of spacings, very slight biases within the primes, etc.. Other data sets are two-dimensional, but we are interested in features beyond trendlines in scatterplots – we are interested in the boundary regions of data sets – mathematicians care about upper and lower bounds and asymptotics.

VISUAL EXPLANATIONS, to use the title of one of Tufte's books,[4] are illustrations meant to aid logical reasoning. For example, we have provided an illustration alongside the proof of the "second supplement" to quadratic reciprocity – the reciprocity law for the squareness of 2. A proof typically covers infinitely many cases, and an illustration can only display a few; but a good illustration depicts a case in "general position" with no unusual features to deceive the reader. Most of our proofs are given with visual explanations; geometric and dynamical proofs are preferred as they work best with illustrations.

[4] "Visual Explanations: Images and quantities, evidence and narrative" by Edward Tufte, published by Graphics Press (1997).

VISUAL MNEMONICS are typically small illustrations to help the reader keep track of a complicated situation, to instill a metaphorical system to aid the digestion of abstract concepts. For example, I have used the terms "lying above" and "lying below" to describe the relationship between Gaussian primes and ordinary primes. There is no inherent "up" or "down" relationship between Gaussian primes and ordinary primes, but mathematicians find it convenient to place primes in particular relative positions. Lakoff and Nuñez[5] argue that such metaphorical systems are central in the human conception of mathematics.

[5] For their detailed argument, see "Where mathematics comes from: how the embodied mind brings mathematics into being", by George Lakoff and Rafael Nuñez, Basic Books (2000).

## For the reader

We have taken care to structure this book in parts, chapters, and in two-page spreads. This means that you can linger with the text open at a point, looking left and right, studying without flipping to other pages. Images and text on one page may complement that on the opposite page.

When reading a theorem (or lemma or proposition or corollary), you must pay equal attention to the *hypothesis* and the *conclusion*. For example, in the theorem "If $x$ and $y$ are integers and $x \neq 0$, then $|xy| \geq |y|$," many readers will remember only the conclusion "$|xy| \geq |x|$." This probably comes from a traditional education of memorizing formulas. But just as important is the hypothesis "If $x$ and $y$ are integers and $x \neq 0$." Without it, the formula has no meaning, and if the word "integers" is changed to "real numbers," the formula is false. All mathematical statements have hypotheses and conclusions – remember both.

THE BEST way to remember a formula is to know why it is true.

## *For the instructor*

I have written this book, in part to demonstrate that a "textbook" can be beautiful and rigorous and interesting. It reflects my experience teaching elementary number theory courses, and I hope that it can be a resource for number theory courses taught by other mathematicians. For those considering adopting this book as a textbook, I make the following remarks and suggestions.

This book has no formal prerequisites beyond high-school algebra and basic coordinate geometry. We do not assume that students remember theorems in Euclidean geometry, e.g. SAS congruence for triangles. However, the Pythagorean theorem will be used freely, along with formulae for areas of basic shapes.

The algebraic prerequisites, while elementary, are probably much more than students encounter in a high school in the United States. We freely introduce new variables, $x$ or $a$ or $p$, even though many students are not so comfortable writing sentences like "Let $a$ be the width of the rectangle." We take care in our language, since beginning students in mathematics need to see academic English language integrated with symbolic algebra. We use essentially no symbolic set theory (e.g. $3 \in \mathbb{N}$).

The exercises at the end of each chapter mix a few drill problems with more interesting mathematical explorations. We trust the instructor to choose exercises appropriate for the students, and to create additional exercises as needed.

For those faculty who are comparing this book to others on the market, I would promote the differences as follows:

1. The proofs in this text have been written, refined, illustrated, refined again and again. Of course, I am not responsible for proving these theorems *first*, but I have gone back to the masters (Euclid, Euler, Gauss, etc.) to gain inspiration. It might be impossible to improve Gauss's *Disquisitiones*, but providing illustrations and modern context can help a student today.

2. I have challenged myself to present complete proofs of all results, without falling back on extended algebraic manipulation. I'm not afraid of a little algebra here and there, but I've tried to use geometric and "dynamical" approaches first or alongside the algebra.

3. The historical notes are not the typical "put a stock-picture in a box next to a Wikipedia-like blurb." I aspire for the history scholarship to reflect the modern state of the art. This includes mathematics inside and outside the Western tradition.

4. The illustrations are meant to aid understanding. I don't think I have included any decorative fluff. There is something to learn from every picture, even those that lie opposite each Chapter heading. I think our students should learn "visual proofs" – not in the sense of producing proofs without words, but in the sense of producing careful diagrams to support carefully-written arguments.

While this text contains mathematical topics that can be found elsewhere, there are some noticeable differences.

1. The text includes a chapter on rational numbers, Ford circles, and Diophantine approximation. This is often absent from beginning textbooks, but the topic can be beneficial to future mathematicians as well as future K-12 teachers.

2. The text introduces the arithmetic of Gaussian and Eisenstein integers. I think this is particularly good for more advanced students – theorems like the uniqueness of prime decomposition are more appreciated by students when they see that it applies in unfamiliar contexts. This can smooth the transition to a later course in algebraic number theory.

3. The text contains a much fuller treatment of binary quadratic forms than most books at this level. This is made possible by Conway's "topograph." With the topograph, one can solve quadratic Diophantine equations and prove the finiteness of class numbers, without abstract algebra or tedious matrix mechanics.

4. For quadratic reciprocity, I have followed Zolotarev's proof. This is not so common, as most texts favor a Gauss sums approach or lattice-point-counting. But I find Zolotarev's proof not only beautiful, but it suggests a "dynamical approach" to modular arithmetic that I find pedagogically advantageous. The dynamical approach can be found, for example, in Gauss' proof of Fermat's Little Theorem, and I think it is interesting to follow the thread from Gauss to Zolotarev in this way. The text includes a full treatment of the sign of a permutation along the way.

## Pathways

There are many pathways through the material:

- For math majors with a strong background in a 12-14 week term, I would recommend Path A,B, or C.

- For math majors in a 12-14 week term, especially those pursuing a career in K-12 education, I would recommend Path A or Path D.

- For math majors in a short 10 week term, I would recommend Path C, excluding Chapter 4, or Path D.

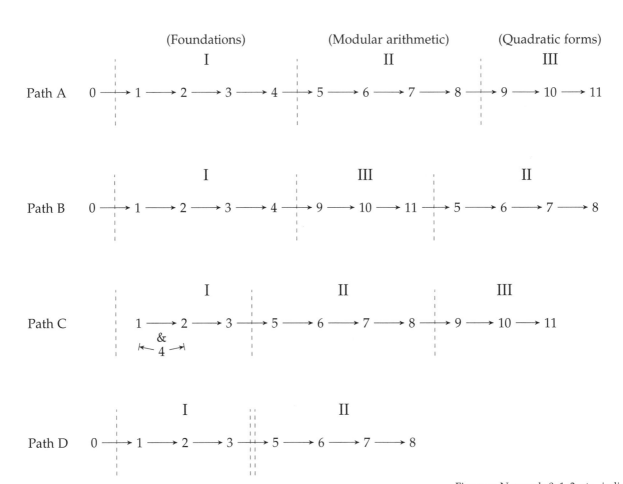

One can safely switch Parts II and III in the book, or exclude Part III entirely. There are few hazards in tackling binary quadratic forms before modular arithmetic, and I have taught the material in this order on many occasions (making quadratic reciprocity the finale).

Figure 1: Numerals 0, 1, 2, etc., indicate Chapter numbers. Numerals I, II, III indicate the parts of the book. Arrows illustrate recommended pathways through the material

## Acknowledgments

I THANK the following students and colleagues for catching typos, big and small: Samit Dasgupta, Erik Carino, Kevin Deutsch, Amanda Springer, Tanya Paraon, Lane Bacchi, Colin Ritchey, Branden Laske, Brenda Minjares, Samantha Orozco, Vanessa Jelmyer, Anshuman Mohan, Nathan Kaplan, Jake Goh Si Yuan, Evangeline Pousson, Seow Yongzhi, Crysta Bullard, Justin Raizes, Thomas Bryant, Liza Mednikov, Miguel Ruiz, Shirlyna Trinh, Ryan Alexiadis, Robert Chung, Suzana Milea, Luke Harrison, Olivia Pratt, Luke Shepherd, Laetitia Cabrol. I thank Stephen DeBacker for feedback on the title, and Ken Ribet for strengthening the writing in many places. I thank Anne Kreps and Jesse Weissman for their sharp eyes and design suggestions.

I DID NOT LIKE number theory when I first encountered it. But a class with Goro Shimura changed my mind, as it brought out connections between number theory and geometry,[6] and I became hooked. I am grateful to Professors Shimura, Conway, Katz, Sarnak, and Wiles for cementing my interest in number theory as an undergraduate, and for their incredible breadth of perspective and generosity. In graduate school, I was a student of Dick Gross, who inspired me with connections between representation theory, geometry, and arithmetic. I hadn't thought much about *elementary number theory* – the topic of this book – until late in graduate school when I was a teaching assistant for Gross and Joe Harris in "The Magic of Numbers" at Harvard University. This book is written for a different audience than their celebrated course, but their influence may be seen in some choices I have made. With this book, I hope to provide an experience in number theory and geometry that is a taste of what these professors provided for me.

[6] Specifically, the analogy between number fields and Riemann surfaces.

FINALLY, I would like to thank my family. My parents were my first teachers and fostered an early interest in mathematics. Pertinax provided lap support and sat on rough drafts to provide judgment. My wife, Anne, sharpened the historical notes and illustrations with her expertise and style. I appreciate her confidence, from the moment I started thinking about the book through its finishing touches.

*Martin H. Weissman, December 2016*

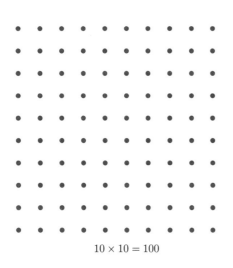
$10 \times 10 = 100$

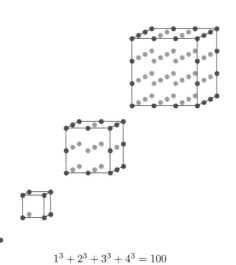
$1^3 + 2^3 + 3^3 + 4^3 = 100$

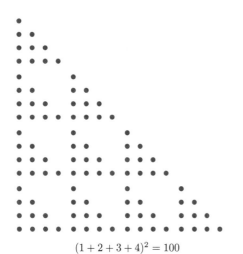
$(1 + 2 + 3 + 4)^2 = 100$

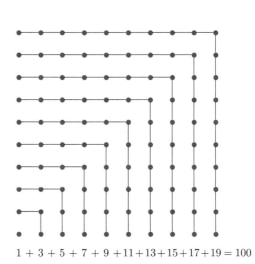
$1 + 3 + 5 + 7 + 9 + 11 + 13 + 15 + 17 + 19 = 100$

# 0
# *Seeing Arithmetic*

THE ORIGINAL numbers are counting numbers. We can use counting numbers to count dots: *two* red dots .•, *five* blue dots ⁙ . Counting numbers are answers to questions that begin "How many...?", asked about things we can see or touch or hear. Mathematicians use the term **natural numbers** to refer to the counting numbers. The natural numbers are zero[1], one, two, three, four, etc..

Rearranging dots does not change their number. Here ∴ are three red dots and here ⋰ are three red dots. "Threeness" persists.

If we use numbers to count dots, then we can see equality of numbers by pairing off the dots. Here ⋮• are some myred dots and here ⁙ are some blue dots. It is hard, at first glance, to see that there are the same number of red dots as blue dots. But arrange the dots in straight lines, each red dot paired with a blue dot, one over the other, ⁞⁞⁞⁞⁞ , and see that they have the same number.

VISUALIZING EQUALITY by "pairing off" is the heart of counting. In some time past (long before 3000BCE), a shepherd paired sheep and pebbles: as the first sheep walked by, he placed down the first pebble; the second sheep walked by and down went the second pebble. At the end of the day, the number of pebbles placed down equaled the number of sheep that walked past. Whether sheep and pebbles, or red dots and blue dots, equality of numbers is defined by pairing off[2] the items of two sets.

To count things, arrange them in a straight line above the positive numbers. For example, to count the primes between 1 and 20, arrange them:

| Primes: | 2 | 3 | 5 | 7 | 11 | 13 | 17 | 19 |
|---|---|---|---|---|---|---|---|---|
|  | $1^{st}$ | $2^{nd}$ | $3^{rd}$ | $4^{th}$ | $5^{th}$ | $6^{th}$ | $7^{th}$ | $8^{th}$ |

There are 8 primes between 1 and 20.

---

[1] Is zero a natural number? Some mathematicians include zero in the set of natural numbers, and some do not. It is a matter of taste. In this text, we declare: **0 is a natural number.**

[2] Such a pairing off is called a **bijection** by mathematicians, a term coined by Nicolas Bourbaki (the pseudonym of a tightly-knit group of mathematicians most active in the mid-twentieth century).

COUNTING will play a central role in this text. Not counting sheep or pebbles, but counting numbers of various flavors. The following examples illustrate how pairing off can help one count with confidence.

**Problem 0.1** How many even numbers are between 1 and 100?

SOLUTION: Arrange the even numbers in a line above the positive numbers.

$$\begin{array}{cccccc} 2 & 4 & 6 & 8 & \cdots & 100 \\ 1^{st} & 2^{nd} & 3^{rd} & 4^{th} & \cdots & ?^{th} \end{array}$$

The top row includes the even numbers between 1 and 100. The bottom row consists of positive numbers, in effect counting the top row. The unknown number ? is the answer to the problem.

Filling in the dots would be tiresome. Instead, observe that each number in the bottom row is precisely half its partner in the top row. 1 is half of 2. 2 is half of 4. 3 is half of 6. Et cetera. The unknown number ? is half of 100 – it is 50. There are 50 even numbers between 1 and 100. ✓

$$\left.\begin{array}{cccccc} 2 & 4 & 6 & 8 & \cdots & 100 \\ 1^{st} & 2^{nd} & 3^{rd} & 4^{th} & \cdots & 50^{th} \end{array}\right\} \text{Divide by two.}$$

**Problem 0.2** How many natural numbers are between[3] 37 and 185?

SOLUTION: Arrange the numbers between 37 and 185 in a line above the positive numbers:

$$\begin{array}{cccccc} 37 & 38 & 39 & 40 & \cdots & 185 \\ 1^{st} & 2^{nd} & 3^{rd} & 4^{th} & \cdots & ?^{th} \end{array}$$

Each number in the bottom row is 36 less than its partner in the top row. Therefore, $? = 185 - 36 = 149$. There are 149 natural numbers between 37 and 185. ✓

$$\left.\begin{array}{cccccc} 37 & 38 & 39 & 40 & \cdots & 185 \\ 1^{st} & 2^{nd} & 3^{rd} & 4^{th} & \cdots & 149^{th} \end{array}\right\} \text{Subtract 36.}$$

**Problem 0.3** How many multiples of three are between 1 and 1000?

SOLUTION: Arrange the multiples of three in a line above the positive numbers:

$$\left.\begin{array}{cccccc} 3 & 6 & 9 & 12 & \cdots & 999 \\ 1^{st} & 2^{nd} & 3^{rd} & 4^{th} & \cdots & ?^{th} \end{array}\right\} \text{Divide by three.}$$

Notice that 999 is the last multiple of three between 1 and 1000. Each number in the bottom row is one-third of its partner in the top row. Therefore, $? = 999 \div 3 = 333$. There are 333 multiples of three between 1 and 1000. ✓

[3] When we write "between," we always mean this in the *inclusive* sense. So "between 37 and 185" means between *and including* 37 and 185.

Many students approach the problem by simply subtracting $185 - 37 = 148$. But 148 is not the correct answer. This exemplifies a common "off-by-one" error that occurs when counting within an interval. Some students are taught to "always add one," to correct this off-by-one error, but that rule does not apply in other circumstances.

It is better to line up the numbers, and never make such mistakes.

ADDITION is the numerical complement to aggregation. One *aggregates* two collections of dots, by putting them together. Aggregating two red dots .* with five blue dots ∷ yields seven dots .*∷. If we care only about the numbers, then we would just write $2 + 5 = 7$.

While addition of two numbers has only one correct result, aggregation of two collections has many correct results – there are many ways of arranging the resulting collection. For example the numerical expression $1 + 3 = 4$ might be realized in any of the following ways:

$$. + .^{.\cdot} = :^{.\cdot} , \text{ or } . + \text{...} = \overset{.}{\text{...}} , \text{ or } . + \text{...} = ::.$$

You might say "who cares," since, in the end, there are still four dots (three blue and one red). But the creative arrangement of dots allows one to solve otherwise difficult addition problems.

**Problem 0.4** Add the odd numbers between 1 and 30. In other words, what is $1 + 3 + 5 + \cdots + 25 + 27 + 29$?

SOLUTION: An odd number of dots can be arranged into a ⌐-shaped configuration. Stacking these configurations of dots yields a square configuration.

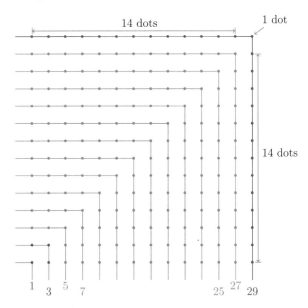

The entire figure contains $1 + 3 + 5 + \cdots + 25 + 27 + 29$ dots, arranged in a pattern of "stacking corners".[4] The square arrangement is 15 dots wide and 15 dots tall, having a total of $15 \times 15 = 225$ dots. Hence $1 + 3 + 5 + \cdots + 25 + 27 + 29 = 225$. ✓

[4] The corner-configuration is the *gnomon* of the square, to use the Pythagorean language.

This method of stacking corners allows one to add the odd numbers between 1 and every natural number.

MULTIPLICATION of two numbers can be represented by the arrangement of dots into a rectangular array. The numerical expression $3 \times 4 = 12$ can be realized by the following figure:

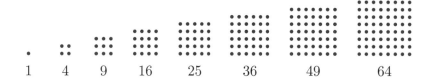

**Square numbers**[5] are those obtained by multiplying a natural number by itself. Below are eight square numbers.

[5] Square numbers are sometimes called "perfect squares". But the word "perfect" has another meaning in number theory, and adds little more than emphasis, so we avoid it.

1   4   9   16   25   36   49   64

For every partition of the square, there is a corresponding addition fact. Some of these facts are otherwise difficult to find. Consider the following partitions of the square number 100.

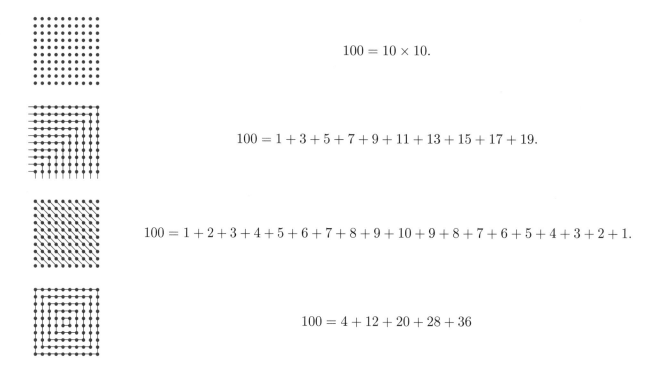

$$100 = 10 \times 10.$$

$$100 = 1 + 3 + 5 + 7 + 9 + 11 + 13 + 15 + 17 + 19.$$

$$100 = 1 + 2 + 3 + 4 + 5 + 6 + 7 + 8 + 9 + 10 + 9 + 8 + 7 + 6 + 5 + 4 + 3 + 2 + 1.$$

$$100 = 4 + 12 + 20 + 28 + 36$$

Figure 1: Four partitions of 100 dots

We challenge the reader to find other interesting partitions of the square, and corresponding addition facts. In the other direction, when the reader encounters a difficult problem of repeated addition, we recommend searching for a geometric solution – a partition of a square or rectangle.

TRIANGULAR arrangements of dots can be duplicated to form rectangular arrangements. This principle plays a role in the solution of some addition problems below.

**Problem 0.5** What is $1 + 2 + 3 + \cdots + 19 + 20$?

SOLUTION: Assemble dots in a right triangle, with one dot in the first row, two dots in the second row, etc., with 20 dots in the 20$^{\text{th}}$ row.

The sum $1 + 2 + 3 + \cdots + 19 + 20$ is the number of dots in this triangular figure. Placing this figure adjacent to a duplicate yields a rectangular[6] figure.

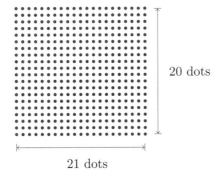

There are $21 \times 20 = 420$ dots in the rectangular figure. Half of the dots are red and half of the dots are blue, so there are $420 \div 2 = 210$ red dots. Hence
$$1 + 2 + 3 + \cdots + 19 + 20 = 210.$$

This method of duplicating a triangular arrangement generalizes.

**Proposition 0.6 (Summation up to N)** *For every positive integer N,*
$$\sum_{i=1}^{N} i = 1 + 2 + 3 + \cdots + N = \frac{N(N+1)}{2}.$$

**Triangular numbers** are numbers obtained by summing the first $n$ numbers for some nonzero natural number $n$. The first five triangular numbers are below.

$$\begin{array}{rcl} 1 & = & 1 \\ 1+2 & = & 3 \\ 1+2+3 & = & 6 \\ 1+2+3+4 & = & 10 \\ 1+2+3+4+5 & = & 15 \end{array}$$

[6] The figure is very slightly but certainly *not* a square. This exhibits a subtle and crucial difference between number theory and geometry. In geometry, one might see the triangle of base 20 and height 20, and compute its area as
$$Area = \frac{1}{2}(20 \times 20) = 200.$$
But when counting dots, the area does not suffice. There are
$$\frac{1}{2}(21 \times 20) = 210 \text{ dots.}$$

For every nonzero natural number $N$, the sum $1 + 2 + 3 + \cdots + N$ equals the number of dots in a right triangle of height and width $N$. Duplicating this triangle yields a rectangle of height $N$ and width $N+1$. There are $N \times (N+1)$ dots in this rectangle. There are half as many dots in the triangle, whence the formula $N \times (N+1) \div 2$ for the triangular number.

PARTNERING is a method of considering a sequence in pairs (allowing a few lonely numbers on occasion). Like duplication, partnering can be used to add numbers in arithmetic progression.

**Problem 0.7 (Revisited)** What is $1 + 2 + 3 + \cdots + 19 + 20$?

SOLUTION: Partner 1 with 20, and partner 2 with 19, and 3 with 18, et cetera. One can see the partnerships in the following diagram.

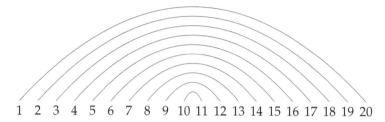

Notice that $1 + 20 = 21$, and $2 + 19 = 21$, and $3 + 18 = 21$; each number together with its partner sums to 21. The 20 numbers become 10 pairs of numbers, and each pair sums to 21. Hence

$$1 + 2 + 3 + \cdots + 18 + 19 + 20 = 10(21) = 210.$$

For more steps, we may write
$$\begin{aligned}
& 1 + 2 + 3 + \cdots + 18 + 19 + 20 \\
= {} & (1 + 20) + \cdots + (10 + 11) \\
= {} & (21) + (21) + \cdots + (21) \\
= {} & 10(21).
\end{aligned}$$

If we impose partnerships on an odd number of things, one thing is always left alone.

**Problem 0.8** What is $1 + 2 + 3 + \cdots + 47 + 48 + 49$?

SOLUTION: Partner 1 with 49, and partner 2 with 48 and 3 with 47, et cetera. Each number, with its partner, sums to 50.

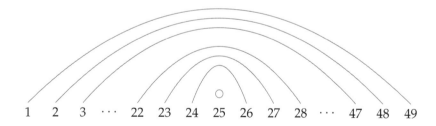

Notice that 24 is partnered with 26 (since they sum to 50). But 25 is left without a partner. In the end, we find 24 pairs – each pair summing to 50 – and 25 left over by itself. Hence

$$1 + 2 + 3 + \cdots + 47 + 48 + 49 = 24(50) + 25 = 1200 + 25 = 1225.$$

Partnering is an effective method of adding arithmetic progressions. While it may seem redundant after the geometric method of duplication, we will find scenarios later in which partnering is crucial and duplication ineffective.

BEYOND PARTNERING and duplicating, one can try to add numbers by putting them into groups of three or more. The problem below applies the rare technique of *triplication*.

**Problem 0.9** Add the first ten square numbers. In other words, what is $1 + 4 + 9 + \cdots + 64 + 81 + 100$?

SOLUTION: We may view each square number as the result of repeated addition: $1 = 1$; $4 = 2 + 2$; $9 = 3 + 3 + 3$; et cetera. So, in this way, the sum of the first ten square numbers equals the sum of the numbers in the triangle in the margin.

*Triplicating* this triangle – overlaying it with itself, but rotated by 120° and 240° – yields the diagram below.

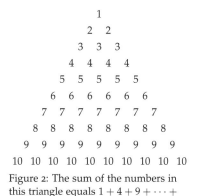

Figure 2: The sum of the numbers in this triangle equals $1 + 4 + 9 + \cdots + 64 + 81 + 100$.

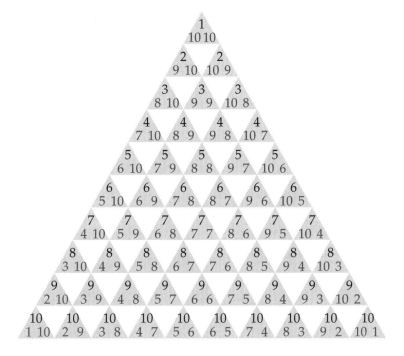

Add only on the black numbers, and you find the answer to the problem. Add only the red numbers, and again you find the answer to the problem. Add only the blue numbers, and again you find the answer to the problem. So, adding all the numbers yields *three times* the answer to the problem.

But each small triangle, consisting of a black, red, and blue number, sums to 21. There are $1 + 2 + 3 + 4 + 5 + 6 + 7 + 8 + 9 + 10 = 55$ small triangles. Hence the sum of *all numbers*, black, red, and blue, equals $55 \times 21 = 1155$. The black numbers sum to $1155 \div 3 = 385$. ✓

Why does each small triangles sum to 21? The top triangle certainly sums to 21 since $10 + 10 + 1 = 21$. Traveling southwest, the blue number in the triangle stays the same, the black increases, and the red decreases, so their sum is unchanged. Traveling southeast, the red number in the triangle stays the same, the black increases, and the blue decreases, so their sum is unchanged. Hence the sum, 21, is unchanged as one travels from triangle to triangle within the large figure.

COUNTING PAIRS requires some cleverness, but most importantly a clear definition. Consider the set

$$S = \{1, 2, 3, 4, 5, 6, 7, 8, 9, 10\}.$$

How many pairs are there, taken from this set? It depends on how one interprets the word "pair"?

An **ordered pair** from $S$ is a pair written $(a, b)$ in which $a \in S$ and $b \in S$. Ordered pairs from $S$ include $(1, 10)$ and $(2, 8)$ and $(8, 2)$ and $(6, 6)$. The adjective "ordered" means that we distinguish $(2, 8)$ from $(8, 2)$.

**Problem 0.10** How many ordered pairs are there from the set $S$?

SOLUTION: Each dot in the 10 by 10 square may be identified by coordinates of the form $(a, b)$ with $a \in S$ and $b \in S$. There are the same number of dots as there are ordered pairs. Hence there are $10 \times 10 = 100$ ordered pairs from the set $S$. ✓

In general, if $S$ is a set with $N$ elements, there are $N \times N$ ordered pairs from $S$.

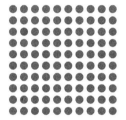

Figure 3: The blue dot has coordinates $(2, 8)$; it is in the 2nd column from the left and 8th row from the bottom. To each dot there corresponds an ordered pair from $S$.

A **subset** of $S$ is an unordered collection of elements of $S$; an element may appear in a subset once or not at all. For example, a two-element subset of $S$ is $\{3, 5\}$. A four-element subset of $S$ is $\{7, 1, 3, 10\}$. A zero-element subset of $S$ is the empty set, denoted $\emptyset$. Order does not matter: $\{1, 5, 6\} = \{6, 1, 5\} = \{6, 5, 1\}$; they are the same subset. Repetition is not allowed: one would never write $\{2, 2\}$.

**Problem 0.11** How many two-element subsets are there in the set $S$?

SOLUTION: Choose a two-element subset $\{a, b\}$ of $S$, so $a \neq b$. There are two ways to order the subset: as $(a, b)$ or as $(b, a)$. The resulting ordered pairs correspond to dots on the square, excluding the diagonal (since $a$ cannot equal $b$).

The number of such ordered pairs is $(10 \times 10) - 10 = 90$, the number of dots in the square, excluding the diagonal. There are twice as many such ordered pairs as two-element subsets; the ordered pairs $(2, 8)$ and $(8, 2)$ correspond to the same two-element subset $\{2, 8\}$. Hence there are $90 \div 2 = 45$ two-element subsets of $S$. ✓

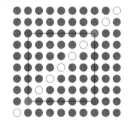

Figure 4: The blue dots have coordinates $(2, 8)$ and $(8, 2)$. These arise from two ways of ordering the same subset $\{2, 8\} = \{8, 2\}$. The red dots are in the 2nd and 8th positions along the diagonal.

The following theorem generalizes the above result.

**Proposition 0.12 (Counting pairs)** *Let $S$ be a set with $N$ elements. The number of ordered pairs from $S$ is $N^2$. The number of two-element subsets[7] of $S$ is equal to*

$$\frac{1}{2}(N \times N - N) = \frac{N^2 - N}{2} = \frac{N(N-1)}{2}.$$

[7] The number of two-element subsets is called "$N$ **choose** 2" and written $\binom{N}{2}$. Thus we find

$$\binom{N}{2} = \frac{N(N-1)}{2}.$$

Two questions have the answer 45:

1. What is $1 + 2 + 3 + 4 + 5 + 6 + 7 + 8 + 9$?

2. What is $\binom{10}{2}$, i.e., how many 2-element subsets are there of $\{1,2,3,4,5,6,7,8,9,10\}$?

Why? The figure below connects these questions.[8]

[8] Similar figures have been published. For example, see Roger B. Nelsen, "Visual Gems of Number Theory," from *Math Horizons*, Feb. 2008, p. 7. I had the benefit of learning about this visualization from Juan Felipe Pérez in the Summer of 2011.

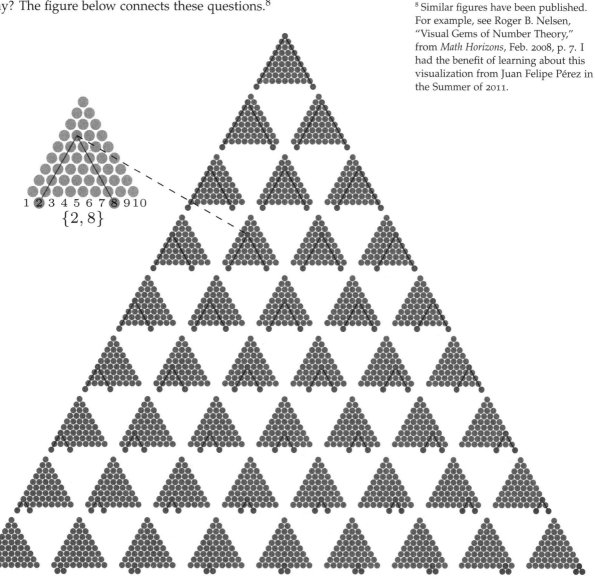

Figure 5: Visualizing 10 choose 2 as the sum of numbers from 1 to 9.

There are $1 + 2 + 3 + \cdots + 8 + 9$ locations in the gray triangle for a blue dot. To each blue dot, there corresponds a pair of red dots, whose locations are form a two-element subset of $\{1,2,3,4,5,6,7,8,9,10\}$.

**Proposition 0.13 (Summation up to $N-1$)** *If $N \geq 2$ then*

$$1 + 2 + 3 + \cdots + (N-1) = \binom{N}{2}.$$

ROUNDING is not usually counted among the fundamental operations. We expect rounding to introduce imprecision, to sacrifice accuracy for convenience. But in number theory, the opposite can be true – an imprecise approximate result can be made correct by rounding.

Since one may round down or up, we use two words and notations. The result of rounding down is called the **floor** and the result of rounding up is called the **ceiling**. The floor of 3.2 is 3. The ceiling of 17.5 is 18. Symbolically, we write this[9]

$$\lfloor 3.2 \rfloor = 3, \quad \lceil 17.5 \rceil = 18.$$

The floor is the nearest integer[10] less than or equal to the number.

$$\lfloor \pi \rfloor = 3, \quad \lfloor -7.2 \rfloor = -8, \quad \lfloor 5 \rfloor = 5.$$

The ceiling is the nearest integer greater than or equal to the number.

$$\lceil \pi \rceil = 4, \quad \lceil -7.2 \rceil = -7, \quad \lceil 5 \rceil = 5.$$

The floor equals the ceiling precisely when the number is an integer.

$$\lfloor 5 \rfloor = \lceil 5 \rceil = 5.$$

[9] This wonderful notation for floor and ceiling was introduced by Kenneth Iverson, in his book "A Programming Language" (John Wiley and Sons, 1962).

[10] The **integers** are the whole numbers, positive and negative and zero. The bold letter $\mathbb{Z}$ is used to denote the set of integers.

$$\mathbb{Z} = \{\ldots, -4, -3, -2, -1, 0, 1, 2, 3, 4, \ldots\}.$$

The letter $\mathbb{Z}$ comes from the German word "Zahl" meaning number.

**Proposition 0.14 (Counting multiples)** *Let N and m be positive integers. The number of multiples of m between 1 and N is equal to $\lfloor N/m \rfloor$.*

PROOF: Arrange the multiples of $m$ on the top row, and positive integers on the bottom.

| $m$ | $2m$ | $3m$ | $\cdots$ | $\lfloor N/m \rfloor m$ | $N$ |
|---|---|---|---|---|---|
| $1^{st}$ | $2^{nd}$ | $3^{rd}$ | $\cdots$ | $\lfloor N/m \rfloor^{th}$ | $N/m$ |

Divide by $m$.

The quotient $N/m$ is not necessarily a natural number; the natural number just beneath it is the floor, $\lfloor N/m \rfloor$. And that is the number of multiples of $m$ between 1 and $N$. ∎

Underappreciated, the floor is most convenient for counting.

**Proposition 0.15 (Counting squares)** *Let N be a positive integer. The number of squares between 1 and N is equal to $\lfloor \sqrt{N} \rfloor$.*

PROOF: Arrange the squares on the top row, and positive integers on the bottom.

| 1 | 4 | 9 | $\cdots$ | $\lfloor \sqrt{N} \rfloor^2$ | $N$ |
|---|---|---|---|---|---|
| $1^{st}$ | $2^{nd}$ | $3^{rd}$ | $\cdots$ | $\lfloor \sqrt{N} \rfloor^{th}$ | $\sqrt{N}$ |

Square root.

The square root $\sqrt{N}$ is not necessarily a natural number; the natural number just beneath it is the floor, $\lfloor \sqrt{N} \rfloor$. And that is the number of squares between 1 and $N$. ∎

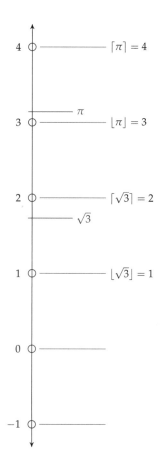

REPRESENTING numbers in decimal is second nature. We learn the base-ten positional system for representing natural numbers, and eventually for all real numbers. According to this decimal system, an expression such as 8039 consists of a units digit (9), tens digit (3), hundreds digit (0), and thousands digit (8). The digits belong to the set $\{0, 1, 2, 3, 4, 5, 6, 7, 8, 9\}$

The expression 8093 is a shorthand for a sum,

$$8093 = (3 \times 10^0) + (9 \times 10^1) + (0 \times 10^2) + (8 \times 10^3).$$

Nestled between $10^3$ and $10^4$; the number 8093 has 4 digits.

| Number: | 1 | 10 | 100 | 1000 | 8093 | 10000 |
|---|---|---|---|---|---|---|
| Digits: | 1 | 2 | 3 | 4 | 4 | 5 |

The number of decimal digits in $10^k$ is $k + 1$. More generally,

**Proposition 0.16 (Counting digits)** *If N is a positive integer, then the number of digits in the decimal representation of N is $\lfloor \log_{10}(N) \rfloor + 1$.*

BINARY representations are more natural than decimal expansions, if we care more about numbers than about how humans[11] represent numbers. In the binary (base-two) system, the only digits are the binary digits (**bits**) 0 and 1.

To avoid confusion, we use a bold **b** to precede a binary representation. For example, an expression such as **b** 1011 consists of a units digit (1), twos digit (1), fours digit (0), eights digit (1). The expression **b** 1011 is a shorthand for a sum,

$$\mathbf{b}\,1011 = 1 \times 2^0 + 1 \times 2^1 + 0 \times 2^2 + 1 \times 2^3 = 1 + 2 + 0 + 8 = 11.$$

Note that **b** 1 = 1 and **b** 10 = 2 and **b** 100 = 4 and **b** 1000 = 8.

Numerically, a bit is a 1 or a 0. But more importantly, a bit is the smallest unit of information. The smallest unit of information is the answer to a yes/no question – yes or no, on or off, 1 or 0. Such an answer is given with a bit. Computers store information in bits, and so computers work in binary.

The number of bits of a number is the amount of information carried by the number, and thus the amount of storage space required to hold the number in memory. Algebraically, the number of bits of a number is related to its base-two logarithm.

**Proposition 0.17 (Counting bits)** *If N is a positive integer, then the number of bits in the binary representation of N is $\lfloor \log_2(N) \rfloor + 1$.*

[11] Humans use a base-ten system as most of us have ten fingers.

| Decimal | Binary |
|---|---|
| 0 | **b** 0 |
| 1 | **b** 1 |
| 2 | **b** 10 |
| 3 | **b** 11 |
| 4 | **b** 100 |
| 5 | **b** 101 |
| 6 | **b** 110 |
| 7 | **b** 111 |
| 8 | **b** 1000 |

Table 1: Counting up to 8, and counting up to **b** 1000.

Rather than calling **b** 10 "ten", it is better to say "one-zero" to avoid confusion.

DIVISION WITH REMAINDER is not a childhood relic, to be forgotten in favor of grown-up decimal division or fractions. Though the word "remainder" suggests unwanted left-overs and "quotient" suggests a terminus, mathematicians respect remainders as much as quotients since they answer different questions of similar importance.

Let's begin by introducing some new notation for division with remainder. In school, many students are taught to write "$23 \div 4 = 5R3$", to mean that "23, divided by 4, is 5 with a remainder of 3". This again suggests that 5 is the answer, and 3 an undesirable left-over. We recommend the following compact notation: $23 = 5(4) + 3$. It is a statement about multiplication and addition. And that is what division with remainder is: a statement about multiplication and addition.

Why do we write $23 = 5(4) + 3$ when it is numerically the same as $23 = 5 \times 4 + 3$? We ask the reader to interpret two notations for multiplication in two visually different ways.

Interpret $5 \times 4$ as a five-by-four array and say "five times four".

Figure 6: $5 \times 4$ suggests a rectangular arrangement.

Interpret $5(4)$ as five groups of four and say "five fours".

Figure 7: $5(4)$ means five fours.

So when we write $23 = 5(4) + 3$, you should think "23 is composed of five groups of four (five fours), with three left over."

This grouping with leftovers *is* division with remainder. Here are some more examples, in both notations:

$$93 \div 80 = 1R13, \quad 93 = 1(80) + 13.$$

$$4 \div 13 = 0R4, \quad 4 = 0(13) + 4.$$

$$1000 \div 4 = 250R0, \quad 1000 = 250(4) + 0.$$

Figure 8: $5(4)$ means five fours. The arrangement within the "fours" is not important.

Figure 9: $23 = 5(4) + 3$.

Sometimes division leaves no remainder, as in $35 \div 7 = 5R0$. This common phenomenon can be expressed in a multitude of ways; most common in schools are the following: "35 is a **multiple** of 7", "7 goes into 35", "7 is a **factor** of 35". In ancient Greece, one might say (translated) "7 **measures** 35"; the interpretation is that given a 7-foot-long rod, one can precisely measure a 35-foot-long distance.[12]

[12] Questions about this type of measurability (divisibility) are studied in the latter half of Euclid's *Elements*.

7 feet

35 feet

ROUNDING is intimately connected to division with remainder. In fact, properties of the floor demonstrate the existence of quotient and remainder.

**Proposition 0.18 (Traditional division with remainder)** *Let $a$ and $b$ be integers, with $b$ positive. Then there exist integers $q$ and $r$ satisfying*
$$a = q(b) + r \text{ and } 0 \leq r < b.$$

PROOF: Since $a/b$ is a real number, it is not far above its floor:
$$0 \leq \frac{a}{b} - \left\lfloor \frac{a}{b} \right\rfloor < 1.$$

Multiply through by $b$ and we find
$$0 \leq b \cdot \left( \frac{a}{b} - \left\lfloor \frac{a}{b} \right\rfloor \right) < b$$

and distribute the $b$ to arrive at
$$0 \leq a - b \left\lfloor \frac{a}{b} \right\rfloor < b.$$

Define $q = \lfloor a/b \rfloor$ and $r = a - q(b)$. These satisfy
$$a = q(b) + r \text{ and } 0 \leq r < b. \qquad \blacksquare$$

Quotient and remainder may be defined *formulaically* using addition, subtraction, multiplication, division, and the floor. Traditionally, dividing $a$ by $b$ yields a remainder beween 0 and $b - 1$. Another convention gives remainders between $-b/2$ and $b/2$.

**Proposition 0.19 (Division with minimal remainder)** *Let $a$ and $b$ be integers, with $b$ positive. Then there exist integers $q$ and $r$ satisfying*
$$a = q(b) + r \text{ and } -\frac{b}{2} \leq r \leq \frac{b}{2}.$$

PROOF: Let $q$ be an integer *closest* to $a/b$. Thus $q = \lfloor a/b \rfloor$ or $q = \lceil a/b \rceil$, or if $a/b$ is halfway between integers, we may choose either floor or ceiling. Thus
$$-\frac{1}{2} \leq \frac{a}{b} - q \leq \frac{1}{2}.$$
Multiply through by $b$ and distribute to find
$$-\frac{b}{2} \leq a - q(b) \leq \frac{b}{2}.$$
Define $r = a - q(b)$ and observe
$$a = q(b) + r \text{ and } -\frac{b}{2} \leq r \leq \frac{b}{2}. \qquad \blacksquare$$

For example, we divide 23 by 4 with remainder. The quotient is
$$q = \lfloor 23 \div 4 \rfloor = \lfloor 5.75 \rfloor = 5.$$
The remainder is what's left:
$$23 = 5(4) + r, \text{ so } r = 3.$$

For example, we divide 23 by 4 with remainder, allowing negative remainders. The quotient $q$ is the integer *closest* to $23/4 = 5.75$, so $q = 6$. The remainder is what's "left":
$$23 = 6(4) + r, \text{ so } r = -1.$$
In this convention, 23 divided by 4 is 6 with a remainder of $-1$.

Figure 10: Using positive remainders, we have $23 = 5(4) + 3$.

Figure 11: Using minimal remainders, we have $23 = 6(4) - 1$.

DIVISIBILITY is the blanket term for when one number "goes into" another. But "goes into" is a bit informal; mathematicians use the verb "**divides**" instead. Examples of this usage of "divides" are in the margin; read them aloud until they becomes second nature. This may be a shift in perspective; no longer do *people divide numbers*. Now *numbers divide numbers*. Synonyms for "divides" are "goes into" and "is a factor of" and, in ancient Greece, "measures."

2 divides 4. 4 divides 12. 3 divides 99. 13 divides 91. 1000 divides 98000. 5 does not divide 13. 3 does not divide 11.

The universally accepted but most unfortunate[13] notation for "divides" is a single vertical line.

$$x \mid y \text{ means that } x \text{ divides } y.$$

[13] One should never use a symmetric symbol for an antisymmetric relation. But overthrowing accepted notation is not the goal of this text. In case of a revolution, $\triangleleft$ would be a nice notation for divides, and $\triangleright$ for "is a multiple of".

Addition endows the integers with a linear order. Multiplication endows the integers with infinite dimensions of complexity.[14]

[14] This is not hyperbole. See Chapter 3.

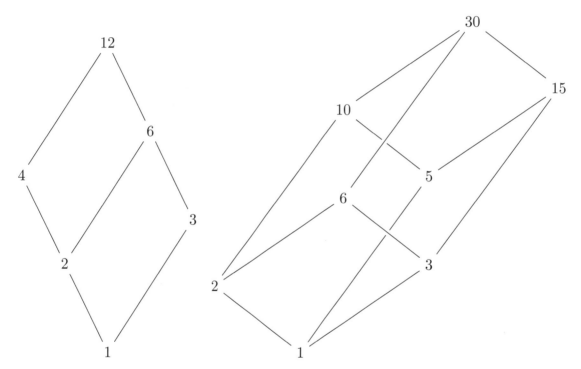

To see the divisors of a number, we display them in a **Hasse diagram**. Each line segment, from $x$ below to $y$ above, expresses a divisibility $x \mid y$. On the left, observe $1 \mid 2$, $1 \mid 3$, $2 \mid 6$, $3 \mid 6$, $2 \mid 4$, $4 \mid 12$, and $6 \mid 12$. Further, every upwards *path* of line segments gives a chain of divisibilities. Since $2 \mid 6$ and $6 \mid 12$, we also see $2 \mid 12$.

Parallel line segments in the same diagram display proportions; for example, the line segment joining 2 to 6 is parallel to the line segment joining 4 to 12 reflecting the fact that $6/2 = 12/4 = 3$.

Figure 12: The Hasse diagrams for the divisors of 12 and 30. The Hasse diagram for 30 is best seen in three dimensions, since 30 has three prime factors: 2, 3, and 5.

To write $x \mid y$ means that $x$ "goes into" $y$, but implicit is a third integer. Namely, $x \mid y$ means that *there exists* an integer $m$ such that $y = mx$. It is crucial to observe that the definition of *divisibility* does not refer at all to the operation *division*. It refers to the existence of an "accesory" integer which satisfies a multiplicative relationship.

**Problem 0.20** Demonstrate that $-3 \mid 12$.

SOLUTION: $-4$ is an integer which satisfies $12 = (-4)(-3)$. ✓

**Problem 0.21** Demonstrate[15] that $0 \mid 0$.

SOLUTION: Every integer $m$ satisfies $0 = m(0)$. ✓

**Problem 0.22** Is it true or false that $0 \mid 12$?

SOLUTION: If $0 \mid 12$ then there would exist an integer $m$ such that $12 = m(0)$. But $m(0) = 0$ and $12 \neq 0$. Hence 0 does not divide 12. ✓

Figure 13: In a Hasse diagram, we express the divisibility $x \mid y$ by a line segment joining $x$ to $y$. To express the accessory integer $m = y/x$, we label the line segment with the number $m$.

[15] The fact that 0 divides 0 is most alarming to novices!

MULTIPLES of a number are evenly spaced on the number line.

**Problem 0.23** Plot the integers $x$ which satisfy $7 \mid x$.

SOLUTION: The multiples of 7 are $\{\ldots, -14, -7, 0, 7, 14, \ldots\}$. ✓

More generally, arithmetic progressions are evenly spaced.

**Problem 0.24** Plot the integers $x$ which satisfy $5 \mid (x-2)$.

SOLUTION: If $5 \mid (x-2)$ then $x-2$ is among the multiples of 5.

Thus $x$ is two more than a multiple of five, lying among the arithmetic progression $\{\ldots, -8, -3, 2, 7, 12, \ldots\}$ plotted above. ✓

**Problem 0.25** Plot the integers $x$ which satisfy $x \mid 12$.

SOLUTION: The positive divisors of 12 are $1, 2, 3, 4, 6, 12$. The negative divisors of 12 are $-1, -2, -3, -4, -6, -12$. ✓

Three properties of | follow below. We leave the reader to consider analagous properties of $\leq$.

**Proposition 0.26 (Reflexive property of |)** *For every integer $x$, $x \mid x$.*

> Every integer is a multiple of itself. To see that $x \mid x$, it suffices to observe that $x = 1 \times x$.

**Proposition 0.27 (Antisymmetric property of |)** *For integers $x, y$, if $x \mid y$ and $y \mid x$ then $x = \pm y$.*

> If two numbers are multiples of each other, then they are equal up to sign.

PROOF: The hypothesis implies $y = mx$ and $x = ny$ for some integers $m$ and $n$. Hence
$$x = ny = n(mx) = (nm)x. \tag{1}$$

If $x = 0$ then $y = mx = 0$, so $x = \pm y$ as claimed.

If $x \neq 0$, then we divide Equation (1) by $x$ to discover that $m$ and $n$ are reciprocals: $nm = 1$. The only integers whose reciprocals are integers are 1 and $-1$. Hence $m = \pm 1$ and $n = \pm 1$, and so $x = \pm y$.

**Proposition 0.28 (Transitive property of |)** *If $x, y, z$ are integers, and $x \mid y$ and $y \mid z$, then $x \mid z$.*

> The relation of "being a multiple" or "divides" is transitive.

PROOF: The hypothesis implies $y = mx$ and $z = ny$ for some integers $m$ and $n$. Substitution yields $z = ny = n(mx) = (nm)x$ and so $x \mid z$. ∎

Divisibility is preserved by addition and multiplication in the following two circumstances.

**Proposition 0.29** *If $d, x,$ and $y$ are integers and $d \mid x$ then $d \mid xy$.*

> The product of a multiple of $d$ with an integer is again a multiple of $d$.

PROOF: If $d \mid x$ then there exists an integer $m$ such that $x = md$. Multiplying both sides by $y$ yields $xy = mdy = (my)d$. Hence $xy$ is a multiple of $d$, i.e., $d \mid xy$. ∎

**Proposition 0.30** *Let $d, x,$ and $y$ be integers. If $d \mid x$ and $d \mid y$, then $d \mid (x+y)$ and $d \mid (x-y)$.*

> The sum or difference of two multiples of $d$ is another multiple of $d$.
>
> Intuitively, if we have two collections, arranged in clumps of $d$, their union can again be arranged in clumps of $d$. This explains the theorem, at least when $d$ and its two multiples are positive.

PROOF: The hypothesis implies $x = md$ and $y = nd$ for some integers $m$ and $n$. Adding or subtracting $x$ and $y$ yields
$$x \pm y = (md) \pm (nd) = (m \pm n)d.$$

Hence $x \pm y$ is a multiple of $d$. ∎

We rephrase the previous theorem as a principle.[16]

**Corollary 0.31 (The two out of three principle for divisibility)** *Let $a, b, c$ be integers, satisfying the equation $a + b = c$. Let $d$ be an integer. If two of the numbers from the set $\{a, b, c\}$ are multiples of $d$, then the third number must also be a mutliple of $d$.*

> [16] This is a rephrasing. Indeed if $a + b = c$ then $b = c - a$ and $a = c - b$. Hence each term is a sum or difference of the other two. Hence if two terms are multiples of $d$, then the third is a multiple of $d$.

**Problem 0.32** Demonstrate that $2\,999\,997$ is a multiple of three.

SOLUTION: Observe that $3\,000\,000$ is a multiple of three; it is three million, or a million threes:
$$3\,000\,000 = 1\,000\,000(3), \text{ and so } 3 \mid 3\,000\,000.$$

Since $3 \mid 3$ and $3 \mid 3\,000\,000$, we find that 3 goes into the difference:
$$3 \mid (3\,000\,000 - 3) = 2\,999\,997. \qquad \checkmark$$

**Problem 0.33** Find all integers $x$ which satisfy $x \mid (x+6)$.

SOLUTION: Note that $x \mid x$ (always). If $x \mid (x+6)$, then $x$ divides the difference:
$$x \mid ((x+6) - x) = 6.$$
Conversely, if $x \mid 6$, then $x$ divides the sum
$$x \mid (x+6).$$
Hence $x \mid (x+6)$ if and only if $x \mid 6$. The solutions to $x \mid (x+6)$ are the divisors of 6: $x \in \{-6, -3, -2, -1, 1, 2, 3, 6\}$. $\qquad \checkmark$

**Problem 0.34** Does the equation $7x^2 + 11 = 21y$ have any integer solutions?

SOLUTION: If there were integer solutions $x$ and $y$ to the equation, then $7x^2$ would be a multiple of 7 and $21y$ would be a multiple of 7. By the two out of three principle, 11 would also be a multiple of 7.

But this is untrue; 11 is *not* a multiple of 7 and hence there can be no integer solutions to the equation $7x^2 + 11 = 21y$. $\qquad \checkmark$

In general, if one is faced with an equation and asked to find integer solutions, one should first check whether there is an obvious *divisibility obstruction*. Perhaps, as the previous problem indicates, no solution can be found because any integer solution would lead to a contradiction of principles of divisibility. Later, we will generalize this to find *congruence obstructions* to solubility of equations.

THIS CHAPTER has introduced the main characters of number theory: the natural numbers and integers, counting, addition, multiplication, rounding, division with remainder, and divisibility. But although these characters may feel like old friends, their mysteries will be exposed in the chapters to follow. When difficulties arise, the reader is advised to return to the techniques of this zeroth chapter.

## Historical Notes

TRIANGULAR NUMBERS, those numbers obtained by adding $1 + 2 + 3 + \cdots$ up to some fixed number, have at least two thousand years of history. Here we present some highlights, tracing sources back in time. Teachers and books often repeat a legend about C.F. Gauss (1777–1855CE) as a child, in which a cruel or frustrated teacher asks the boy to sum the numbers between 1 and 100. According to this legend, Gauss quickly obtained 5050, the correct answer. This legend, embellished like any good fishing story, has only a remote connection to textual source, which in turn has questionable connection to fact. The textual source is the 1856 Gauss memorial volume, written by Gauss's colleague at Göttingen, Wolfgang Sartorus. This source mentions an arithmetic problem given to the seven-year-old Gauss, but nothing about the problem except that he solved it quickly, and was confident in the correctness of his solution.[17]

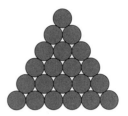

Figure 14: The sixth triangular number, $1 + 2 + 3 + 4 + 5 + 6 = 21$.

Alcuin of York, an English scholar and teacher of the late 8th century CE, presents the problem of adding the numbers from 1 to 100 as Problem 42 in his "Problems to sharpen the young"; we have a copy of his text[18] from the late 9th century. The author presents the reader with a staircase of 100 steps; one pigeon stands on the first step, two pigeons on the second, et cetera, and the question is posed "How many pigeons are there altogether?" The solution given is similar to, but not entirely the same as, our solution by partnering. Alcuin partners 1 and 99, and 2 and 98, etc.., leaving 50 and 100. See the annotated translation[19] by J. Hadley and D. Singmaster for a thorough treatment of this medieval document.

But such problems certainly predate Alcuin. An example is found in a Jewish legal text from before 500CE, the Babylonian Talmud, Menachot folio 106a, in which the numbers from 1 to 60 are correctly summed; the method of summation is not explained, but the commentaries (Tosafot) written between 1000CE and 1400CE explain precisely the method of pairing off we have described in this chapter.[20]

[17] For a critical examination of the Gauss legend, see "Gauss's Day of Reckoning," by Brian Hayes, in *American Scientist*, May-June 2006, Vol. 94, No. 3, p. 200.

[18] "Propositiones ad Acuendos Juvenes," by Alcuin of York, c. 800CE. This problem is in XLII, titled "Propositio de scala habente gradus centum," which reads "Est scala una habens gradus C. In primo gradu sedebat columba una: in secundo duae; in tertio tres; in quarto IIII; in quinto V. Sic in omni gradu usque ad centesimum. Dicat, qui potest, quot columbae in totum fuerunt?"

[19] "Problems to Sharpen the Young," by John Hadley and David Singmaster, in *The Mathematical Gazette,* March 1992, Vol. 76, No. 475, pp. 102–126.

[20] For more about this finding in the Babylonian Talmud, see "A Mediaeval Derivation of the Sum of an Arithmetic Progression," by Martin D. Stern, in *The Mathematical Gazette,* June 1990, Vol. 74, No. 468, pp. 157–159.

TO ADD PERFECT SQUARES, $1 + 4 + 9 + \cdots$ up to a given square $N^2$, one may use the formula

$$\sum_{i=1}^{N} i^2 = \frac{(N)(N+1)(2N+1)}{6},$$

provable by induction. The result of this summation is called the $N^{\text{th}}$ square pyramidal number. Such a formula occurs in the Gaṇipāda of Āryabhata (India, 499CE), which reads

The product of the three quantities, the number of terms plus one [$N+1$], the same increased by the number of terms [$2N+1$], and the number of terms [$N$], when divided by six, gives the sum of squares of natural numbers (*varga-citighana*).[21]

A millenium after Āryabhata, Nīlakaṇṭha (c. 1500CE) of the Kerala school in India demonstrates a "stacking-corners" style argument for this formula, using a solid of dimensions $N$ by $N+1$ by $2N+1$. We leave this to the exercises, and refer to K. Plofker's translation and exposition[22] for more details.

In the "Introduction to Arithmetic" by Nicomachus of Gerasa (Roman Syria, now Jordan, c. 100CE), we find the following treatment of cubes:

But all the products of a number multiplied twice into itself, that is, the cubes... 1,8,27,64,125, and 216, and those that go on analogously, in a simple, unvaried progression as well. For when the successive odd numbers are set forth indefinitely beginning with 1, observe this: The first one makes the potential cube; the next two added together, the second; the next three, the third; the four next following, the fourth; the succeeding five, the fifth; the next six, the sixth; and so on.[23]

The term "Nicomachus's Theorem" today often refers to the result that the sum of the first $N$ cubes equals the square of the $N^{\text{th}}$ triangular number.[24] This can be deduced from Nicomachus's observation above: see the exercises.

EUCLID introduces divisibility and proportion in Book V of the *Elements*. The Definitions of Book V begin with

Μέρος ἐστὶ μέγεθος μεγέθους τὸ ἔλασσον τοῦ μείζονος, ὅταν καταμετρῇ τὸ μεῖζον.

Translated,

A magnitude is a part of a(nother) magnitude, the lesser of the greater, when it measures the greater.[25]

Factors of a number are the "parts" for Euclid, and while we might say "goes into" or "divides", Euclid would write "measures."

[21] Translation from "Development of Calculus in India," by K. Ramasubramanian and M.D. Srinivas, in *Studies in the History of Indian Mathematics*, ed. C. S. Seshadri, published by Hindustan Book Agency, 2010.

[22] "Aryabhatiyabhasya of Nilakantha," translated by K. Plofker and H. White, in Chapter 4 of *The Mathematics of Egypt, Mesopotamia, China, India, and Islam, A Sourcebook*, ed. Victor J. Katz, Princeton University Press, 2007.

[23] From Book 2, Chapter XX, of "Introduction to Arithmetic", by Nicomachus of Gerasa, translated by Martin Luther D'Ooge, published by the Macmillan Company, New York, 1926.

Nicomachus was aware of the following, expressed in a modern table:

| | | |
|---|---|---|
| 1 | = | 1 |
| 8 | = | 3 + 5 |
| 27 | = | 7 + 9 + 11 |
| 64 | = | 13 + 15 + 17 + 19 |
| 125 | = | 21 + 23 + 25 + 27 + 29 |

[24] In other words,
$$\sum_{i=1}^{N} i^3 = \left(\sum_{i=1}^{N} i\right)^2.$$

[25] The Greek text here comes from the 1885 edition of J.L. Heiberg, and the English translation is by Richard Fitzpatrick. Many editions of Euclid, including Fitzpatrick's and Byrne's remarkable illustrated edition of Books I-VI, are freely available online.

## Exercises

1. How many multiples of 7 are between 10 and 500?

2. How many numbers between 1 and 100 are not multiples of three?

3. How many numbers between 1 and 100 are mulitples of 2 or[26] multiples of 3? Caution: do not double-count the multiples of 6.

4. Add the even numbers between 1 and 100.

5. Add the numbers in the series $3 + 11 + 19 + 27 + \cdots + 395 + 403$.

6. How many square numbers are between 100 and 10000?

7. Given integers $x$ and $y$, with $x > 0$ and $y < 0$ and $x - y = 5$. What are the possible values of $x$ and $y$? Reason geometrically.

8. Carry out division with *positive* remainder and division with minimal (positive or negative) remainder for the following: $27 \div 7$, $30 \div 6$, $100 \div 3$, $90 \div 13$. Express your answers as equalities of the form $a = q(b) + r$.

9. It is possible to fit three congruent pentagons – all equilateral and equiangular – around a single point without overlapping. For which positive integers $(A, B)$ is it possible to fit $A$ congruent shapes – each equilateral and equiangular with $B$ sides – around a point without overlapping? (This is the crux of Euclid's argument that there are five Platonic solids.)

10. If $x \leq y$ then $x + 5 \leq y + 5$. Is the same true, with $\leq$ replaced by $|$? Prove it or find a counterexample.

11. If $x \leq y$ then $5x \leq 5y$. Is the same true, with $\leq$ replaced by $|$? Prove it or find a counterexample.

12. Draw Hasse diagrams for the positive divisors of 7, 15, 18, and 105.

13. Plot all integers $x$ which satisfy $(x + 1) \mid 14$.

14. Plot integers $x$ which satisfy $4 \mid (x + 1)$.

15. Plot all integers $x$ which satisfy $(x + 4) \mid 2x$.

16. Graph the functions $y = \lfloor x \rfloor$ and $y = \lceil 2x \rceil$.

17. How many bits does the binary expansion of $10^{500}$ have?

18. Describe the effect of the functions $f(x) = \lfloor 10x \rfloor / 10$ and $g(x) = \lfloor 100x \rfloor / 100$.

[26] In mathematical contexts, "or" means the "inclusive or". So "$P$ or $Q$" means "$P$ or $Q$ or both". So the first few numbers which are multiples of 2 or 3 are
$$2, 3, 4, 6, 8, 9, 10, 12, \ldots.$$

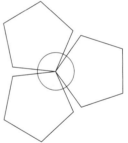

Figure 15: 3 regular 5-gons fit around a point, with a bit of room to spare. So $(3, 5)$ is a possible ordered pair.

19. Write down a formula with one variable $x$, using basic arithmetic operations and the floor function, whose output is the ones digit of $x$ in base 10. Challenge: write down a formula[27] whose output is the *leading* digit of $x$, i.e., if $x = 372.5$ then $f(x) = 3$.

[27] You may use basic arithmetic operations, exponents and logarithms, and the floor function.

20. Prove that
$$\lim_{x \to \infty} \frac{\lfloor x \rfloor}{x} = \lim_{x \to \infty} \frac{\lceil x \rceil}{x} = 1.$$

21. Consider a square chessboard, with $N$ rows and $N$ columns, whose bottom-left corner is black. Find a formula (using floor or ceiling) for the number of black squares on the chessboard.

Figure 16: A chessboard with 6 rows and 6 columns.

22. Prove that if $x \mid (x^2 + 1)$ then $x = \pm 1$.

23. Prove that $6 \mid (x^3 - x)$ for every integer $x$. Hint: among three consecutive integers, one must be a multiple of 3.

24. Let $T(N) = \sum_{i=1}^{N} i$ be the $N^{\text{th}}$ triangular number. When is $T(N)$ even? Prove your answer.

25. Draw a spiral to demonstrate that
$$100 = 10 + 2(9) + 2(8) + 2(7) + 2(6) + 2(5) + 2(4) + 2(3) + 2(2) + 2(1).$$

26. (Challenge) Let $S(N) = \sum_{i=1}^{N} i^2$ be the sum of the first $N$ perfect squares. Prove that $6N = (N)(N+1)(2N+1)$ by a "stacking corners" argument in a rectangular solid of dimensions $N \times (N+1) \times (2N+1)$.

27. Let $C(N) = \sum_{i=1}^{N} i^3$ be the sum of the first $N$ cubes. Let $T(N) = \sum_{i=1}^{N} i$ be the $N^{\text{th}}$ triangular number. Prove that $C(N) = T(N)^2$ by induction. Can you find a proof using Nicomachus's observation (see the *Notes*) earlier in the chapter? Can you find a four-dimensional stacking-corners proof?

28. Demonstrate that the equation $3x^{17} + 1111 = 27y + 15z$ has no solution in which $x$, $y$, and $z$ are integers.

29. Let $N$ be a positive integer. Let $\sigma_0(N)$ denote the number of positive divisors of $N$. Prove that $\sigma_0(N)$ is odd if and only if $N$ is a square number. Prove that the *product* of all positive divisors of $N$ equals $N^{\sigma_0(N)/2}$. Hint: Use partnering arguments.

30. Let $A(N)$ be the average of $\sigma_0(1), \sigma_0(2), \sigma_0(3), \ldots, \sigma_0(N)$ – the average number of positive divisors among the numbers up to $N$. Demonstrate that $A(N)$ is approximately $\log(N)$, with error bounded by 1. Hint: Relate $NA(N)$ to the number of dots in a shaded region such as the one on the right.

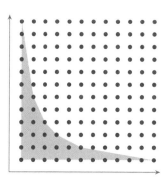

# Part I

# Foundations

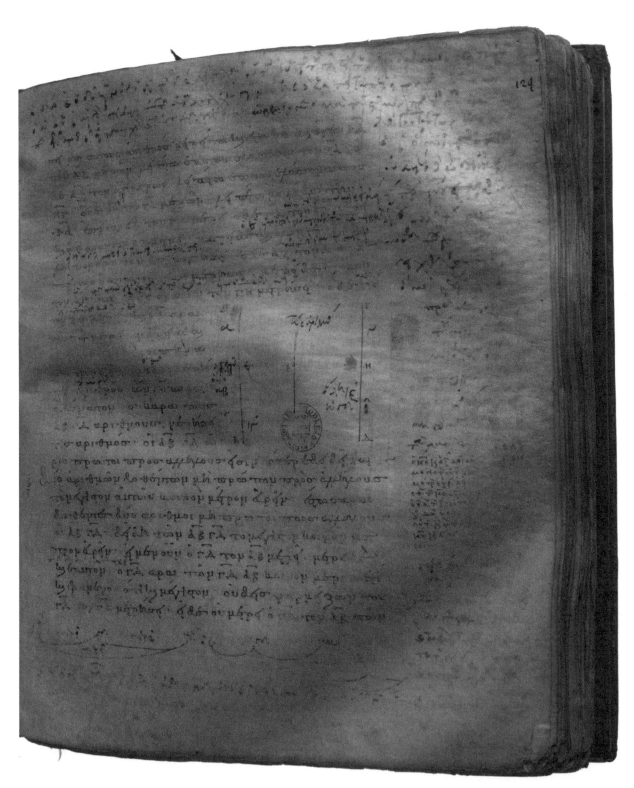

Figure 17: *Elements*, Book VII, Proposition 1, in MS D'Orville 301, 888CE. Image courtesy of the Clay Mathematics Institute.

# 1
# The Euclidean Algorithm

Euclid wrote or assembled the *Elements* around 300BCE. The first four books of Euclid's *Elements* develop plane geometry, and books five and six discuss proportion and similarity. At the beginning of Book VII, we find "anthyphairesis," now called "the Euclidean algorithm," soon followed by the most important implications. The Euclidean algorithm, subject of this chapter, is the deductive root of number theory – without it, the later chapters of this text would have no logical legs to stand on. Anthyphairesis begins our number theory as much as it began Euclid's number theory.

In China, we find anthyphairesis under a different name, in a different context. In 1983, archeologists found a large collection of Chinese writings in a tomb that had been sealed around 186BCE. Among these are 190 readable bamboo strips, referred to as the Suàn shù shū or "Writings on reckoning". On strips 17 and 18, we find the "method for simplifying parts":

> Take the numerator from the denominator; in turn take the denominator from the numerator. When the numbers, the numerator and denominator, are equal to one another, then you can go on to simplify.[1]

To introduce this algorithm, we present a problem here.

**Problem 1.1 (1-Dimensional Hop and Skip)** Imagine you are standing on the number line. You can hop, moving 133 units to the right or left, and you can skip, moving 85 units to the right or left.

Using only hops and skips, and beginning at 0, where on the number line can you travel? Can you travel from 0 to 1?

[1] Translation from p.18 of "The Suàn shù shū, 'Writings on reckoning': Rewriting the history of early Chinese mathematics in the light of an excavated manuscript," by C. Cullen, in *Historia Mathematica* **34** (2007).

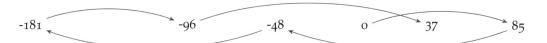

We urge the reader to consider this problem for a few minutes before turning the page to its solution.

Figure 1.1: Beginning at zero, you can skip to the right, hop twice to the left, skip to the right, then hop to the right, ending at 37.

SOLUTION: If one can hop 133 units and skip 85 units, then one can combine these to jump 48 units. Indeed, a hop in one direction followed by a skip in the other direction can be thought of as a jump. The jump is a **compound move**; we write: jump = hop − skip.

We now have three ways to move along the number line: hopping, skipping, and jumping. We carry on in this fashion, defining leaps and bounds, pounces and stretches, and finally a step.

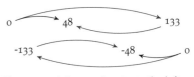

Figure 1.2: A jump, 48 units to the left or right, is built out of a hop and a skip.

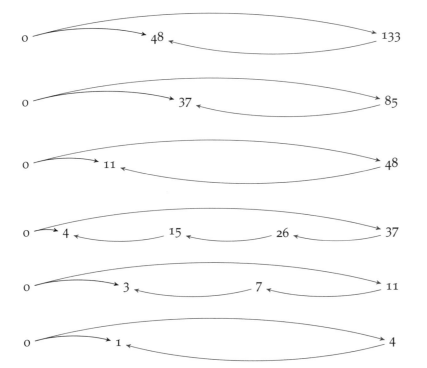

jump = hop − skip.

leap = skip − jump.

bound = jump − leap.

pounce = leap − 3 bounds.

stretch = bound − 2 pounces.

step = pounce − stretch.

If one can hop and skip, then one can jump; hence one can leap, bound, pounce, stretch, and step. And so, by hopping and skipping, one can step 1 unit to the left or right. By hopping and skipping, one can travel to every *integer* by stepping left and right. ✓

The compound moves are *remainders*:

$$\begin{aligned} 133 &= 1(85) + 48, \\ 85 &= 1(48) + 37, \\ 48 &= 1(37) + 11, \\ 37 &= 3(11) + 4, \\ 11 &= 2(4) + 3, \\ 4 &= 1(3) + 1. \end{aligned}$$

When written in this fashion, the procedure of repeated division with remainder is called the **Euclidean algorithm**.

**Proposition 1.2** *Suppose one can hop a units and skip b units, and $a > b$. Then if r is a remainder after dividing a by b, i.e., $a = q(b) + r$, then one can jump r units.*

PROOF: A jump of $r$ units can be built by hopping once to the right, and skipping $q$ times to the left: $r = a - q(b)$. ∎

We illustrate this with one more example.

**Problem 1.3** Imagine you are standing on the number line again, and you can hop 91 units to the right or left and skip 49 units to the right or left. Using only hops and skips, and beginning at zero, where on the number line can you travel?

SOLUTION: Consider the results of repeated division with remainder:

$$\begin{aligned} 91 &= 1(49) + 42, \\ 49 &= 1(42) + 7, \\ 42 &= 6(7) + 0. \end{aligned}$$

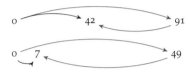

Figure 1.3: A jump of 42 units is built out of a hop and a skip. A leap of 7 units is built out of a skip and a jump.

The first division with remainder implies that we can jump to 42 (a hop minus a skip). The second division with remainder implies that we can leap to 7 (a skip minus a jump). The last division with remainder does not help, except to tell us that one cannot continue this process.

By leaping 7 units, we may travel to every multiple of 7. In fact, we can never travel anywhere else. Each hop is a multiple of 7 (note that $91 = 13(7)$) and each skip is a multiple of 7 (note that $49 = 7(7)$). By moving in multiples of 7, beginning at 0, one cannot travel anywhere except to multiples of 7.

Hence by hops of 91 and skips of 49, one can move to all multiples of 7 and nowhere else. ✓

HIGH SCHOOL ALGEBRA has limited utility in solving these problems. Given the ability to hop 91 units and skip 49 units, where can we travel? A combination of $x$ hops and $y$ skips takes us from 0 to $91x + 49y$. So the following two questions are the same:

1. Can we hop by 91 and skip by 49, to travel from zero to one?

2. Does the equation $91x + 49y = 1$ have a solution in which $x$ and $y$ are *integers*?

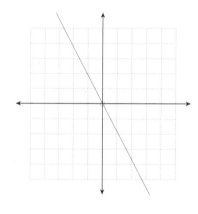
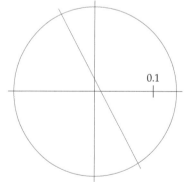

Figure 1.4: The graph of $91x + 49y = 1$. It comes very close to the grid-points, where $x$ and $y$ are integers, but never passes through them. A close zoom shows that the line does not pass through the origin.

High school algebra might allow you to solve the equation $91x + 49y = 1$ (the solution set is a line). But it will not tell you where on the line to find the *integer* solutions, or even whether such a pair of integers $(x, y)$ exists.

THE EUCLIDEAN ALGORITHM is the process of repeated division with remainder, beginning with two natural numbers.[2] We have already seen the algorithm carried out twice, but it is worth some examination and practice.

Begin with two natural numbers, $a$ and $b$. Assume that $a$ is the larger of the two numbers. Divide with remainder, to find a quotient $q$ and remainder $r$. The algebraic relationship between these four quantities is

$$a = q(b) + r.$$

To use words instead of symbols,

$$\text{Dividend} = \text{quotient}(\text{divisor}) + \text{remainder}.$$

For now, we use the traditional division with remainder, as in Proposition 0.18, so $0 \leq r \leq b - 1$.

If the remainder is nonzero, proceed to the next step in the Euclidean algorithm. In this next step, *ignore the quotient* and divide $b$ by $r$ with remainder. In haiku,

Divisor to front,
remainder to divisor,
ignore the quotient.

Repeat this process, carrying out division with remainder repeatedly, until one finds a remainder 0. When the remainder is zero, STOP.

**Problem 1.4** Carry out the Euclidean algorithm on 99 and 51.[3]

SOLUTION: We divide with remainder,

$$99 = 1(51) + 48.$$

The divisor (51) moves to the front, the remainder (48) becomes the divisor, ignore the quotient (1), and divide with remainder again:

$$51 = 1(48) + 3.$$

Now the divisor (48) becomes the dividend and the remainder (3) becomes the divisor, and we divide again:

$$48 = 16(3) + 0.$$

The remainder is zero, and we STOP. ✓

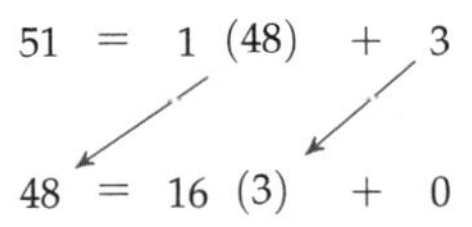

[2] Generalizations apply to integers, to the Gaussian and Eisenstein integers, to polynomials with rational or real or complex coefficients, and to many other settings. The term "Euclidean domain" is used to describe settings in which division with remainder, and hence the Euclidean algorithm, is possible.

[3] For the moment, the algorithm is just a sequence of actions we carry out. But it might help to think of 99 as a hop and 51 as a skip. Successive remainders are the compound moves.

$$51 = 1\ (48) + 3$$
$$48 = 16\ (3) + 0$$

Figure 1.5: Two pieces of information get carried to the next step: the divisor and the remainder. These become the dividend and divisor in the next step.

THE Euclidean algorithm cannot go on forever. We quantify this, and estimate how quickly the algorithm finishes

**Problem 1.5** Carry out the Euclidean algorithm on 123987 and 34762.

SOLUTION: Repeatedly divide with remainder,

$$\begin{aligned} 123987 &= 3(34762) + 19701, \\ 34762 &= 1(19701) + 15061, \\ 19701 &= 1(15061) + 4640, \\ 15061 &= 3(4640) + 1141, \\ 4640 &= 4(1141) + 76, \\ 1141 &= 15(76) + 1, \\ 76 &= 76(1) + 0. \end{aligned}$$

Keep in mind that this problem tells us something not so obvious. If we can hop 123987 units and skip 34762 units, then the Euclidean algorithm demonstrates that we can step from 0 to 1. The remainders – the numbers 19701, 15061, 4640, 1141, 76, 1 – are the compound moves.

One can also perform the Euclidean algorithm while using minimal remainders[4] as discussed in Proposition 0.19:

$$a = q(b) + r \text{ with } -\frac{b}{2} \le r \le \frac{b}{2}.$$

The previous problem is solved anew:

[4] When using negative remainders, we still use positive dividends and divisors. At each step, the divisor becomes the dividend, and the *absolute value* of the remainder becomes the divisor.

(Step 1)   $123987 = 4(34762) - 15061$,   $15061 \le 1/2 \cdot 34762$.
(Step 2)   $34762 = 2(15061) + 4640$,   $4640 \le 1/2 \cdot 15061$.
(Step 3)   $15061 = 3(4640) + 1141$,   $1141 \le 1/2 \cdot 4460$.
(Step 4)   $4640 = 4(1141) + 76$,   $76 \le 1/2 \cdot 1141$.
(Step 5)   $1141 = 15(76) + 1$,   $1 \le 1/2 \cdot 76$.
(Step 6)   $76 = 76(1) + 0$.   $0 \le 1/2 \cdot 1$.

If $x$ is a positive integer, write $\text{bits}(x)$ for the number of bits in the binary expansion of $x$. Recall that $\text{bits}(x) = \lfloor \log_2(x) \rfloor + 1$; it follows that $\text{bits}(\lfloor x/2 \rfloor) \le \text{bits}(x) - 1$.

**Proposition 1.6 (Euclidean algorithm runtime)** *Let $a$ and $b$ be positive integers with $a > b$. Let $\text{bits}(b)$ be the number of bits in the binary expansion of $b$. Then the Euclidean algorithm on $a$ and $b$, performed with minimal remainders, stops after at most $\text{bits}(b)$ steps.*

PROOF: The first remainder is bounded by $b/2$, and so the first remainder has at most $\text{bits}(b) - 1$ bits (if it is nonzero). The second remainder is bounded by half the first remainder, and so the second remainder has at most $\text{bits}(b) - 2$ bits. Et cetera – in step $s$, the remainder has at most $\text{bits}(b) - s$ bits. Therefore, the Euclidean algorithm must terminate in no more than $\text{bits}(b)$ steps.

In practice, the Euclidean algorithm is very very fast on a computer; it can be carried out in a split-second even for numbers with thousands of digits.

The Euclidean algorithm leads to the greatest common divisor.

**Theorem 1.7 (Euclid's algorithm gives the GCD)** *Let a and b be natural numbers, not both zero.*[5] *Carrying out the Euclidean algorithm on a and b, the final nonzero remainder is the greatest common divisor of a and b.*

[5] We leave the case $a = b = 0$ for the next page.

Why does the Euclidean algorithm give the greatest common divisor of two numbers? An example illustrates the reason. Let us apply the Euclidean algorithm to 221 and 182.

$$221 = 1(182) + 39,$$
$$182 = 4(39) + 26,$$
$$39 = 1(26) + 13,$$
$$26 = 2(13) + 0.$$

Consider a common divisor[6] $d$ of 221 and 182; our reasoning below follows the Euclidean algorithm from the top downwards, applying the two-out-of-three principle (Corollary 0.31) at each step.

[6] This means that $d$ is a divisor of 221 and $d$ is a divisor of 182. Symbolically, this is written $d \mid 221$ and $d \mid 182$.

| | |
|---|---|
| $(d \mid 221$ and $d \mid 182)$ implies $d \mid 39$ | since $221 = 1(182) + 39$. |
| $(d \mid 182$ and $d \mid 39)$ implies $d \mid 26$ | since $182 = 4(39) + 26$. |
| $(d \mid 39$ and $d \mid 26)$ implies $d \mid 13$ | since $39 = 1(26) + 13$. |

Hence every common divisor of 221 and 182 is a divisor of 13.

Now we follow the Euclidean algorithm from the bottom up.

| | |
|---|---|
| $(13 \mid 0$ and $13 \mid 13)$ implies $13 \mid 26$ | since $26 = 2(13) + 0$. |
| $(13 \mid 13$ and $13 \mid 26)$ implies $13 \mid 39$ | since $39 = 1(26) + 13$. |
| $(13 \mid 26$ and $13 \mid 39)$ implies $13 \mid 182$ | since $182 = 4(39) + 26$. |
| $(13 \mid 39$ and $13 \mid 182)$ implies $13 \mid 221$ | since $221 = 1(182) + 39$. |

To summarize, we find that

1. Every common divisor of 221 and 182 is a divisor of 13.

2. 13 itself is a common divisor of 221 and 182.

The conjunction of these two conclusions is that not only is 13 a common divisor of 221 and 182, but moreover every common divisor of 221 and 182 is a divisor of 13. Hence every common divisor is less than or equal to 13; thus 13 is the *greatest* common divisor of 221 and 182.

The best modern definition of "greatest common divisor" is a bit stronger than the definition taught in school.

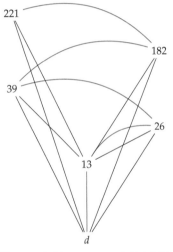

Figure 1.6: Lines represent divisibility, from smaller on bottom to larger on top. Blue arcs trace the steps in the Euclidean algorithm. $13 \mid 221$ and $13 \mid 182$. If another number $d$ is a common divisor of 221 and 182, then (as indicated in red) $d \mid 13$ as well.

**Definition 1.8** Let $a$ and $b$ be integers. A natural number $g$ is called the[7] **greatest common divisor** of $a$ and $b$ if it satisfies the following two properties:

1. $g$ is a **common divisor** of $a$ and $b$. Symbolically, $g \mid a$ and $g \mid b$.

2. If $d$ is a common divisor of $a$ and $b$, then $d \mid g$.

This is *stronger* than the school definition, since it requires not that $g$ is the *greatest* of the common divisors, but that $g$ is a *multiple* of the common divisors. A greatest common divisor in our strong sense of the word is a greatest common divisor in the school sense of the word. We write $g = \text{GCD}(a,b)$ to mean that the greatest common divisor of $a$ and $b$ exists[8] and equals $g$.

**Lemma 1.9** *If $a$ is a natural number, then* $\text{GCD}(a,0) = a$.

PROOF: Since $a \mid a$ and $a \mid 0$, we find that $a$ is a common divisor of $a$ and 0. If $d \mid a$ and $d \mid 0$, then in particular $d \mid a$. Every common divisor of $a$ and 0 is a divisor of $a$. ∎

Note that, according to the definition above, $\text{GCD}(0,0) = 0$. Indeed, $0 \mid 0$ and a number which divides both 0 and 0 divides 0. This technicality will not play a great role in what follows.

The next lemma states that throughout the Euclidean algorithm, the greatest common divisor remains *invariant*.

**Lemma 1.10** *Suppose that $a,b,q,r$ are integers and $a = q(b) + r$. Let $g$ be a natural number. Then $g = \text{GCD}(a,b)$ if and only if $g = \text{GCD}(b,r)$.*

PROOF: If $g = \text{GCD}(a,b)$ then $g \mid a$ and $g \mid b$, and thus $g \mid r$ by the two-out-of-three principle. Hence $g$ is a common divisor of $b$ and $r$. If $d \mid b$ and $d \mid r$, then $d \mid a$ by the two-out-of-three principle, and hence $d \mid g$. Thus $g = \text{GCD}(b,r)$ as well.

Since $r = -q(b) + a$, the same argument demonstrates that if $g = \text{GCD}(r,b)$ then $g = \text{GCD}(a,b)$. ∎

Now we can prove Theorem 1.7: that $\text{GCD}(a,b)$ (in the strong sense) is the final nonzero remainder in the Euclidean algorithm applied to $a$ and $b$.

PROOF: Carry out the Euclidean algorithm on $a$ and $b$. In the last line, we find an expression of the form

$$a_{\text{final}} = q_{\text{final}}(r_{\text{final}}) + 0,$$

where $r_{\text{final}}$ is the final nonzero remainder.

Lemma 1.9 states that $\text{GCD}(r_{\text{final}}, 0) = r_{\text{final}}$. Applying Lemma 1.10 to each line of the Euclidean algorithm, from the bottom up, we find that $\text{GCD}(a,b) = r_{\text{final}}$. ∎

[7] There can be at most one greatest common divisor of $a$ and $b$. Indeed, if $g_1$ and $g_2$ are greatest common divisors of $a$ and $b$, then we have $g_1 \mid g_2$ and $g_2 \mid g_1$. But since $g_1$ and $g_2$ are *natural* numbers, this implies $g_1 = g_2$.

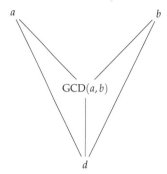

Figure 1.7: $\text{GCD}(a,b) \mid a$ and $\text{GCD}(a,b) \mid b$. If another number $d$ is a common divisor of $a$ and $b$, then (as indicated in red) $d \mid \text{GCD}(a,b)$ as well.

[8] It is not yet clear that the greatest common divisor exists, in the strong sense of the above definition! But indeed it does, as the proof below demonstrates.

$$a = q\,(b) + r$$

Figure 1.8: $\text{GCD}(a,b) = \text{GCD}(b,r)$.

$$221 = 1\,(182) + 39$$
$$182 = 4\,(39) + 26$$
$$39 = 1\,(26) + 13$$
$$26 = 2\,(13) + 0$$

Figure 1.9: Dividends are paired with divisors, and divisors are paired with remainders. Every pair has the same GCD by Lemma 1.10.

$$\text{GCD}(221,182) = \text{GCD}(182,39);$$
$$\text{GCD}(182,39) = \text{GCD}(39,26);$$
$$\text{GCD}(39,26) = \text{GCD}(26,13);$$
$$\text{GCD}(26,13) = \text{GCD}(13,0) = 13.$$

FACTORIZATION is not required when finding greatest common divisors! In school, most students are taught to find greatest common factors by factoring numbers and finding commonalities. This method is entirely unnecessary, deprecated[9] about 2300 years ago by Euclid's algorithm. Factoring large numbers is very time-consuming, while the Euclidean algorithm is reliably fast.

[9] In computer science, an algorithm is **deprecated** when a better algorithm is found.

**Problem 1.11** What is the greatest common divisor of 2059 and 1711?

SOLUTION: Perform the Euclidean algorithm on 2059 and 1711.

$$
\begin{aligned}
2059 &= 1(1711) + 348, \\
1711 &= 4(348) + 319, \\
348 &= 1(319) + 29, \\
319 &= 11(29) + 0.
\end{aligned}
$$

The final nonzero remainder is the greatest common divisor.

$$\text{GCD}(2059, 1711) = 29. \quad \checkmark$$

**Problem 1.12** Does $713x + 851y = 10$ have an integer solution?

SOLUTION: Apply the Euclidean algorithm.

$$
\begin{aligned}
851 &= 1(713) + 138, \\
713 &= 5(138) + 23, \\
138 &= 6(23) + 0.
\end{aligned}
$$

We find that $\text{GCD}(713, 851) = 23$. It follows that 713 is a multiple of 23 and 851 is a multiple of 23. If we could find integers $x$ and $y$ for which $713x + 851y = 10$, then 10 would also be a multiple of 23. But 10 is not a multiple of 23, so the equation $713x + 851y = 10$ has no integer solutions. $\quad \checkmark$

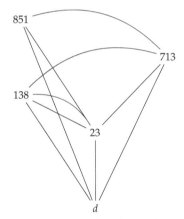

Figure 1.10: $23 = \text{GCD}(851, 713)$.

**Problem 1.13** Does $71x + 83y = 2$ have an integer solution?

SOLUTION: Apply the Euclidean algorithm.

$$
\begin{aligned}
83 &= 1(71) + 12, \\
71 &= 5(12) + 11, \\
12 &= 1(11) + 1, \\
11 &= 11(1) + 0.
\end{aligned}
$$

If one can hop by 83 and skip by 71, then one can jump by 12, leap by 11, and step by 1. Hence by stepping twice, we can step to 2. Using hops and skips, we can travel to 2, and therefore $71x + 83y = 2$ has an integer solution. $\quad \checkmark$

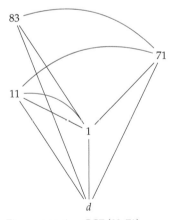

Figure 1.11: $1 = \text{GCD}(83, 71)$.

The previous two problems exhibit the following general result, determining which **linear Diophantine equations**[10] can be solved.

**Theorem 1.14 (Solubility of linear Diophantine equations)** *Let $a$, $b$, and $c$ be integers. The equation $ax + by = c$ has an integer solution $(x, y)$ if and only if $c$ is a multiple of $\text{GCD}(a, b)$.*

PROOF: If $x$ and $y$ are integers, then $ax + by$ is a multiple of $\text{GCD}(a, b)$. Hence for $ax + by = c$ to have an integer solution, it is *necessary* that $c$ be a multiple of $\text{GCD}(a, b)$.

Conversely, suppose that $c$ is a multiple of $\text{GCD}(a, b)$. Apply the Euclidean algorithm to the natural numbers $|a|$ and $|b|$. Considering hops of $|a|$ and skips of $|b|$, the remainders provide a sequence of compound moves, culminating in a move by $\text{GCD}(a, b)$ – let us call this move a step. By steps, one can travel to every multiple of $\text{GCD}(a, b)$. Hence by steps, one can travel to $c$. As steps are compound moves, one can hop and skip to travel to $c$.

Therefore, for $ax + by = c$ to have an integer solution, it suffices that $c$ be a multiple of $\text{GCD}(a, b)$. ∎

The above theorem is an *existence* theorem – it claims the existence of a solution without actually providing it. But fortunately, the Euclidean algorithm can be used to construct solutions to equations of the form $ax + by = c$. To construct solutions, one must carefully track the compound moves throughout.

**Problem 1.15** Find integers $x$ and $y$ satisfying $71x + 83y = 2$.

SOLUTION: Apply the Euclidean algorithm, and name the moves along the way. Begin with $hop = 83$ and $skip = 71$.

$$83 = 1(71) + 12 \qquad 12 = jump = hop - skip$$
$$71 = 5(12) + 11 \qquad 11 = leap = skip - 5(jumps)$$
$$12 = 1(11) + 1 \qquad 1 = step = jump - leap.$$

Now, we express our *step* in terms of hops and skips.[11]

$$1 = 12 - 1(11) = 12 - 1(71 - 5(12))$$
$$= 6(12) - 1(71) = 6(83 - 1(71)) - 1(71)$$
$$= 6(83) - 7(71).$$

We find that $1 = 6(83) - 7(71)$. Doubling each side yields

$$2 = 12(83) - 14(71).$$

This exhibits a solution $x = -14$, $y = 12$, to the given Diophantine equation $71x + 83y = 2$. ✓

[10] A Diophantine equation is an equation involving integers, variables, and the elementary operations of addition, subtraction, and multiplication. Solving a Diophantine equation means finding *integer* solutions. A Diophantine equation is called *linear* if variables are never multiplied with variables. For example, $2x + 3y = 17$ is a linear Diophantine equation, whereas $2x^2 - 7xy + y^2 = 17$ is a quadratic (nonlinear) Diophantine equation.

Diophantine equations are named for the Greek mathematician Diophantus; we refer to the historical notes at the end of this chapter, and at the end of Chapter 9 for more on the history of these equations.

[11] It is very easy to make mistakes in this process. The process consists of substitutions (like changing a jump to a hop minus a skip) and tallying (like recognizing 1 jump plus 5 jumps equals 6 jumps). The key is following the steps of the Euclidean algorithm, from bottom to top, alternately substituting and tallying – it takes a bit of practice.

Three questions arise in the study of Diophantine equations. For linear Diophantine equations, and in order of increasing difficulty, these questions are

1. Do there exist integers $x$ and $y$ such that $ax + by = c$?

2. Find a pair of integers $(x, y)$ such that $ax + by = c$.

3. Find *all* pairs of integers $(x, y)$ such that $ax + by = c$.

The first has a quick answer: "yes" when $c$ is a multiple of $\text{GCD}(a, b)$ and "no" otherwise. The Euclidean algorithm quickly gives the GCD, and so answers the first question.

The second is a bit more difficult. To *find one* integer solution $(x, y)$, one performs the Euclidean algorithm, and carefully back-tracks, substituting and tallying, until a solution $(u, v)$ is found to the equaions $au + bv = \text{GCD}(a, b)$. Then one scales $(u, v)$ to find the solution to $ax + by = c$

The third question requires a study of common multiples. But when the numbers are small, the third question can sometimes be answered with a coarser and elementary method.

**Problem 1.16** Find all integer solutions to the equation $4x + 6y = 10$.

SOLUTION: One solution is easy to guess; when $x = 1$ and $y = 1$, we find that $4x + 6y = 10$. Now we attempt to find more solutions by incrementing $x$, solving for $y$, and tracking the results in a table. Note that if $4x + 6y = 10$, then $y = (10 - 4x)/6$.

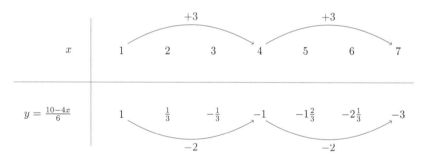

We find a new solution every time that $x$ increases by 3; this gives solutions when $x = 1, 4, 7, \ldots$. Each time $x$ increases by 3, $y$ decreases by 2. Thus, beginning from our solution $(x, y) = (1, 1)$, we may add $3n$ to $x$ and subtract $2n$ from $y$ to find a new solution:

$$x = 1 + 3n, \quad y = 1 - 2n$$

is a solution to the Diophantine equation for every integer $n$. These are all of the integer solutions. ✓

We find that there are an *infinite number* of solutions to the linear Diophantine equation $4x + 6y = 10$. To describe the infinite set of solutions, we need to introduce an auxiliary parameter – this is the integer $n$. For each integer value of $n$, we identify a solution.

THE LEAST COMMON MULTIPLE is defined below; the definition is a mirror image of the definition of greatest common divisor.

**Definition 1.17** Let $a$ and $b$ be integers. A natural number $\ell$ is called the[12] **least common multiple** of $a$ and $b$ if it satisfies both of the following.

1. $\ell$ is a **common multiple** of $a$ and $b$, i.e., $a \mid \ell$ and $b \mid \ell$.

2. If $m$ is a common multiple of $a$ and $b$, then $\ell \mid m$.

The school definition of least common multiple is that it is the smallest of the common multiples. Our stronger definition is that the least common multiple must divide every other common multiple. We write $\ell = \mathrm{LCM}(a,b)$ to mean that the least common multiple of $a$ and $b$ exists[13] and equals $\ell$.

**Lemma 1.18** *Let $a$ and $b$ be integers. If $m$ is a common multiple of $a$ and $b$, and $n$ is a common multiple of $a$ and $b$, then $\mathrm{GCD}(m,n)$ is a common multiple of $a$ and $b$.*

PROOF: Since $a \mid m$ and $a \mid n$, we find that $a \mid \mathrm{GCD}(m,n)$. Similarly, since $b \mid m$ and $b \mid n$, we find that $b \mid \mathrm{GCD}(m,n)$. Hence $\mathrm{GCD}(m,n)$ is a common multiple of $a$ and $b$. ∎

**Proposition 1.19** *Let $a$ and $b$ be nonzero integers. Let $\ell$ be the **smallest** positive integer which is a common multiple of $a$ and $b$. Then $\ell$ divides every common multiple of $a$ and $b$ and so $\ell = \mathrm{LCM}(a,b)$.*

PROOF: Let $m$ be a common multiple of $a$ and $b$. Since $\ell$ is a common multiple of $a$ and $b$, we find that $\mathrm{GCD}(\ell,m)$ is a common multiple of $a$ and $b$. Since $\mathrm{GCD}(\ell,m) \mid \ell$ and $\ell$ is a positive integer, we find that $\mathrm{GCD}(\ell,m) \leq \ell$. But since $\ell$ was the smallest positive common multiple of $a$ and $b$, we find that $\mathrm{GCD}(\ell,m) = \ell$. Therefore $\ell \mid m$. ∎

Two MANTRAS remind us about common multiples and divisors.

**Every common divisor is a divisor of the greatest common divisor.**

**Every common multiple is a multiple of the least common multiple.**

Now we use properties of the least common multiple to solve linear Diophantine equations in the strongest sense – finding *all* solutions.

[12] There can be only one least common multiple.

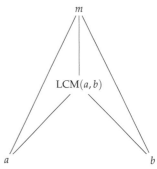

Figure 1.12: $a \mid \mathrm{LCM}(a,b)$ and $b \mid \mathrm{LCM}(a,b)$. If another number $m$ is a common multiple of $a$ and $b$, then (as indicated in red) $\mathrm{LCM}(a,b) \mid m$ as well.

[13] The existence of the least common multiple, in the strong sense, is proven in the proposition below.

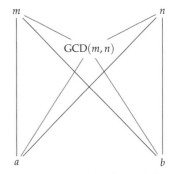

Figure 1.13: The hypothesis of the lemma is displayed in black. The conclusion of the lemma is displayed in red.

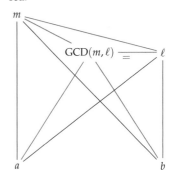

Figure 1.14: In the proof of the Theorem, we deduce that $\ell$ equals $\mathrm{GCD}(m,\ell)$ and thus $\ell \mid m$.

Consider a linear Diophantine equation $ax + by = c$. Suppose we have *two* solutions, meaning two ordered pairs $(x_1, y_1)$ and $(x_2, y_2)$:

$$(1) \qquad ax_1 + by_1 = c,$$
$$(2) \qquad ax_2 + by_2 = c.$$

Subtracting the second equation from the first yields

$$(1) - (2) \qquad a(x_1 - x_2) + b(y_1 - y_2) = 0.$$

Define $u = x_1 - x_2$ and $v = y_1 - y_2$. Then

$$(1) - (2) \qquad au + bv = 0.$$

This is called a **homogeneous**[14] linear Diophantine equation (in the variables $u$ and $v$). The key to solving homogeneous linear Diophantine equations is the least common multiple.

[14] A linear equation is called homogeneous if there are no nonzero constant terms; for example $2x + 3y = 7$ is not homogeneous, but the equation $2u + 3v = 0$ is homogeneous.

**Proposition 1.20** *Let $a$ and $b$ be integers. Then for every integer $n$, there is an integer solution $(u, v)$ of the equation $au + bv = 0$ given by*

$$u = n \cdot \frac{\text{LCM}(a,b)}{a}, \quad v = -n \cdot \frac{\text{LCM}(a,b)}{b}.$$

*All solutions of $au + bv = 0$ are obtained in this way.*

PROOF: If $au + bv = 0$, then $au = -bv$. The left side is evidently a multiple of $a$ and the right side is evidently a multiple of $b$; since they are equal, they equal a *common multiple* of $a$ and $b$. Every common multiple of $a$ and $b$ is a multiple of $\text{LCM}(a, b)$. Therefore, for some integer $n$,

$$n \cdot \text{LCM}(a, b) = au = -bv.$$

Dividing through by $a$ or by $-b$ yields

$$u = n \cdot \frac{\text{LCM}(a,b)}{a}, \quad v = -n \cdot \frac{\text{LCM}(a,b)}{b}.$$

From this we can solve the general linear Diophantine equation.

**Theorem 1.21 (General solution of linear Diophantine equations)**
*Let $a$, $b$, and $c$ be integers. Suppose[15] that $a \neq 0$ or $b \neq 0$. If $c$ is not a multiple of $\text{GCD}(a, b)$ then the equation $ax + by = c$ has no integer solutions. On the other hand, if $c$ is a multiple of $\text{GCD}(a, b)$ then the equation $ax + by = c$ has infinitely many solutions. Given one solution $(x_0, y_0)$, all solutions have the form*

$$x = x_0 + n \cdot \frac{\text{LCM}(a,b)}{a}, \quad y = y_0 - n \cdot \frac{\text{LCM}(a,b)}{b}.$$

[15] If $a = 0$ and $b = 0$, then the equation $ax + by = c$ is the equation $0x + 0y = c$; this technically has infinitely many solutions if $c = 0$ and no solutions otherwise.

PROOF: We have already seen that the equation $ax + by = c$ has a solution if and only if $c$ is a multiple of $\gcd(a,b)$; we are left to check that all solutions can be obtained by the formulae of the theorem.

So suppose that $(x_0, y_0)$ is one solution to the equation, so $ax_0 + by_0 = c$. Given integers $(x, y)$, we find that $ax + by = c$ if and only if[16]

$$a(x - x_0) + b(y - y_0) = 0.$$

Therefore $ax + by = c$ if and only if $u = (x - x_0)$ and $v = (y - y_0)$ are solutions to the homogeneous equation $au + bv = 0$. We know the solutions to the homogeneous equation by the previous proposition; thus $ax + by = c$ if and only if

$$x - x_0 = n \cdot \frac{\operatorname{LCM}(a,b)}{a}, \quad y - y_0 = -n \cdot \frac{\operatorname{LCM}(a,b)}{b}$$

for some integer $n$. Adding $x_0$ or $y_0$ to both sides yields

$$x = x_0 + n \cdot \frac{\operatorname{LCM}(a,b)}{a}, \quad y = y_0 - n \cdot \frac{\operatorname{LCM}(a,b)}{b}.$$

These are all of the solutions to the equation $ax + by = c$. ∎

[16] If $ax + by = c$, then subtracting $ax_0 + by_0 = c$ yields $a(x - x_0) + b(y - y_0) = 0$. Conversely, if $a(x - x_0) + b(y - y_0) = 0$, then adding $ax_0 + by_0 = c$ yields $ax + by = c$.

**Problem 1.22** Find all solutions to the equation $71x + 83y = 2$.

SOLUTION: In Problem 1.15, we found one solution; we call this solution $(x_0, y_0)$:

$$x_0 = -14, \quad y_0 = 12.$$

The least common multiple of 71 and 83 is 5893; this is *not obvious* but we will soon see how to compute it. Thus the general solution to the equation $71x + 83y = 2$ is given by:

$$x = -14 + n \cdot \frac{5893}{71}, \quad y = 12 - n \cdot \frac{5893}{83}.$$

We can simplify this further.

$$x = -14 + 83n, \quad y = 12 - 71n$$

is a solution for every integer $n$. ✓

To understand this better, consider what happens when $n$ increments from 0 to 1 to 2, etc.. At each step, $x$ is increased by 83 and $y$ is decreased by 71. So the quantity $71x + 83y$ is increased by $71 \times 83$ and decreased by $83 \times 71$. As we increment $n$, the quantity $71x + 83y$ does not change; if we begin at $n = 0$, with $x = x_0$ and $y = y_0$, then $71x + 83y = 2$ no matter how many times we increment $n$.

The previous problem required the computation of a least common multiple, LCM(71, 83). The general solution of linear Diophantine equations seems to require computation of an LCM. Fortunately, LCMs are not harder to compute than GCDs; these are relatively easy thanks to the Euclidean algorithm.

The octahedral Hasse diagram below displays the intricate divisibility relationships between $a$, $b$, GCD$(a,b)$ and LCM$(a,b)$, whenever $a$ and $b$ are positive integers.

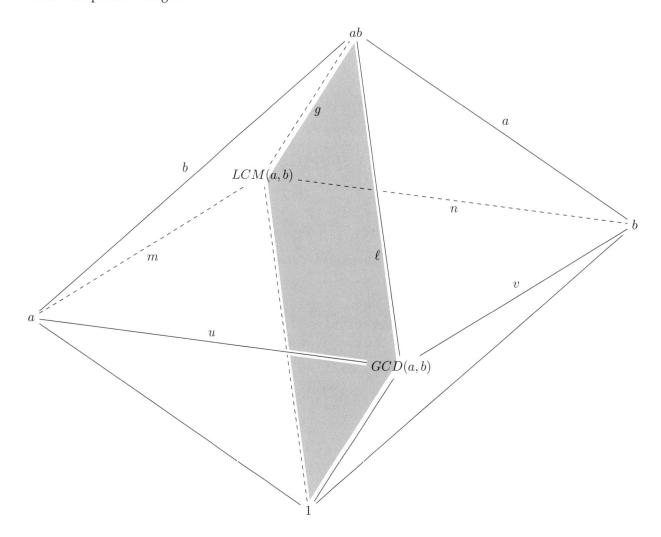

Figure 1.15: Divisibility relationships between $a$ and $b$, greatest common divisor and least common multiple.

The quantities $u$, $v$, $m$, $n$, $g$, and $\ell$ are defined by the properties

$$a = u \cdot \text{GCD}(a,b); \qquad b = v \cdot \text{GCD}(a,b);$$
$$\text{LCM}(a,b) = m \cdot a; \qquad \text{LCM}(a,b) = n \cdot b;$$
$$ab = g \cdot \text{LCM}(a,b). \qquad ab = \ell \cdot \text{GCD}(a,b).$$

**Theorem 1.23 (GCD-LCM product formula)** *Let a and b be positive integers. Then*
$$\mathrm{GCD}(a,b) \times \mathrm{LCM}(a,b) = a \times b.$$

PROOF: We refer to the diagram on the opposite page throughout. Every triangle in the diagram expresses a multiplication fact.

First, observe that $b = mg$ and $a = ng$. Hence $g$ is a common divisor of $a$ and $b$. It follows that

$$g \mid \mathrm{GCD}(a,b). \tag{1.1}$$

Second, observe that $\ell = av$ and $\ell = bu$. Hence $\ell$ is a common multiple of $a$ and $b$. It follows that

$$\mathrm{LCM}(a,b) \mid \ell. \tag{1.2}$$

∎

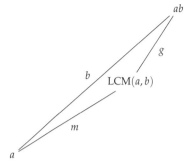

Figure 1.16: This triangle in the Hasse diagram demonstrates that $b = g \cdot m$.

Putting together the divisibilities of (1.1) and (1.2), we find

$$ab = g \cdot \mathrm{LCM}(a,b) \mid \mathrm{GCD}(a,b) \cdot \mathrm{LCM}(a,b) \mid \mathrm{GCD}(a,b) \cdot \ell = ab.$$

But since $ab = ab$, it follows by a squeeze[17] that all intermediate terms (being positive integers) are equal:

$$ab = g \cdot \mathrm{LCM}(a,b) = \mathrm{GCD}(a,b) \cdot \mathrm{LCM}(a,b) = \mathrm{GCD}(a,b) \cdot \ell = ab.$$

[17] Here is the squeeze argument. If one has a tower of divisibilities of positive integers, like $w \mid x \mid y \mid z$, and $w = z$, then we must have $w = x = y = z$. The equality $w = z$ leaves no room for $x$ and $y$ in between!

**Problem 1.24** Compute $\mathrm{LCM}(1711, 2059)$.

SOLUTION: The Euclidean algorithm, applied to 1711 and 2059, demonstrates that
$$\mathrm{GCD}(1711, 2059) = 29.$$

The previous theorem demonstrates that

$$29 \times \mathrm{LCM}(1711, 2059) = 1711 \times 2059.$$

Hence
$$\mathrm{LCM}(1711, 2059) = \frac{1711 \times 2059}{29} = 121481. \quad \checkmark$$

Theorem 1.23 is stated for *positive* integers $a$ and $b$ only. But for *all* integers $a$ and $b$, $\mathrm{LCM}(a,b) = \mathrm{LCM}(|a|, |b|)$, so we need not consider negative integers. Moreover, $\mathrm{LCM}(0, b) = 0$ for every integer $b$, so zeros do not present any difficulty.

With the ability to compute least common multiples, we have all of the tools we need to solve every linear Diophantine equation.

Theorem 1.23 states that $\text{GCD}(a,b) \times \text{LCM}(a,b) = ab$, when $a$ and $b$ are positive integers. From this, we deduce

$$\frac{\text{LCM}(a,b)}{a} = \frac{b}{\text{GCD}(a,b)}, \quad \frac{\text{LCM}(a,b)}{b} = \frac{a}{\text{GCD}(a,b)}. \quad (1.3)$$

This gives a practical simplification for solving linear Diophantine equations, a new version of Theorem 1.21.

**Corollary 1.25 (General solution of linear Diophantine equations)**
*Suppose that $(x_0, y_0)$ is a solution to the linear Diophantine equation $ax + by = c$. Then the general solution to this Diophantine equation is*

$$x = x_0 + n \cdot \frac{b}{\text{GCD}(a,b)}, \quad y = y_0 - n \cdot \frac{a}{\text{GCD}(a,b)}.$$

PROOF: Theorem 1.21 states the general solution as

$$x = x_0 + n \cdot \frac{\text{LCM}(a,b)}{a}, \quad y = y_0 - n \cdot \frac{\text{LCM}(a,b)}{b}.$$

Equations (1.3) give the result. ∎

**Problem 1.26** Find all solutions to the linear Diophantine equation

$$782x + 253y = 92.$$

SOLUTION: To begin we find the greatest common divisor $\text{GCD}(782, 253)$ with the Euclidean algorithm. At the same time, we name our moves; we call 782 a hop and 253 a skip.

$$782 = 3(253) + 23, \quad jump = 23 = hop - 3(skips).$$
$$253 = 11(23) + 0.$$

Thus $23 = \text{GCD}(782, 253)$. Since $23 \mid 92$, the Diophantine equation has a solution.[18] Since $92 = 4(23)$, we may find a solution by four jumps:

$$92 = 4(jumps) = 4(hops) - 12(skips) = 4(782) - 12(253).$$

This gives one solution to the equation $782x + 253y = 92$:

$$x_0 = 4, \quad y_0 = -12.$$

The general solution to the Diophantine equation is now given by

$$x = 4 + n \cdot \frac{253}{\text{GCD}(782, 253)}, \quad y = -12 - n \cdot \frac{782}{\text{GCD}(782, 253)}.$$

As $\text{GCD}(782, 253) = 23$, and $253/23 = 11$ and $782/23 = 34$, we find the general solution to the Diophantine equation:

$$x = 4 + 11n, \quad y = -12 - 34n. \quad \checkmark$$

[18] At this point, one could also divide through by 23. The equation

$$782x + 253y = 92$$

is equivalent to the equation

$$34x + 11y = 4$$

after dividing through by 23.

It is always good to check one's solution. To check, just plug into the original Diophantine equation; for every integer $n$ we find

$$782x + 253y$$
$$= 782(4 + 11n) + 253(-12 - 34n)$$
$$= 3128 + 8602n - 3036 - 8602n$$
$$= 3128 - 3036$$
$$= 92.$$

The following scaling property of GCDs and LCMs is useful computationally and theoretically in what follows.

**Proposition 1.27 (Scaling property of GCD and LCM)** *If $g$, $a$, and $b$ are integers, and $g > 0$, then*

$$\text{GCD}(ga, gb) = g \cdot \text{GCD}(a, b), \text{ and } \text{LCM}(ga, gb) = g \cdot \text{LCM}(a, b).$$

PROOF: When $a$ or $b$ is zero, one may check the result directly. Since $\text{GCD}(a, b) = \text{GCD}(|a|, |b|)$, we may assume $a$ and $b$ are positive in what follows.

To see the scaling property of the GCD, consider the Euclidean algorithm applied to $a, b$ alongside $ga, gb$.

$$a = qb + r \qquad ga = qgb + gr.$$

If $r$ is the (natural) remainder after dividing $a$ by $b$, then $gr$ is the remainder after dividing $ga$ by $gb$. Indeed, if $0 \leq r < b$ then $0 \leq gr < gb$. Thus $\text{GCD}(ga, gb) = \text{GCD}(gb, gr)$. In this way, every step of the Euclidean algorithm for $ga, gb$ is obtained by scaling a step of the Euclidean algorithm for $a, b$. Scaling the last nonzero remainders, we find that

$$g \cdot \text{GCD}(a, b) = \text{GCD}(ga, gb).$$

For the LCM, we apply Theorem 1.23 to find

$$\text{LCM}(ga, gb) = \frac{ga \cdot gb}{\text{GCD}(ga, gb)} = \frac{g^2 \cdot ab}{g \cdot \text{GCD}(a, b)} = g \cdot \text{LCM}(a, b).$$

Now, when computing GCDs and LCMs, one can pull out obvious common divisors.

**Problem 1.28** Compute $\text{GCD}(9100, 4900)$.

SOLUTION: Since both 9100 and 4900 have 100 as an obvious common divisor, we can use the Theorem above.

$$\text{GCD}(9100, 4900) = 100 \cdot \text{GCD}(91, 49).$$

With the Euclidean algorithm we find that $\text{GCD}(91, 49) = 7$. Hence $\text{GCD}(9100, 4900) = 700$. ✓

**Corollary 1.29** *If $a$ and $b$ are integers, not both zero,[19] and $g = \text{GCD}(a, b)$, then*

$$\text{GCD}(a/g, b/g) = 1.$$

PROOF: By the scaling property, we have

$$g \cdot \text{GCD}(a/g, b/g) = \text{GCD}(ga/g, gb/g) = \text{GCD}(a, b) = g.$$

Dividing both sides by $g$,

$$\text{GCD}(a/g, b/g) = 1.$$

$91 = 1(49) + 42$
$49 = 1(42) + 7$
$42 = 6(7) + 0.$

[19] If both $a = 0$ and $b = 0$, then $g = 0$ and we cannot divide by $g$.

## Historical Notes

THE EUCLIDEAN ALGORITHM is found in Propositions VII.1 and VII.2, of Euclid's *Elements*. Many editions of the *Elements*, including the edition from Constantinople (888CE) and the first printed edition (1482CE), include illustrations akin to our "hop and skip" diagrams.

We use the designation "Euclidean algorithm," despite historical problems. Euclid uses the word *anthyphairesis*[20] for the process of continual subtraction of lesser from greater – the process that forms the basis of his algorithm. David Fowler argues[21] that anthyphairesis played a central role in the concept of proportion, at least in the early years of Plato's Academy. One could use the term anthyphairesis instead of "the Euclidean algorithm" to avoid attributing the process to any particular individual.

Still, using the term anthyphairesis suggests that the process belongs solely to the Greek Mediterranean. But the process can be found in texts from China of nearly the same time period. In the *Suàn shù shū*[22] (c. 186BCE) we have mentioned the "method for simplifying parts" – a mutual subtraction algorithm equivalent to Euclid's algorithm, at least from the modern algebraic standpoint.

Also a product of Han dynasty China is the centerpiece of the Chinese mathematical canon[23]: the *Jiǔzhāng Suànshù*, translated as *Nine Chapters on the Mathematical Practice*, or simply the *Nine Chapters*. In Liu Hui's commentary (c. 263CE) to the *Nine Chapters*, we find a tutorial on fractions. Chapter 1, Problem 6 displays a ternary question-answer-method format. The question asks the reader to divide 91/49 (more literally, 91 split into 49). The answer is 13/7 (13 split into 7). The method described is anthyphairesis: to repeatedly subtract the smaller number from the larger, until "equal numbers" (*deng shu*) are obtained. We follow the suggested method below:

$$\frac{91}{49} \to \frac{42}{49} \to \frac{42}{7} \to \frac{35}{7} \to \frac{28}{7} \to \frac{21}{7} \to \frac{14}{7} \to \frac{7}{7}.$$

The number 7 is the "equal number"; in our language, $7 = \mathrm{GCD}(49, 91)$, and therefore one reduces 91/49 by dividing top and bottom by 7:

$$\frac{91}{49} = \frac{91 \div 7}{49 \div 7} = \frac{13}{7}.$$

LINEAR DIOPHANTINE EQUATIONS can be found scattered in many texts over the past 2000 years. Examples arise in the remainder problems of Sunzi's *suan jing* (c. 300CE), the *Aryabhatiya* of Āryabhata (499CE), tavern problems (Zechrechnen) in Germany (c. 1500), and elsewhere.

[20] In Proposition VII.1 of the *Elements* we find the word ἀνθυφαιρουμένου. Here ἀνθ is a prefix meaning "repeatedly", and ὑφαιρέω is a verb meaning "to diminish"

[21] See "Ratio in Early Greek Mathematics" by D. H. Fowler, in the *Bulletin of the American Mathematical Society* 1 **6** (1979).

[22] See "The Suán shù shū, 'Writings on reckoning': Rewriting the history of early Chinese mathematics in the light of an excavated manuscript," by C. Cullen, in *Historia Mathematica* **34** (2007).

[23] See "A Chinese canon in mathematics and its two Layers of commentaries: Reading a collection of texts as shaped by actors" by Karine Chemla, in *Looking at it from Asia*, Boston Studies in the Philosophy and History of Science, 2010, vol. 265, part 2, pp.169–210. The term *jing* refers to canonical texts in China, and the term was applied to the *Nine Chapters* – the second of *Ten Canons* described by Li Chunfeng (c. 656CE).

For more on the history of such "remainder problems," see "Modular arithmetic before C.F. Gauss: Systematizations and discussions on remainder problems in 18th century Germany," by Maarten Bullynck, *Historia Mathematica* **36** (2009) pp.48–72.

In the work of Bachet de Méziriac, whose 1621 translation and adaptation of Diophantus's *Arithmetica*[24] was famously annotated with Fermat's "Last Theorem", we find a systematic solution to linear Diophantine equations. A special case of our Theorem 1.14 is the following Proposition from Bachet's book of recreational math problems:

> Deux nombres premiers entre eux estant donnez, treuuer le moindre multiple de chascun d'iceux, surpassant de l'unité un multiple de l'autre.[25]

Translated, the Proposition begins:

> Given two numbers, prime to each other;[26] find a multiple of one number which exceeds by one a multiple of the other number.

In symbolic algebra, the passage studies the linear Diophantine equation $ax - by = 1$ whenever $\mathrm{GCD}(a,b) = 1$. About 150 years later, Étienne Bézout generalized Bachet's Proposition, to study the linear Diophantine equation $ax + by = c$ – not only in the context of integers but also in the context of polynomials. In his *Cours de Mathématiques* (1766) [27] Bézout writes algebra in notation much like ours today, with letters standing for numbers. Bézout begins his algebra text (p. 4) by telling the reader that the sum of $a$ with itself is $2a$, that the sum of $a$ and $b$ will be written $a + b$, etc.. By page 13, Bézout writes that $a^3 b^2 c$ is equivalent to $aaabbc$, and reminds his reader that $a^2$ should not be confused with $2a$. On page 118, we find the problem

> On demande en combien de manieres on peut 542 livres[28], en donnant des pieces de 17 liv. & recevant en échange des pieces de 11 livres.

This problem is expressed soon after as an algebraic equation; Bézout writes $17x - 11y = 542$. He notes that solving such equations – while demanding the solutions be positive integers – is not always so easy. But Bézout gives a solution, using a strategy of compound moves akin to the Euclidean algorithm.

The fundamental relationship between GCD and LCM is that $\mathrm{GCD}(a,b) \times \mathrm{LCM}(a,b) = ab$. This will be proven in the next chapter using prime decomposition. The proof using prime decomposition is easier, but by placing it in this chapter we have given a complete treatment of linear Diophantine equations. In addition, the proof given in this chapter exhibits the intricacy of divisibilities between $ab$, $a$, $b$, $\mathrm{GCD}(a,b)$, and $\mathrm{LCM}(a,b)$. The squeeze method occurs across fields of mathematics (e.g. in calculus, to prove that $\lim_{x \to 0} \sin(x)/x = 1$).

---

[24] The dating of Diophantus is debated, but to be conservative, it is certainly between 150 BCE and 320 CE. Wilbur Knorr places Diophantus at the time of Heron, around 80 CE; P. Tannery (1896) placed Diophantus later, around 240 CE.

[25] Proposition XVIII of "Problèmes plaisants et délectables, qui se font par les nombres," by Bachét de Méziriac, 1624. This is a later edition; the earlier edition from 1612 contains a similar variety of recreational mathematics problems, but fewer Theorems and Propositions in the introduction. Bachet includes his own Propositions in the introduction which he finds useful for solving the problems he gives later. The style of Bachet is similar to Greek works by Euclid and Diophantus, in sharp contrast to the later algebraic style of Bézout.

[26] Two numbers $x$ and $y$ are a said to be "prime to each other" if $\mathrm{GCD}(x,y) = 1$.

[27] See "Cours de mathématiques à l'usage des gardes du pavillon et de la marine. Troisième partie: L'algèbre" by E. Bézout, 1766

[28] The *livre* was the French currency, in the time of Bézout. A translation of this problem is found in the exercises.

Notably, Bézout places such equations in his algebra textbook after traditional linear equations, but before any study of quadratic equations and roots. Today, students do not encounter the Euclidean algorithm until long after they encounter the quadratic formula!

## Exercises

1. Perform the Euclidean algorithm, using positive remainders, and then using minimal (positive or negative) remainders, on the numbers 527 and 340.

2. Compute GCD(300, 111), GCD(289, 323), and GCD(3939, 10403).

3. Compute LCM(300, 111), LCM(289, 323), and LCM(3939, 10403).

4. If you can hop 289 units left and right and skip 323 units left and right, where can you travel (beginning at zero)?

5. If you can hop 35 units left and right, skip 55 units left and right, and jump 77 units left and right, where can you travel?

6. Find all solutions to the Diophantine equation $17x + 19y = 100$.

7. Find all solutions to the Diophantine equation $3939x + 10403y = 909$.

8. You have a bag of silver coins, each coin weighing 17 ounces of silver. A merchant has a bag of silver coins, each weighing 11 ounces. You wish to pay the merchant 542 ounces of silver. How many 17-ounce coins should you give the merchant, and how many 11-ounce coins should the merchant give to you?[29]

    What is the smallest amount of coins that can be exchanged in this transaction?

    [29] This is based on a problem of Étienne Bézout, from 1766. See the *Historical Notes* for a description.

9. Prove that if $n$ is an integer, then $GCD(n, n+1) = 1$.

10. Prove that if $a$ and $b$ are positive integers and $GCD(a,b) = LCM(a,b)$, then $a = b$.

11. Let $a$, $b$, and $c$ be positive integers. Let $g = GCD(a, GCD(b,c))$. Prove that $g$ satisfies the following two properties:

    (a) $g \mid a$ and $g \mid b$ and $g \mid c$.
    (b) If $d \mid a$ and $d \mid b$ and $d \mid c$, then $d \mid g$.

    (For this reason, it is fair to write $GCD(a,b,c)$ for $GCD(a, GCD(b,c))$ and call it the greatest common divisor of $a$, $b$, and $c$.)

12. Write and prove the statements analagous to the previous theorem, for *LCM* instead of *GCD*.

13. Describe all solutions to the Diophantine equation $15x + 35y + 21z = 1$.

14. Prove that $\text{GCD}(a,b) = 1$ if and only if $\text{GCD}(a^2, b^2) = 1$. Use this and the scaling property to prove that $\text{GCD}(u^2, v^2) = \text{GCD}(u,v)^2$ and $\text{LCM}(u^2, v^2) = \text{LCM}(u,v)^2$ for all integers $u, v$.

    Hint: Cube $ax + by = 1$. Compare to Euclid's *Elements*, VII.24 and VII.25.

15. Use the Euclidean algorithm to find $\text{GCD}(221, 85)$. Draw the Hasse diagram displaying all divisibilities among the numbers 1, 85, 221, $\text{GCD}(85, 221)$, $\text{LCM}(85, 221)$, and $85 \times 221$.

16. Perform the Euclidean algorithm, using only positive remainders, on two consecutive Fibonacci numbers[30] (e.g. 13 and 21 or 21 and 34). What do you notice? State and prove your observations about the remainders that arise.

    [30] The Fibonacci numbers begin with
    $$1, 1, 2, 3, 5, 8, 13, 21, 34, \ldots.$$
    The first two Fibonacci numbers are 1 and 1. Afterwards, each Fibonacci number is defined to be the sum of the two previous, e.g. $2 = 1 + 1$ and $3 = 1 + 2$ and $5 = 2 + 3$, etc..

17. Perform the Euclidean algorithm, using minimal (positive or negative) remainders, on two consecutive Fibonacci numbers. State and prove your observations.

18. Consider the diagrams below. Why can they *not* be the Hasse diagrams for *all* positive divisors of a number $N$?

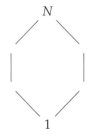
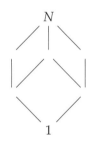

19. Let $N$ be a positive integer, and consider the Hasse diagram containing all positive divisors of $N$. Why does the Hasse diagram look (geometrically) the right-side-up as it does up-side-down?

20. (A tavern problem from 1544) A group of 20 persons, men, women and children, drink in a tavern. Together they spend 18 Thaler. The men drink for 3 Thaler a person, the women for 2 Thaler, and the children for half a Thaler. How many men, women, and children were in the drinking party?[31]

    [31] This translation of a 1544 exercise by Adam Riese is adapted from "Modular arithmetic before C.F. Gauss: Systematizations and discussions on remainder problems in 18th century Germany," by Maarten Bullynck, *Historia Mathematica* **36** (2009).

21. A tortoise and hare jog around a circular track. First the tortoise begins from the starting line; he jogs at a steady pace, completing one lap every 45 minutes. The hare begins from the starting line 30 minutes after the tortoise, travelling in the same direction; the hare completes one lap every 10 minutes. How often do the tortoise and hare cross paths?

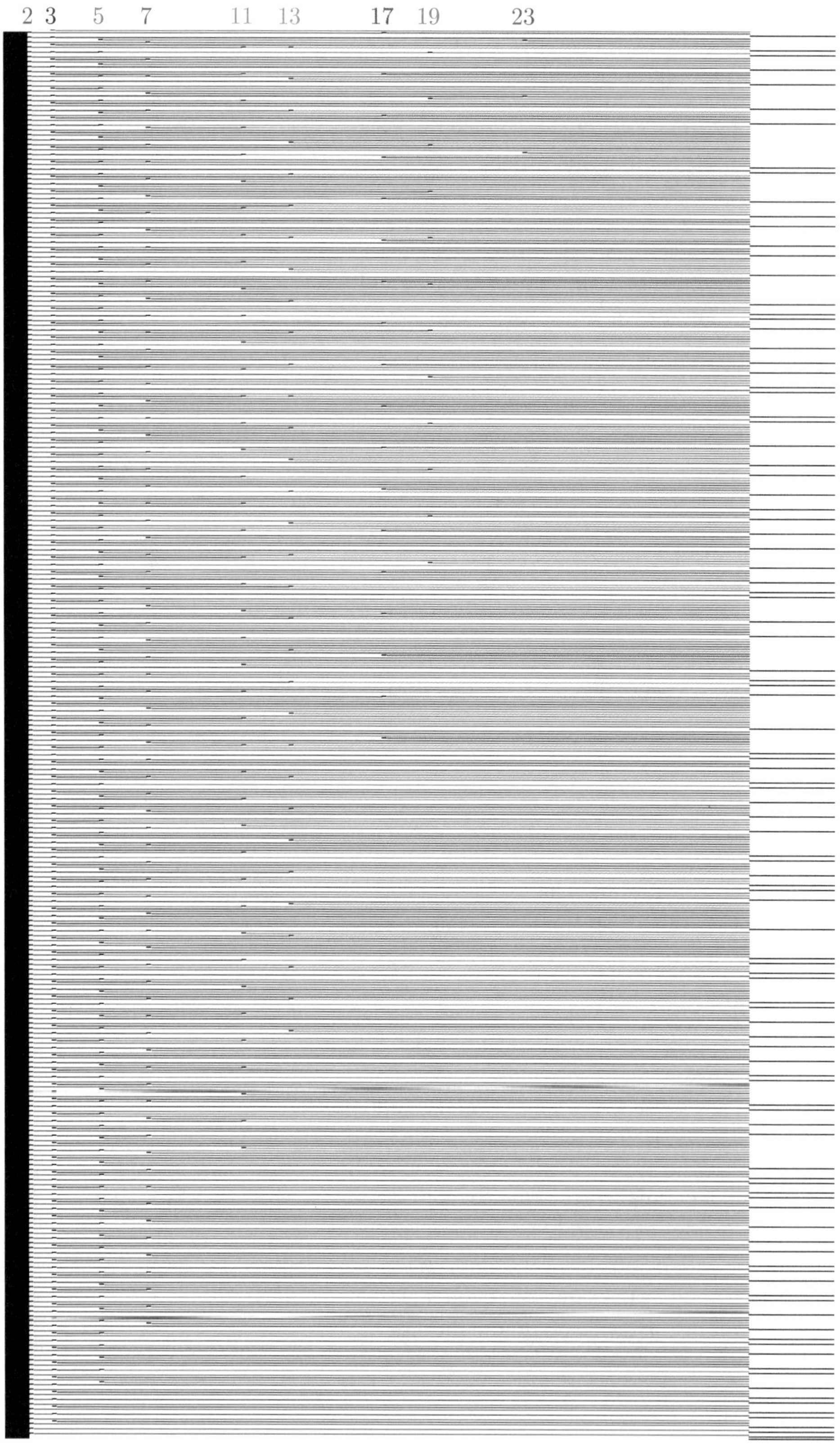

Figure 1.17: After multiples of 2, 3, 5, 7, 11, 13, 17, 19, and 23 are sieved, the prime numbers between 2 and 577 remain.

# 2
# *Prime Factorization*

PRIME NUMBERS are the building blocks of the positive integers, when the verb "building" is taken to mean "multiplying". All positive integers may be obtained by adding 1 to itself, over and over again: $2 = 1+1$ and $3 = 1+1+1$ and $4 = 1+1+1+1$, et cetera. But when multiplication is used instead of addition, the number 1 does not get us very far: $1 \cdot 1 = 1$ and $1 \cdot 1 \cdot 1 = 1$ and $1 \cdot 1 \cdot 1 \cdot 1 = 1$.

By multiplying prime numbers, one can build all positive integers. As[1] a molecule of water is built out of two hydrogen atoms and one oxygen atom, 12 is built (multiplied) out of two 2s and one 3.

$$\text{Water} = HOH, \quad 12 = 2 \cdot 3 \cdot 2.$$

From school experience, we might take prime factorization for granted. We learn factorization as a rote procedure, breaking numbers into smaller bits until primes are found. In this chapter we explain why this works, why factorization – a procedure with many paths – leads to a reliable result. Then we demonstrate how factorization can be used to solve some difficult problems.

On the opposite page, you can see the first 106 prime numbers as thin black lines. The first prime numbers are 2 and 3, adjacent black lines at the bottom. The $106^{\text{th}}$ prime number is 577, at the top of the stack. We produced the list by *sieving*; we filtered out multiples of 2, beginning with 4. Then we filtered out multiples of 3, beginning with 9. Then we filtered out multiples of 5, beginning with 25. We continued to filter, until all that remained were primes. This procedure, the **Sieve of Eratosthenes**, can produce lists of primes extremely quickly.

The Sieve of Eratosthenes can produce long lists of prime numbers. But questions about primes remain: How long a list can be obtained? How are the primes spread out among the numbers? How much space is between the primes? How can one determine whether a number is prime?

[1] The analogy between molecules and numbers is imperfect, since atoms can be configured in complicated geometric ways, while multiplication of primes does not depend on order or grouping: $12 = 2 \times 2 \times 3 = 2 \times 3 \times 2 = 3 \times 2 \times 2$ by the commutativity and associativity of multiplication.

PRIME NUMBERS are the atoms of number theory. A **prime number** is a positive integer, *not* equal to one, whose only positive factors are 1 and itself. Thus, when $p$ is a prime number, $1 \mid p$ and $p \mid p$, but no other positive integer divides $p$.

**Composite** numbers are positive integers, *not* equal to one, and which are *not* prime. So 4 is composite since $4 = 2 \times 2$ and 6 is composite since $6 = 2 \times 3$, and 120 is composite since $120 = 6 \times 20$.

**One** is its own sort of number. It is a **unit**, and we call it neither prime nor composite. It is special because its reciprocal is also an integer; no other positive integer has this property.

**Factoring** a number means expressing the number as a product of one² or more prime number(s). Every positive integer $n$, if it is greater than one, can be factored into prime numbers. Why? Either $n$ is prime and no factoring is necessary, or else $n$ is composite. If $n$ is composite, then $n$ has a positive factor $d$ besides 1 and $n$. Then we can break up $n$ as a product of smaller positive integers: $n = d \times e$.

Repeat the process on $d$ and $e$, et cetera, breaking numbers into smaller and smaller pieces until prime numbers are obtained. This process must end, using a fundamental property of positive integers: every decreasing sequence of positive integers must eventually terminate! The positive integer pieces cannot get smaller forever.

So when one is given a positive number, like 490, one can factor it like one learns in school.

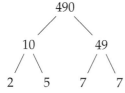 490 is the product $2 \times 5 \times 7 \times 7$.

[2] What is the product of one number? This is not a Zen question (what is the sound of one hand clapping?); by convention the product of one number $p$ is defined to be $p$ itself.

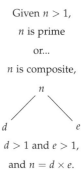

The numbers are factored like stones broken into pebbles. Sometimes numbers are cleft evenly and sometimes only a small factor is chipped off. There are many roads one can take to factorization.

490 is the product of $7 \times 7 \times 2 \times 5$.

It *should* be surprising that these two different roads lead to the same result – the same factors, each the same number of times. But the reader is probably not surprised, since years of experience remove one's healthy skepticism. But the proof of the *unique decomposition* into primes is difficult – it requires the Euclidean algorithm.

Consider only the numbers in the arithmetic progression $1, 4, 7, 10, 13, 16, \ldots$. Products of numbers in this progression stay within this progression. We may factor elements of this progression into smaller elements of this progression; for example, $16 = 4 \times 4$, but 10 is "prime" within this progression. But "prime factorization" in this system is not unique!

$$4 \times 25 = 10 \times 10.$$

PRIME NUMBERS are important because of their atomic role in factorization. So it is worthwhile to develop ways to recognize a prime number – these are called primality tests[3]. In this chapter, our best primality tests require some brute force.

**Problem 2.1** Is 599 a prime number?

SOLUTION: If 599 were composite, then $599 = d \times e$ for two positive integers $d$ and $e$, with $d, e > 1$. This implies that $d \leq \sqrt{599}$ or $e \leq \sqrt{599}$; indeed, if both were greater than $\sqrt{599}$ then $d \times e$ would be greater than $\sqrt{599} \times \sqrt{599} = 599$. But 599 is not greater than itself!

Hence if 599 were composite, then 599 would have a factor between 2 and $\lfloor\sqrt{599}\rfloor$. That factor, in turn, would have some *prime* factor $p$, which also would be a factor of 599.

Therefore, if 599 were composite, then 599 would have a *prime* factor between 2 and $\lfloor\sqrt{599}\rfloor$. Now, to compute $\lfloor\sqrt{599}\rfloor$, we note that $24 \times 24 = 576$ and $25 \times 25 = 625$. Hence $\lfloor\sqrt{599}\rfloor = 24$. If 599 were composite, then 599 would have a prime factor between 1 and 24. The prime numbers between 1 and 24 are

$$2, 3, 5, 7, 11, 13, 17, 19, \text{ and } 23.$$

- 2 does not go into 599 (the last digit is not even).
- 3 does not go into 599 (the sum of the digits is not divisible by 3).
- 5 does not go into 599 (the last digit is neither 0 nor 5).
- 7 does not go into 599.
- 11 does not go into 599 ($5 - 9 + 9$ is not a multiple of 11).[4]
- 13 does not go into 599.
- 17 does not go into 599.
- 19 does not go into 599.
- 23 does not go into 599.

By checking these nine facts, we find that 599 has no prime factor between 1 and 24. Hence 599 cannot be composite. 599 is prime. ✓

This solution demonstrates a general technique.

**Proposition 2.2** *Let $N$ be a positive integer. If $N$ is composite then $N$ has a prime factor between 2 and $\lfloor\sqrt{N}\rfloor$. Thus, if $N > 1$ and $N$ has **no** prime factors between 2 and $\lfloor\sqrt{N}\rfloor$ then $N$ must be prime.*

Our first primality test uses this theorem; to determine whether $N$ is prime, look for prime factors between 1 and $\lfloor\sqrt{N}\rfloor$. If you don't find any, $N$ is prime.

[3] Results of primality tests are sometimes called primality certificates! When we verify the primality of a number, we will issue a certificate like the one below:

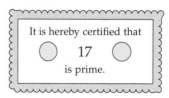

[4] The trick to check divisibility by 11 will be explained in a later chapter.

EUCLID, in Proposition IX.20 of the *Elements*, demonstrated the infinitude of primes. There is no biggest prime number[5]. Before presenting Euclid's demonstration, it is important to observe that the infinitude of the positive integers does *not* imply the infinitude of the prime numbers. Even with only one prime number, 2, one can build an infinite collection of counting numbers by multiplication:

$$2 = 2, \quad 4 = 2 \times 2, \quad 8 = 2 \times 2 \times 2, \quad 16 = 2 \times 2 \times 2 \times 2, \text{ etc..}$$

[5] This fact does not stop some people from competing to find larger and larger prime numbers. Read about the Great Internet Mersenne Prime Search at http://www.mersenne.org/

How can one prove the infinitude of any set? One way is to directly enumerate a set. Why are there infinitely many even positive numbers? We can put them into a never-ending list, indexed by *all* positive integers.

$$\begin{array}{cccc} 2 & 4 & 6 & 8 & \cdots \\ 1^{\text{st}} & 2^{\text{nd}} & 3^{\text{rd}} & 4^{\text{th}} & \cdots \end{array} \quad \text{Divide by two.}$$

So if one believes there are infinitely many positive integers, then there are infinitely many even positive integers. But this sort of list-making doesn't suffice for prime numbers since it is not clear how to find, for example, the millionth prime number. Identifying the millionth even number is easy – it's just 2 million. The millionth prime number is harder to find.

Euclid suggests a less direct method to prove the infinitude of the primes. He convinces his reader that, given a finite list of prime numbers, he can produce a prime not on that list. Hence there cannot be a finite and all-inclusive list of prime numbers.

**Theorem 2.3 (Infinitude of primes)** *There are infinitely many prime numbers.*

PROOF: Suppose you have a finite list of prime numbers. Multiply all the prime numbers in your list together, and call the result $N$. So $N$ is a positive integer. As $N + 1$ is again a positive integer, bigger than 1, $N + 1$ has a prime factor $p$.

Remember, you began with a list of prime numbers! If $p$ were in your list, then $p \mid N$ (since $N$ was the product of primes in your list) and $p \mid N + 1$ (we chose $p$ to be a factor of $N + 1$). By the two-out-of-three principle,

$$p \mid ((N+1) - N), \text{ so } p \mid 1.$$

But no prime number is a factor of 1! Hence the hypothesis ("if $p$ were in the list") cannot be true. The prime number $p$ is a "new prime", a prime not in the list.

Since no finite list of prime numbers contains all the prime numbers, there are infinitely many prime numbers. ∎

For example, imagine your list is $\{3, 5, 11\}$. Then

$$N = 3 \times 5 \times 11 = 165,$$
$$N + 1 = 166.$$

The prime factors of 166 are 2 and 83. Both of these are "new primes", primes not in the original list. So enlarge the list to $\{2, 3, 5, 11, 83\}$, including the new primes. Multiply these together:

$$N = 2 \times 3 \times 5 \times 11 \times 83 = 27390.$$
$$N + 1 = 27391.$$

The prime factors of 27391 are 7 and 7 and 13 and 43; these are "new primes". So enlarge the prime list to

$$\{2, 3, 5, 7, 11, 13, 43, 83\}.$$

Continue if you dare!

PRIME NUMBERS inspire feelings of reverence, colorful and musical metaphors, fiction books, not-so-fiction books, dictionaries, TV episodes, songs, and monetary prizes. Authors attach adjectives to prime numbers like "mysterious" and "solitary". What's the big deal? For mathematicians, one appeal of prime numbers is their remarkably precise macroscopic behavior and entirely unpredictable microscopic wobbling.

What do we know about prime numbers? What do number theorists wish they knew? Below are some highlights.

We know there are infinitely many prime numbers. So what about looking for prime numbers of special kinds? In fact, there are infinitely many prime numbers within the arithmetic progressions $1, 4, 7, 10, 13, 16, \ldots$ and $2, 5, 8, 11, 14, 17, \ldots$ and $1, 5, 9, 13, 17, 21, \ldots$ and $3, 7, 11, 15, 19, 23, \ldots$. In 1837, Dirichlet proved that there are infinitely many prime numbers within most arithmetic progressions:[6]

**Theorem 2.4 (Infinitude of primes in arithmetic progressions)** *If $a, b$ are positive integers, and $\mathrm{GCD}(a, b) = 1$, then there are infinitely many prime numbers in the progression*

$$a, a+b, a+2b, a+3b, \ldots.$$

The proof is beyond the scope of this text, though we will take care of a few specific cases of $a$, $b$, using variants of Euclid's method.

Instead of looking for primes within an arithmetic progression, one can look for an arithmetic progression within the primes! For example, $3, 5, 7$ is an arithmetic progression of primes, and so is $251, 257, 263, 269$ and so is $9843019, 9843049, 9843079, 9843109, 9843139$. These progressions have lengths 3,4,5, respectively. In 2008, Ben Green and Terence Tao proved[7]

**Theorem 2.5 (Green-Tao Theorem)** *The set of prime numbers contains infinitely many arithmetic progressions of length k, for all positive k.*

How about looking for prime numbers within other sequences? Take the sequence $1, 2, 5, 10, 17, 26, 37, \ldots$; these are the squares, plus one. It seems (experimentally) that there are an infinite number of primes therein, but nobody knows how to prove such a fact.

How about the sequence $3, 7, 15, 31, 63, 127, 255$; these are the powers of two, minus one. The primes in this sequence are called **Mersenne primes**. How many Mersenne primes are there? Nobody knows, though number theorists suspect an infinite number.

The next two pages display some statistical information about prime numbers, mostly macroscopic. How densely are the prime numbers packed within the integers? And how closely can we estimate the number of prime numbers in a range, for example between 1 and 5 million?

[6] The arithmetic progressions $0, 3, 6, 9, 12, \ldots$ and $2, 6, 10, 14, 18, \ldots$ cannot have an infinite number of primes! In the first progression, all the numbers are multiples of 3, so 3 is the only prime number in the first progression. In the second progression, all the numbers are even, so only 2 is prime. If $\mathrm{GCD}(a, b) \neq 1$, there cannot be infinitely many primes in the progression $a, a+b, a+2b, \ldots$.

[7] See "The primes contain arbitrarily long arithmetic progressions," in *Annals of Mathematics* **167** (2008).

**Conjecture 2.6** *There are infinitely many primes of the form $x^2 + 1$.*

**Conjecture 2.7** *There are infinitely many Mersenne primes (primes of the form $2^x - 1$).*

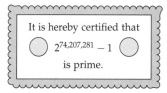

Figure 2.1: The largest known (in 2016) prime number is the monstrous Mersenne prime above.

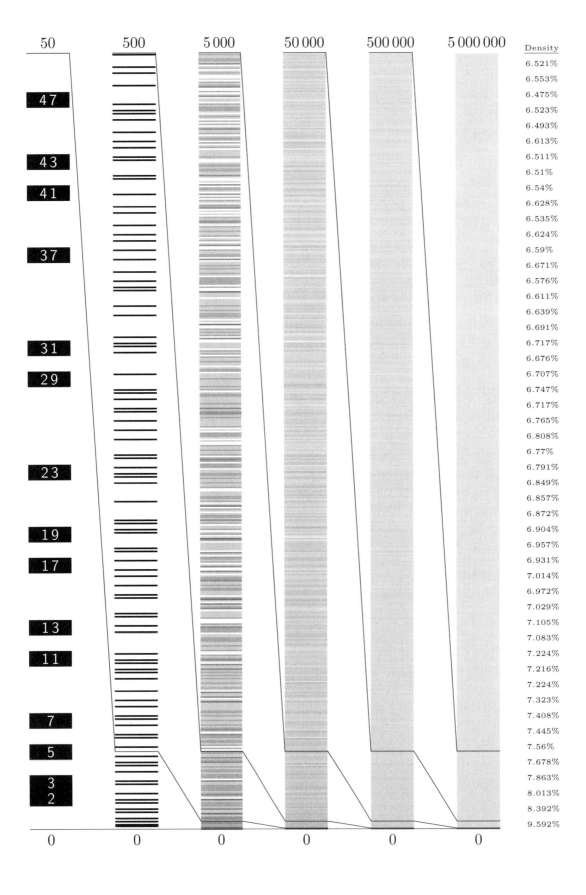

# PRIME FACTORIZATION

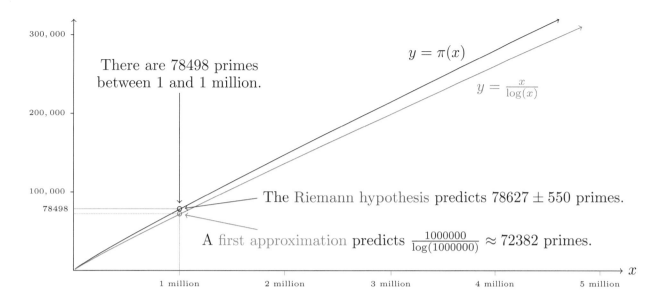

The opposite page displays the distribution of prime numbers, between 1 and 50, between 1 and 500, between 1 and 5000, etc.. In the leftmost column you can see individual primes, from 2 and 3, up to 47. In the next column, you can see the individual primes between 1 and 500 – sharp lines, more frequent at the bottom than top, otherwise spaced irregularly. Between 1 and 5000, we can no longer see the individuals, but some regions are denser than others. In the rightmost column, resolving individual primes is impossible, so density is indicated in ranges of 100 000. The primes gradually spread apart, viewed macroscopically.

Above, the black curve indicates the number of primes between 1 and $x$; this is called $\pi(x)$. The green curve is the approximation Gauss used when he was 15 years old: $\pi(x) \approx x/\log(x)$.

A much better approximation is given by the logarithmic integral:

$$\pi(x) \approx \mathrm{Li}(x) = \int_2^x \frac{1}{\log t} dt.$$

The BIG question in number theory is the **Riemann hypothesis**, stating that the approximation $\pi(x) \approx \mathrm{Li}(x)$ is very good, with error bounded by a constant times $\sqrt{x}\log(x)$.[8]

**Conjecture 2.8 (Riemann hypothesis, prime error-term formulation)**
*For all $x \geq 2657$,*

$$|\pi(x) - \mathrm{li}(x)| \leq \frac{\sqrt{x} \cdot \log(x)}{8\pi}.$$

The Riemann hypothesis very accurately predicts the macroscopic distribution of prime numbers.

Figure 2.2: Graphs of $y = \pi(x)$ (the number of primes between 1 and $x$) and the nearby graph $y = x/\log(x)$. Here and elsewhere, $\log(x)$ denotes the natural logarithm: $\log(x) = \int_1^x \frac{dt}{t}$.

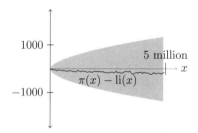

Figure 2.3: The black line plots the "error term" $\pi(x) - \mathrm{li}(x)$, which lies within the red region predicted by the Riemann hypothesis. Note the different scaling in the axes!

[8] The formulation we give here comes from "Sharper bounds for the Chebyshev functions $\theta(x)$ and $\psi(x)$," by L. Schoenfeld, *Mathematics of Computation*, vol. 30, **134** (1976). The functions $\mathrm{li}(x)$ and $\mathrm{Li}(x)$ differ only by a small constant, $\mathrm{li}(x) = \mathrm{Li}(x) + 1.045\ldots$. The common formulation involves the zeros of the Riemann zeta function.

The opposite page displays the distribution of **prime gaps**, for primes up to 1 million. The leftmost column exhibits the density of "twin primes" – pairs of primes like $(3,5)$ and $(5,7)$ and $(11,13)$ which differ by 2. As the column grows lighter from bottom to top, it appears that twin primes become more rare. There are 8169 twin prime pairs among the primes up to 1 million. The following conjecture is widely believed.

**Conjecture 2.9 (Twin prime conjecture)** *There are infinitely many twin primes. In other words, there are infinitely many prime numbers $p$ such that $p + 2$ is also prime.*

Stronger, and further out of reach, is the following conjecture.

**Conjecture 2.10** *Let $k$ be a positive even number. Then there are infinitely many prime numbers $p$ such that the next prime number is $p + k$.*

The best results in this direction were obtained recently (2013) by Yitang Zhang, and refined (the original constant was 70 million rather than 246) by a grand collaboration, giving the following result.

**Theorem 2.11 (Zhang-Maynard bounded gaps theorem)** *There exist infinitely many pairs of consecutive prime numbers $(p, q)$ such that $0 < q - p \leq 246$.*

Infinitude aside, it appears that some gaps are more frequent than others. Among the primes up to 1 million, gaps of 6 are most frequent, occurring among 13549 pairs of primes. Gaps of 18 and 24 and 30 seem more frequent than gaps of 20,22,26,28. In fact, it is expected that gaps of 30 eventually supersede gaps of 6 in frequency, if one looks at a *much* larger range of primes, and later 30 is superseded by 210, and 210 by 2310, etc..[9]

Red dots on the opposite page represent the trailblazers – the first prime pairs for each gap. For example, the first pair of consecutive primes with gap 100 is $(396733, 396833)$.

HEURISTICALLY, prime numbers are scattered as if every natural number $x$ has a probability $1/\log(x)$ of being prime.[10] But "probability" is a strange word to use – 6 never had a chance of being prime! – and so this is a heuristic rather than a statement about randomness. Still this heuristic supports the macroscopic distribution of primes we see. But it is not enough – such a heuristic does not predict the dominance of certain gaps (6, 30, etc.) over others. Deeper heuristics are needed to explain the distribution of prime gaps. And much deeper insights will be needed to prove conjectures like the twin prime conjecture.

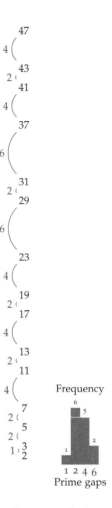

Figure 2.4: Frequency of prime gaps, among primes up to 50.

[9] This is the sequence of "primorials" – products of the first primes. $6 = 2 \cdot 3$ and $30 = 2 \cdot 3 \cdot 5$ and $210 = 2 \cdot 3 \cdot 5 \cdot 7$, etc.. The conjecture is due to Odlyzko, Rubinstein, and Wolf, "Jumping Champions," in *Experiment. Math.* **8** no. 2 (1999).

[10] This heuristic is sometimes called Cramér's model; Harald Cramér used this heuristic to make predictions about prime gaps. See "On the order of magnitude of the difference between consecutive prime numbers," *Acta Arithmetica* **2** (1936).

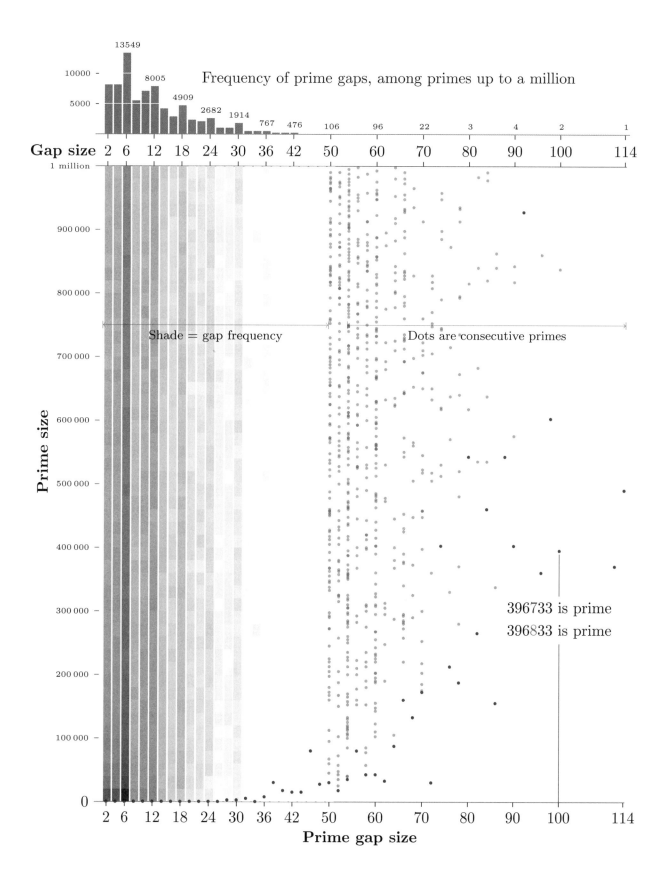

Having surveyed the landscape of prime numbers, we turn our attention to the way in which positive integers are built from primes. The **prime decomposition** of a positive integer is the result of factorization, ordering primes, and grouping. We carry out the prime decomposition of 490 here; its factorization depends on choices, but one result might be:

$$490 = 7 \cdot 5 \cdot 2 \cdot 7.$$

Ordering primes, we have $490 = 2 \cdot 5 \cdot 7 \cdot 7$. Grouping with exponents,

$$490 = 2 \cdot 5 \cdot 7^2.$$

This expression is the prime decomposition of 490, and we say that the primes 2, 5, and 7 occur. Every[11] positive integer $N$ has such a prime decomposition, an expression of the form

$$N = 2^{e_2} 3^{e_3} 5^{e_5} 7^{e_7} 11^{e_{11}} \cdots ,$$

where the exponents (the numbers $e_2$, $e_3$, $e_5$, etc.) are natural numbers which describe how many times each prime occurs. Finitely many of the exponents are nonzero, since factorization yields a *finite* product of prime numbers.

But why does factorization – a process with many possible paths – always lead to the same prime decomposition? Or, put another way, why can one number not have two prime decompositions?[12] This fact will be proven on the next page, following a few key results. The first depends fundamentally on the Euclidean algorithm.

**Lemma 2.12 (Euclid's Lemma)** *Let $a$, $b$, and $c$ be integers. If $a \mid bc$, and $GCD(a, b) = 1$, then $a \mid c$.*

PROOF: Since $GCD(a, b) = 1$, Theorem 1.14 implies that there exist integers $x$ and $y$ such that $ax + by = 1$. Multiplying both sides by $c$,

$$axc + byc = c.$$

Observe that $a \mid axc$, and $a \mid byc$ since $a \mid bc$. Therefore $a \mid c$. ∎

An important special case occurs when $a$ is a prime number.

**Corollary 2.13 (Euclid's Lemma for primes)** *Let $p$, $b$, and $c$ be integers, and suppose $p$ is a prime number. If $p \mid bc$ then $p \mid b$ or $p \mid c$.*

PROOF: Consider $GCD(p, b)$. It is a positive divisor of $p$, and since $p$ is prime, we find $GCD(p, b) = 1$ or $GCD(p, b) = p$. If $GCD(p, b) = p$, then $p = GCD(p, b) \mid b$, so $p \mid b$. If $GCD(p, b) = 1$, then Lemma 2.12 implies that $p \mid c$. ∎

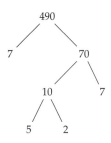

Figure 2.5: One plausible way to factor 490 into primes.

We say that 7 occurs twice in the prime decomposition of 490.

[11] Even 1 has a prime decomposition of this sort. Namely,

$$1 = 2^0 3^0 5^0 7^0 11^0 \cdots .$$

[12] The numbers $23, 73, 17, 97$ are prime. Could it be true that

$$23 \times 73 = 17 \times 97?$$

Why not? Why can a number *not* be factored into two primes in two different ways?

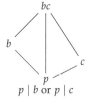

This corollary generalizes to longer products. For example, if $p$ is prime, and $a, b, c$ are integers, and $p \mid abc$, then we find that $p \mid a(bc)$. Hence $p \mid a$ or $p \mid bc$. In the latter case, $p \mid b$ or $p \mid c$. Hence $p \mid abc$ implies that $p \mid a$ or $p \mid b$ or $p \mid c$. In this way, we find a general result.

**Corollary 2.14** *If a prime number divides a product of integers, then $p$ divides at least one term in the product.*

We can now prove the uniqueness of prime decomposition.

**Theorem 2.15 (Uniqueness of prime decomposition)** *Suppose that $e_2, e_3, e_5, \ldots$ and $f_2, f_3, f_5, \ldots$ are natural numbers, and only finitely many of them are nonzero. If*

$$2^{e_2} 3^{e_3} 5^{e_5} 7^{e_7} \cdots p^{e_p} \cdots = 2^{f_2} 3^{f_3} 5^{f_5} 7^{f_5} \cdots p^{f_p} \cdots \quad (2.1)$$

*then the exponents are equal:*

$$e_2 = f_2, \quad e_3 = f_3, \quad e_5 = f_5, \quad e_7 = f_7, \ldots.$$

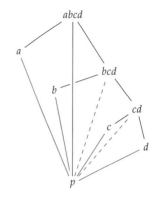

Figure 2.6: If $p \mid abcd$ then $p \mid a$ or $p \mid (bcd)$; in the latter case, $p \mid b$ or $p \mid cd$; in the latter case $p \mid c$ or $p \mid d$.

PROOF: Suppose that $N$ is a positive integer, and

$$N = 2^{e_2} 3^{e_3} 5^{e_5} 7^{e_7} \cdots p^{e_p} \cdots = 2^{f_2} 3^{f_3} 5^{f_5} 7^{f_5} \cdots p^{f_p} \cdots \quad (2.2)$$

Let $p$ be a prime. We wish to show that $e_p = f_p$. As $e_p \leq f_p$ or $f_p \leq e_p$, it suffices by symmetry to consider the case with $e_p \leq f_p$. In this case we may define a natural number $d_p = f_p - e_p$, and consider the positive integer $N/p^{e_p}$ with prime decompositions.

$$N/p^{e_p} = 2^{e_2} 3^{e_3} 5^{e_5} 7^{e_7} \cdots p^0 \cdots = 2^{f_2} 3^{f_3} 5^{f_5} 7^{f_5} \cdots p^{d_p} \cdots. \quad (2.3)$$

If $d_p$ were positive, then $p$ would divide the right side of (2.3), hence would divide the left side of (2.3). By Corollary 2.14, $p$ would divide a term (a prime!) on the left side. Since primes divide primes only when they are equal, we find a contradiction – the prime $p$ does not occur on the left side as its exponent is zero.

Hence $d_p = 0$. Since $d_p = f_p - e_p$, we find that $e_p = f_p$. ∎

We now have the *existence* and *uniqueness* of prime decomposition. If $N$ is a positive integer, then there *exist unique* natural numbers $e_2, e_3, e_5, e_7, \ldots$ such that

$$N = 2^{e_2} 3^{e_3} 5^{e_5} 7^{e_7} \cdots.$$

We began this chapter by introducing prime number as the building blocks of the positive integers. The existence and uniqueness of prime decomposition makes this building-block metaphor precise.

PRIME DECOMPOSITION turns multiplication problems into addition problems.

**Problem 2.16** What is the prime decomposition of $2100 \times 490$?

SOLUTION: The prime decompositions[13] of these numbers are:

$$2100 = 2^2 3^1 5^2 7^1 \text{ and } 490 = 2^1 5^1 7^2.$$

[13] Keep in mind that primes not occurring have implicit exponent zero.

To find the prime decomposition of their product, add the exponents.

$$2100 \times 490 = 2^{2+1} 3^{1+0} 5^{2+1} 7^{1+2} = 2^3 3^1 5^3 7^3.$$

On the other hand, prime decompositions are not particularly useful if you need to add numbers. The prime decomposition of $15 + 77$ bears no discernable[14] relation to the prime decomposition of 15 and that of 77:

$$15 = 3^1 5^1, \quad 77 = 7^1 11^1, \quad 92 = 2^2 23^1.$$

[14] The *abc-conjecture* suggests a subtle statistical relationship between the prime decompositions of coprime integers $a, b$ and that of their sum $c$. For an expositon on this conjecture and its implications, see "The ABC's of Number Theory," by N. Elkies, in the *Harvard College Mathematics Review*, vol. 1, 1 (Spring 2007), pp. 57–76.

What if you add two prime numbers? If $p$ and $q$ are prime, and $p \neq 2$ and $q \neq 2$, then $p + q$ is an even number bigger than 4. What even numbers arise as sums of primes? Some examples are below.

$$4 = 2+2, \quad 6 = 3+3, \quad 8 = 3+5, \quad 10 = 5+5 = 3+7, \quad 12 = 5+7, \ldots.$$

In correspondence with Euler, Goldbach conjectured the following.

**Conjecture 2.17 (Goldbach's conjecture)** *Every even number starting with* 4 *can be expressed as the sum of two prime numbers.*

This conjecture seems far out of reach. But in 1973, Chen Jingrun proved[15] the following partial result.

**Theorem 2.18 (Chen's theorem)** *There exists a positive number N such that if $n > N$ and $n$ is even, then $n$ can be expressed as a sum $n = p + s$ in which $p$ is prime and $s$ is a prime or a product of two primes.*

[15] See "On the representation of a larger even integer as the sum of a prime and the product of at most two primes" in *Sci. Sinica* **16** (1973).

Harald Helfgott has recently (2013, 2015) announced a proof[16] of another variant of Goldbach's conjecture.

[16] See "The ternary Goldbach conjecture is true", on the ArXiv at https://arxiv.org/abs/1501.05438.

**Theorem 2.19 (Ternary Goldbach)** *Every odd number starting at 7 can be expressed as the sum of three primes.*

Despite the deep conjectures and theorems about sums of prime numbers, the primes do not seem tailored for addition. They are the multiplicative atoms, and so we focus on multiplication here.

PRIME DECOMPOSITION is convenient, when the only operation we care about is multplication. When we know the prime decomposition of two numbers, the prime decomposition of their *product* is found by *adding* exponents. It follows that prime decomposition turns statements about divisibility into statements about inequality.

**Proposition 2.20** *Let a and b be positive integers with prime decompositions:*

$$a = 2^{e_2} 3^{e_3} 5^{e_5} \cdots,$$
$$b = 2^{f_2} 3^{f_3} 5^{f_5} \cdots.$$

*Then $a \mid b$ if and only if $e_p \leq f_p$ for every prime number $p$.*

PROOF: By definition (and positivity), $a \mid b$ if and only if there is a positive integer $c$ satisfying $b = a \cdot c$. Such a positive integer $c$ has a prime decomposition also:

$$c = 2^{g_2} 3^{g_3} 5^{g_5} \cdots.$$

We find that $a \mid b$ if and only if there are natural numbers $g_2, g_3, g_5,$ etc., such that

$$(2^{e_2} 3^{e_3} 5^{e_5} \cdots) \cdot (2^{g_2} 3^{g_3} 5^{g_5} \cdots) = 2^{f_2} 3^{f_3} 5^{f_5} \cdots$$

But by the uniqueness of prime decomposition, Theorem 2.15, this occurs if and only if the exponents of every prime number on the left side equal the exponents of every prime number on the right side:

$$e_2 + g_2 = f_2, \quad e_3 + g_3 = f_3, \quad e_5 + g_5 = f_5, \ldots.$$

Finding a positive integer $c$, for which $b = a \cdot c$, is equivalent to finding natural numbers $g_2, g_3, g_5$, etc., which satisfy the above equations. Finding such natural numbers $g_2, g_3, g_5$, etc., is possible if and only if $e_2 \leq f_2$, $e_3 \leq f_3$, $e_5 \leq f_5$, etc..

Hence $a \mid b$ if and only if $e_p \leq f_p$ for every prime number $p$. ∎

For example, the fact that 30 divides 2340 reflects the fact that all of the exponents in the prime decomposition of 30 are less than or equal to the corresponding exponents in the decomposition of 2340.

| Prime $p$ | 2 | 3 | 5 | 7 | 11 | 13 | 17 | $\cdots$ |
|---:|---|---|---|---|---|---|---|---|
| Exponent of $p$ in 30 | 1 | 1 | 1 | 0 | 0 | 0 | 0 | $\cdots$ |
| Exponent of $p$ in 2340 | 2 | 2 | 1 | 0 | 0 | 1 | 0 | $\cdots$ |

The prime decompositions of 30 and 2340 are given by

$$30 = 2^1 3^1 5^1 7^0 11^0 13^0 17^0 \cdots,$$
$$2340 = 2^2 3^2 5^1 7^0 11^0 13^1 17^0 \cdots.$$

**Problem 2.21** How many positive divisors does 2100 have?

SOLUTION: To count the divisors of 2100, one could tediously search through the numbers 1, 2, 3, all the way up to 2100. But instead, recall the prime decomposition:

$$2100 = 2^2 3^1 5^2 7^1.$$

If $d$ is a divisor of 2100, then the prime decomposition of $d$ is

$$d = 2^{e_2} 3^{e_3} 5^{e_5} 7^{e_7} \cdots,$$

where, by the previous theorem,

$$e_2 \leq 2, \quad e_3 \leq 1, \quad e_5 \leq 2, \quad e_7 \leq 1,$$

and all other exponents must be zero. Conversely, every choice[17] of natural numbers $e_2$, $e_3$, $e_5$, $e_7$, satisfying these inequalities, yields a divisor of 2100.

[17] For example, choosing $e_2 = 1$ and $e_3 = 0$ and $e_5 = 2$ and $e_7 = 1$ yields the divisor
$$2^1 3^0 5^2 7^1 = 350.$$

As the exponents are natural numbers, the possible exponents are:

$$e_2 \in \{0, 1, 2\}, \quad e_3 \in \{0, 1\}, \quad e_5 \in \{0, 1, 2\}, \quad e_7 \in \{0, 1\}.$$

Since there are 3 choices for $e_2$, 2 choices for $e_3$, 3 choices for $e_5$, and 2 choices for $e_7$, there are $3 \times 2 \times 3 \times 2 = 36$ divisors of 2100. ✓

When $N$ is a positive integer, we write $\sigma_0(N)$ to denote the number of positive divisors of $N$. So $\sigma_0(2100) = 36$ by the previous exercise. Prime decomposition yields a general formula for $\sigma_0(N)$.

**Corollary 2.22** *Let $N$ be a positive integer. For each prime number $p$, let $e_p$ denote the exponent of $p$ in the prime decomposition of $N$. Then*[18]

$$\sigma_0(N) = (e_2 + 1) \cdot (e_3 + 1) \cdot (e_5 + 1) \cdot (e_7 + 1) \cdots.$$

[18] This is not an infinite product. Only a finite number of the exponents $e_p$ are nonzero; hence all but finitely many of the quantities $(e_p + 1)$ equal one. We only multiply the quantities $(e_p + 1)$ when $e_p \neq 0$ and $e_p + 1 \neq 1$.

PROOF: To choose a divisor of $N$, it is equivalent to choose a power of 2 between 0 and $e_2$, to choose a power of 3 between 0 and $e_3$, a power of 5 between 0 and $e_5$, etc.. The number of choices equals the product of $(e_2 + 1)$, $(e_3 + 1)$, $(e_5 + 1)$, etc.. ∎

**Problem 2.23** How many divisors does 1 000 000 have?

SOLUTION: Note that $1\,000\,000 = 10^6$ and $10 = 2 \times 5$, so the prime decomposition of one million is

$$1\,000\,000 = 2^6 5^6.$$

We can count the divisors of one million using the previous theorem:

$$\sigma_0(1\,000\,000) = (6+1) \cdot (6+1) = 7 \times 7 = 49.$$

The number 1 000 000 has exactly 49 positive divisors. ✓

EVERY SENTENCE about divisibility ($\mid$) may be rephrased as a sentence about inequality ($\leq$), after prime decomposition. Let $a$ and $b$ be positive integers with prime decompositions

$$a = 2^{e_2}3^{e_3}5^{e_5}7^{e_7}\cdots, \quad b = 2^{f_2}3^{f_3}5^{f_5}7^{f_7}\cdots.$$

To say $a \mid b$, it is equivalent to say that $e_p \leq f_p$ for every prime number $p$. A consequence is the following.

**Corollary 2.24** $\text{GCD}(a,b)$ *has the prime decomposition*

$$\text{GCD}(a,b) = 2^{\min(e_2,f_2)}3^{\min(e_3,f_3)}5^{\min(e_5,f_5)}7^{\min(e_7,f_7)}\cdots.$$

*Similarly,* $\text{LCM}(a,b)$ *has the prime decomposition*

$$\text{LCM}(a,b) = 2^{\max(e_2,f_2)}3^{\max(e_3,f_3)}5^{\max(e_5,f_5)}7^{\max(e_7,f_7)}\cdots.$$

PROOF: The greatest common divisor of $a$ and $b$ is characterized by

1. $\text{GCD}(a,b) \mid a$ and $\text{GCD}(a,b) \mid b$;

2. if $d$ is a positive integer and $d \mid a$ and $d \mid b$, then $d \mid \text{GCD}(a,b)$.

Let $g_p$ (for $p$ prime) denote the exponents in the prime decomposition of $\text{GCD}(a,b)$. Let $v_p$ (for $p$ prime) denote the exponents in the prime decomposition of a positive integer $d$. Then the above two properties can be rephrased as

1. $g_p \leq e_p$ and $g_p \leq f_p$ for every prime number $p$;

2. if $v_p \leq e_p$ and $v_p \leq f_p$ then $v_p \leq g_p$, for every prime number $p$.

But these two properties precisely characterize $g_p$ as the *minimum* of $e_p$ and $f_p$ – the first says $g_p$ is less than or equal to $e_p$ and $f_p$, and the second says $g_p$ is as large as possible among numbers less than equal to $e_p$ and $f_p$. Hence $g_p = \min(e_p,f_p)$, for all prime numbers $p$.

The proof for LCM is the same as the proof for GCD, swapping inequalities and changing min to max when appropriate. ∎

**Corollary 2.25** *If* $\text{GCD}(a,b) = 1$ *then* $\text{GCD}(a^m,b^n) = 1$ *for all positive integers* $m, n$.

PROOF: Since $\text{GCD}(a,b) = 1$, we have $\min(e_p,f_p) = 0$ for all primes $p$. Hence $\min(m \cdot e_p, n \cdot f_p) = 0$ for all primes $p$. But $m \cdot e_p$ and $n \cdot f_p$ are the exponents of $p$ in $a^m$ and $b^n$, respectively; therefore $\text{GCD}(a^m,b^n) = 1$. ∎

Every property of min and max can be translated to a property of GCD and LCM. For example, min and max have the property $\min(e,f) + \max(e,f) = e + f$. This gives a new proof of Theorem 1.23: $\text{GCD}(a,b) \times \text{LCM}(a,b) = a \times b$.

---

In this result, we use the **minimum** $\min(e,f)$ and **maximum** $\max(e,f)$ of natural numbers $e$ and $f$. These are nearly self-explanatory. $\min(e,f)$ denotes either $e$ or $f$, whichever is smaller. If $e = f$, then we just define $\min(e,f)$ to be equal to their common value; for example, $\min(3,5) = 3$ and $\min(6,6) = 6$. The maximum is defined similarly, so $\max(3,5) = 5$ and $\max(6,6) = 6$.

$$\text{GCD}(a,b) = 2^{g_2}3^{g_3}5^{g_5}\cdots,$$
$$d = 2^{v_2}3^{v_3}5^{v_5}\cdots.$$

Another proof is the following: if $\text{GCD}(a,b) = 1$, then $a$ and $b$ share no common prime factor. Hence $a^m$ and $b^n$ share no common prime factor; therefore $\text{GCD}(a^m,b^n) = 1$ as well.

**Problem 2.26** Tabulate the prime decompositions of 30, 78, GCD(30, 78), LCM(30, 78) and 30 × 78.

SOLUTION: Their prime decompositions are tabulated below.

| Prime $p$ | 2 | 3 | 5 | 7 | 11 | 13 | 17 | $\cdots$ | $p$ |
|---|---|---|---|---|---|---|---|---|---|
| Exponent of $p$ in 30 | 1 | 1 | 1 | 0 | 0 | 0 | 0 | $\cdots$ | $e_p$ |
| Exponent of $p$ in 78 | 1 | 1 | 0 | 0 | 0 | 1 | 0 | $\cdots$ | $f_p$ |
| Exponent of $p$ in 6 = GCD(30, 78) | 1 | 1 | 0 | 0 | 0 | 0 | 0 | $\cdots$ | $\min(e_p, f_p)$ |
| Exponent of $p$ in 390 = LCM(30, 78) | 1 | 1 | 1 | 0 | 0 | 1 | 0 | $\cdots$ | $\max(e_p, f_p)$ |
| Exponent of $p$ in 2340 = 30 × 78 | 2 | 2 | 1 | 0 | 0 | 1 | 0 | $\cdots$ | $e_p + f_p$ |

When two integers $a$ and $b$ satisfy $\text{GCD}(a, b) = 1$, they are called **relatively prime**, or **coprime**, for short. For example, 147 and 202 are coprime. In fact, most pairs of positive integers are coprime.

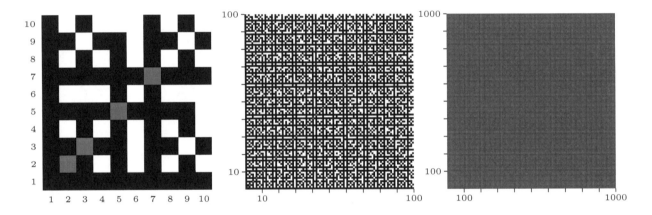

The black squares in these figures have coprime coordinates. When $p$ is a prime number, the coordinate $(p, p)$ along the diagonal is highlighted with a red square; note in this case $\text{GCD}(a, p) = 1$ if $1 \leq a < p$. Like the prime numbers, the coprime pairs exhibit macroscopic regularity. As we zoom out, the uneven patterns of black and white squares become a more uniform shade of gray. The shade is about 60% black, 40% white, according to the following theorem.[19]

**Theorem 2.27 (Probability of coprimality)** *Let $N$ be a positive integer, and let $C(N)$ be the probability that two integers $a$ and $b$, chosen independently and at random from $\{1, \ldots, N\}$, are coprime. Then*

$$\lim_{N \to \infty} C(N) = \frac{6}{\pi^2} \approx 0.6079.$$

*In this sense, a pair of randomly chosen positive integers is coprime with probability about 60.79%*

Figure 2.7: Coprime pairs, shaded in black, at three different scales.

[19] The proof of this theorem is beyond the scope of this text. For a sketch, note that two numbers are coprime if they do not have 2 has a common factor, they do not have 3 as a common factor, they do not have 5 as a common factor, etc. (for each prime $p$). The probability that two numbers do not have 2 as a common factor is 3/4. The probability that they do not have 3 as a common factor is 8/9. Etc.. These events are independent, and so the probability that two numbers are coprime is the product

$$\frac{3}{4} \cdot \frac{8}{9} \cdot \frac{24}{25} \cdot \frac{48}{49} \cdots.$$

In 1737, Euler discovered that this product equals $6/\pi^2$.

When $a$ and $b$ are coprime, the divisors of $ab$ come from multiplying divisors of $a$ with divisors of $b$. For example, consider $ab = 2100$, $a = 21$, and $b = 100$. The multiplication table below generators the divisors of $ab$ from those of $a$ and those of $b$.

|  |  | \multicolumn{4}{c}{Divisors of 21} |  |  |  |
|---|---|---|---|---|---|
|  | × | $3^0 7^0 = 1$ | $3^1 7^0 = 3$ | $3^0 7^1 = 7$ | $3^1 7^1 = 21$ |
|  | $2^0 5^0 = 1$ | 1 | 3 | 7 | 21 |
|  | $2^1 5^0 = 2$ | 2 | 6 | 14 | 42 |
|  | $2^2 5^0 = 4$ | 4 | 12 | 28 | 84 |
| Divisors | $2^0 5^1 = 5$ | 5 | 15 | 35 | 105 |
| of | $2^1 5^1 = 10$ | 10 | 30 | 70 | 210 |
| 100 | $2^2 5^1 = 20$ | 20 | 60 | 140 | 420 |
|  | $2^0 5^2 = 25$ | 25 | 75 | 175 | 525 |
|  | $2^1 5^2 = 50$ | 50 | 150 | 350 | 1050 |
|  | $2^2 5^2 = 100$ | 100 | 300 | 700 | 2100 |

Table 2.1: Divisors of 2100 arise from divisors of 100 and divisors of 21. It follows that

$$\sigma_0(2100) = \sigma_0(100) \cdot \sigma_0(21).$$

**Theorem 2.28** *Let $a$ and $b$ be coprime positive integers. If $u \mid a$ and $v \mid b$, then $uv \mid ab$. This gives a one-to-one correspondence between divisors of $ab$ and ordered pairs $(u,v)$ consisting of a divisor of $a$ and a divisor of $b$.*

PROOF: For the first assertion, note that if $a = um$ and $b = vn$ then $ab = (mn)(uv)$. To see that this gives a one-to-one correspondence, consider the prime decompositions of $a$ and $b$. Since $\text{GCD}(a,b) = 1$, they have no primes in common! Thus we can write

$$a = p_1^{e_1} \cdots p_s^{e_s} \text{ and } b = q_1^{f_1} \cdots q_t^{f_t}$$

for disjoint lists of prime numbers $\{p_1, \ldots, p_s\}$ and $\{q_1, \ldots, q_t\}$. The prime decomposition of $ab$ is then

$$ab = p_1^{e_1} \cdots p_s^{e_s} \cdot q_1^{f_1} \cdots q_t^{f_t}.$$

Giving a divisor $w \mid ab$ is the same as giving exponents $e'_i \leq e_i$ and $f'_j \leq f_j$ for all $1 \leq i \leq s$ and all $1 \leq j \leq t$, so that

$$w = p_1^{e'_1} \cdots p_s^{e'_s} \cdot q_1^{f'_1} \cdots q_t^{f'_t} \mid ab.$$

This is the same, in turn, as giving a pair of divisors

$$u = p_1^{e'_1} \cdots p_s^{e'_s} \mid a \text{ and } v = q_1^{f'_1} \cdots q_t^{f'_t} \mid b.$$

**Corollary 2.29** *If $a$ and $b$ are coprime positive integers, then $\sigma_0(ab) = \sigma_0(a) \times \sigma_0(b)$.*

MULTIPLICATIVE FUNCTIONS are those that satisfy the conclusion of the previous theorem. Let $f$ be a function from the set of positive integers to the set of all (integer, real, complex, whatever) numbers. We say that $f$ is **multiplicative** if $f(ab) = f(a) \cdot f(b)$ whenever $a$ and $b$ are coprime.[20] Corollary 2.29 can be restated: the divisor-counting function $\sigma_0$ is multiplicative.

[20] If $f$ is multiplicative then $f(6) = f(2) \cdot f(3)$ and $f(100) = f(4) \cdot f(25)$. But it is not necessarily true that $f(4) = f(2) \cdot f(2)$. Nor is it likely true that $f(24) = f(6) \cdot f(4)$. Do not forget about the coprime hypothesis in the identity $f(ab) = f(a) \cdot f(b)$.

Multiplicative functions distribute over a prime decomposition.

**Proposition 2.30** *Let $f$ be a multiplicative function. If $N$ is a positive integer with prime decomposition $N = 2^{e_2} 3^{e_3} 5^{e_5} 7^{e_7} \cdots$, then*

$$f(N) = f(2^{e_2}) \cdot f(3^{e_3}) \cdot f(5^{e_5}) \cdot f(7^{e_7}) \cdots.$$

PROOF: Since $\operatorname{GCD}(2^{e_2}, 3^{e_3} 5^{e_5} 7^{e_7} \cdots) = 1$, we have

$$f(N) = f(2^{e_2}) \cdot f(3^{e_3} 5^{e_5} 7^{e_7} \cdots).$$

Since $\operatorname{GCD}(3^{e_3}, 5^{e_5} 7^{e_7} \cdots) = 1$, we have

$$f(N) = f(2^{e_2}) \cdot f(3^{e_3}) \cdot f(5^{e_5} 7^{e_7} \cdots).$$

Carrying on in this manner leads to the result. ∎

Multiplicative functions arise naturally. Some elementary examples of multiplicative functions are $f(x) = x^k$, for every number $k$. Indeed, for this function,

$$f(ab) = (ab)^k = a^k b^k = f(a) \cdot f(b),$$

whether or not $\operatorname{GCD}(a,b) = 1$. Another example is the function $\chi$ given by the rule: $\chi(n) = 1$ if $n$ is odd, and $\chi(n) = 0$ if $n$ is even. The reader can verify that $\chi(ab) = \chi(a)\chi(b)$, again regardless of whether or not $\operatorname{GCD}(a,b) = 1$.

More interesting, and generalizing the divisor-counting function $\sigma_0$, are the **divisor-sum functions** $\sigma_k(n)$. When $k$ is any[21] number and $n$ is a positive integer, let $\sigma_k(n)$ denote the sum of the $k^{\text{th}}$ powers of the positive divisors of $n$. For example,

[21] We will primarily be interested in cases where $k$ is a natural number. But the quantity $n^k$ makes sense, when $n$ is a positive integer, and $k$ is a complex number!

$$\sigma_0(10) = 1^0 + 2^0 + 5^0 + 10^0 = 1 + 1 + 1 + 1 = 4;$$
$$\sigma_1(10) = 1^1 + 2^1 + 5^1 + 10^1 = 1 + 2 + 5 + 10 = 18;$$
$$\sigma_2(10) = 1^2 + 2^2 + 5^2 + 10^2 = 1 + 4 + 25 + 100 = 130.$$

Divisor-sum functions arise across number theory in a surprising variety of contexts. For example, Jacobi proved that every odd positive integer $n$ can be expressed as the sum of four squares, in $8\sigma_1(n)$ ways, and as a sum of eight squares, in $16\sigma_3(n)$ ways.

**Theorem 2.31 (Divisor-power-sums are multiplicative)** *If $a$ and $b$ are positive and coprime,[22] then $\sigma_k(ab) = \sigma_k(a) \cdot \sigma_k(b)$. In other words, $\sigma_k$ is multiplicative.*

PROOF: We begin by illustrating this theorem for $a = 4$, $b = 15$, and $ab = 60$. Consder the table of divisors of $ab$, as they are built from divisors of $a$ and $b$ by Theorem 2.28.

|  |  | Divisors of 4 | | |
|---|---|---|---|---|
|  | × | $2^0 = 1$ | $2^1 = 2$ | $2^2 = 4$ |
|  | $3^0 5^0 = 1$ | 1 | 2 | 4 |
| Divisors of | $3^1 5^0 = 3$ | 3 | 6 | 12 |
| 15 | $3^0 5^1 = 5$ | 5 | 10 | 20 |
|  | $3^1 5^1 = 15$ | 15 | 30 | 60 |

Divisors of 60 arise from divisors of 15 and divisors of 4. Now we multiply $\sigma_k(4)$ and $\sigma_k(15)$, using the distributive law.

$$\begin{aligned}\sigma_k(4) \times \sigma_k(15) =\ & (1^k + 2^k + 4^k) \times (1^k + 3^k + 5^k + 15^k) \\ =\ & (1^k + 2^k + 4^k) \times 1^k \\ & + (1^k + 2^k + 4^k) \times 3^k \\ & + (1^k + 2^k + 4^k) \times 5^k \\ & + (1^k + 2^k + 4^k) \times 15^k \\ =\ & (1^k + 2^k + 4^k) \\ & + (3^k + 6^k + 12^k) \\ & + (5^k + 10^k + 20^k) \\ & + (15^k + 30^k + 60^k) \\ =\ & \sigma_k(60).\end{aligned}$$

This method generalizes, to the extent that one can produce a table of divisors of $ab$ by multiplying divisors of $a$ with divisors of $b$. Hence the result holds when $a$ and $b$ are coprime. ∎

With these results, we can now evaluate a divisor-sum function by using prime decomposition.

**Problem 2.32** Evaluate $\sigma_3(100)$.

SOLUTION: The prime decomposition $100 = 2^2 5^2$ implies that $\sigma_3(100) = \sigma_3(2^2) \cdot \sigma_3(5^2)$. We compute

$$\sigma_3(2^2) = 1^3 + 2^3 + 4^3 = 1 + 8 + 64 = 73;$$
$$\sigma_3(5^2) = 1^3 + 5^3 + 25^3 = 1 + 125 + 15625 = 15751.$$

Hence $\sigma_3(100) = 73 \times 15751 = 1\,149\,823$. ✓

---

[22] If $a$ and $b$ are not coprime, this result certainly does not hold. For example,

$$\sigma_k(2 \cdot 2) = \sigma_k(4) = 1^k + 2^k + 4^k,$$

and

$$\begin{aligned}\sigma_k(2)\sigma_k(2) &= (1^k + 2^k)(1^k + 2^k) \\ &= 1^k + 2^k + 2^k + 4^k.\end{aligned}$$

Since $2^k$ never equals 0, we find $\sigma_k(4)$ never equals $\sigma_k(2)\sigma_k(2)$!

Here is a more typical proof, general and formal, and based on the same ideas:

$$\begin{aligned}\sigma_k(ab) &= \sum_{w|ab} w^k \\ &= \sum_{u|a}\sum_{v|b}(uv)^k \\ &= \sum_{u|a}\sum_{v|b}u^k \cdot v^k \\ &= \left(\sum_{u|a}u^k\right) \cdot \left(\sum_{v|b}v^k\right) \\ &= \sigma_k(a) \cdot \sigma_k(b).\end{aligned}$$

Most important is the transition from first line to second, where we use the fact that every divisor $w \mid ab$ can be written uniquely as $uv$ where $u \mid a$ and $v \mid b$.

PRIME DECOMPOSITION reduces the computation of multiplicative functions to the evaluation of mutiplicative functions on prime powers. For example, to compute $\sigma_3(100)$, we still needed to compute $\sigma_3(2^2)$ and $\sigma_3(5^2)$. This step can be accelerated, using the all-important geometric series formula.[23]

**Proposition 2.33 (Geometric series formula)** *Let x be any number except*[24] *1. Let e be a natural number. Then*

$$1 + x + x^2 + \cdots + x^e = \frac{x^{e+1} - 1}{x - 1}.$$

PROOF: Consider the product

$$(x - 1) \cdot (1 + x + x^2 + \cdots + x^e).$$

We evaluate this using the distributive law:

$$(x-1)\cdot(1+x+x^2+\cdots+x^e) = x\cdot(1+x+x^2+\cdots+x^e)$$
$$- 1\cdot(1+x+x^2+\cdots+x^e),$$
$$= (x+x^2+\cdots+x^e+x^{e+1})$$
$$- (1+x+x^2+\cdots+x^e),$$
$$= -1 + x^{e+1};$$
$$(x-1)\cdot(1+x+x^2+\cdots+x^e) = x^{e+1} - 1.$$

If $x \neq 1$, then we can safely divide both sides by $x - 1$ to obtain the desired result. ∎

**Corollary 2.34** *If p is a prime number, and k is a positive*[25] *integer, and e is a positive integer, then*

$$\sigma_k(p^e) = \frac{p^{ke+k} - 1}{p^k - 1}.$$

PROOF: When $p$ is a prime number, the divisors of $p^e$ are $1, p, p^2, p^3, \ldots, p^e$. The divisor sum can be evaluated explicitly.

$$\sigma_k(p^e) = 1^k + p^k + (p^2)^k + (p^3)^k + \cdots + (p^e)^k$$
$$= 1 + (p^k) + (p^k)^2 + (p^k)^3 + \cdots + (p^k)^e$$
$$= \frac{(p^k)^{e+1} - 1}{(p^k) - 1}$$
$$= \frac{p^{ke+k} - 1}{p^k - 1}.$$

∎

[23] This result should be compared to Euclid's *Elements*, Proposition IX.35.

[24] If $x = 1$, then the formula holds in a limiting sense. By l'Hôpital's rule, one can compute

$$\lim_{x \to 1} \frac{x^{e+1} - 1}{x - 1} = e + 1.$$

Indeed, $e + 1$ is the value of the geometric series if $x = 1$.

The key step is the fortuitous cancellation of all terms except those highlighted in red.

[25] When $k = 0$, the quotient does not make sense. But fortunately, when $k = 0$, we have already seen that

$$\sigma_0(p^e) = e + 1.$$

Here we use the identity $(p^a)^k = p^{ak} = (p^k)^a$ and the penultimate step uses the geometric series formula above, with $x = (p^k)$; note that $x \neq 1$ since $k \neq 0$.

**Problem 2.35** Compute $\sigma_2(2100)$.

SOLUTION: The prime decomposition of 2100 is $2100 = 2^2 3^1 5^2 7^1$, so

$$\sigma_2(2100) = \sigma_2(2^2)\sigma_2(3^1)\sigma_2(5^2)\sigma_2(7^1).$$

These can be evaluated using the previous corollary.

$$\sigma_2(2^2) = \frac{2^{2\cdot 2+2}-1}{2^2-1} = \frac{63}{3} = 21;$$

$$\sigma_2(3^1) = \frac{3^{2\cdot 1+2}-1}{3^2-1} = \frac{80}{8} = 10;$$

$$\sigma_2(5^2) = \frac{5^{2\cdot 2+2}-1}{5^2-1} = \frac{15624}{24} = 651;$$

$$\sigma_2(7^1) = \frac{7^{2\cdot 1+2}-1}{7^2-1} = \frac{2400}{48} = 50.$$

Notice that all these fractions equal natural numbers. This is not coincidence – remember that $\sigma_k(n)$ is always a natural number, since it is a sum of powers of natural numbers by definition.

Hence
$$\sigma_2(2100) = 21 \times 10 \times 651 \times 50 = 6\,835\,500. \quad \checkmark$$

WHY should we care about divisor-sum functions? Counting divisors might be natural enough, but this is only $\sigma_0$. Adding divisors – the function $\sigma_1$ – has a long history, and we study some more interesting parts here. Notice that $\sigma_1(n)$ is the sum of the divisors of $n$. For example,

$$\sigma_1(1) = 1, \quad \sigma_1(2) = 1+2 = 3, \quad \sigma_1(6) = 1+2+3+6 = 12.$$

A **perfect** number is a positive integer $n$ which satisfies $\sigma_1(n) = 2n$. More traditionally, a perfect number is one which equals the sum of its **proper divisors** (divisors except itself, or **aliquot parts**). The first perfect number is 6:

$$6 = 1+2+3, \quad \sigma_1(6) = 1+2+3+6 = 12 = 2 \cdot 6.$$

The next perfect number is 28:

$$28 = 1+2+4+7+14, \quad \sigma_1(28) = 1+2+4+7+14+28 = 2 \cdot 28.$$

The next perfect number is 496 and the following one is 8128, and after that comes 33 550 336. Perfect numbers belong to the Greek tradition, arising in Euclid's *Elements* and in more elaborate mystical contexts in the work of Nicomachus (the Pythagorean) of Gerasa (c. 100CE). As an application of prime decomposition, we study a surprising connection between perfect numbers and Mersenne primes.

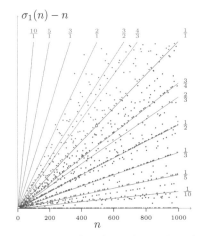

Figure 2.8: A plot of $n$ on the x-axis and the sum of proper divisors $\sigma_1(n) - n$ on the y-axis. Points tend to cluster around lines of rational slope – some lines are indicated above with their slopes. Points on the red diagonal line correspond to perfect numbers. Above are the abundant numbers and below are the deficient numbers.

MERSENNE PRIMES are prime numbers which are one less than a power of 2. The first few Mersenne primes are

$$2^2 - 1 = 3, \quad 2^3 - 1 = 7, \quad 2^5 - 1 = 31, \quad 2^7 - 1 = 127.$$

Observe that the exponents of 2 are prime, in these first few examples. This is not a coincidence!

**Proposition 2.36** *If $n$ is a positive integer, and $2^n - 1$ is a prime number, then $n$ is a prime number.*

PROOF: We prove that if $n$ is not prime, then $2^n - 1$ cannot be prime. As a first case, if $n = 1$, then $2^n - 1 = 1$, which is not prime.

If $n$ is composite, then $n = de$ for two integers $d, e$ with $d > 1$ and $e > 1$. It follows, using the techniques from Proposition 2.33 with $x = 2^d$, that

$$2^n - 1 = 2^{de} - 1 = (2^d - 1)(1 + 2^d + 2^{2d} + 2^{3d} + \cdots + 2^{(e-1)d}).$$

Since $d > 1$, $2^d - 1 > 1$ and $1 + 2^d + \cdots + 2^{(e-1)d} > 1$. Hence the above equality demonstrates that $2^n - 1$ is composite. ∎

The converse of this proposition is FALSE. If $n$ is prime, $2^n - 1$ is not necessarily prime. The first example is

$$2^{11} - 1 = 2047 = 23 \cdot 89.$$

Proposition 2.33 implies a strange property of powers of two:

$$\sigma_1(2^{n-1}) = 1 + 2 + 2^2 + \cdots + 2^{n-1} = \frac{2^n - 1}{2 - 1} = 2^n - 1. \quad (2.4)$$

In Proposition IX.36 of the *Elements*, Euclid states that

> If any multitude whatsoever of numbers is set out continuously in a double proportion, from a unit, until the whole sum added together becomes prime, and the sum multiplied into the last number makes some number, then the number so created will be perfect.[26]

[26] From *Euclid's Elements of Geometry*, edited, and provided with a modern English translation by Richard Fitzpatrick.

In this way Euclid linked Mersenne primes to perfect numbers. We give a modern statement and modern proof below.

**Theorem 2.37 (Mersenne primes yield perfect numbers)** *If the quantity $q = 1 + 2 + 2^2 + \cdots + 2^{n-1}$ is prime, then $2^{n-1}q$ is a perfect number.*

PROOF: Let $q = 1 + 2 + 2^2 + \cdots + 2^{n-1} = 2^n - 1$, a prime number[27] by hypothesis. Let $N = 2^{n-1}(2^n - 1) = 2^{n-1}q$. We compute

$$\begin{aligned}
\sigma_1(N) &= \sigma_1(2^{n-1})\sigma_1(q), && \text{(by Theorem 2.31)} \\
&= (2^n - 1) \cdot (q + 1), && \text{(by Equation (2.4))} \\
&= q \cdot (2^n - 1 + 1), && \text{(Since } q = 2^n - 1\text{)} \\
&= q \cdot 2 \cdot 2^{n-1} \\
&= 2(2^{n-1})q = 2N
\end{aligned}.$$

[27] This implies that $q$ is a Mersenne prime, and that $n$ is prime by the previous Theorem. But these facts are unnecessary in the proof here.

Hence $N$ is a perfect number. ∎

Euclid's result tells us that, beginning from a Mersenne prime $q = 2^p - 1$, one can construct a perfect number $2^{p-1}q$. Two millennia later, Euler proved[28] that every *even* perfect number arises from Euclid's construction.

[28] See "Tractatus de numerorum doctrina capita sedecim quae supersunt" by Leonhard Euler, in *Commentationes arithmeticae* 2, 1849, pp. 503-575, reprinted in *Opera Omnia*, Series I, v. 5, p. 182–283. Available as E792 at the Euler Archive, http://www.eulerarchive.org/.

**Theorem 2.38 (Even perfect numbers arise from Mersenne primes)**
*Suppose N is an even perfect number. Then there exist prime numbers p and q, for which $q = 2^p - 1$ and $N = 2^{p-1} \cdot q$.*

PROOF: Begin with the prime decomposition of $N$:

$$N = 2^{e_2} 3^{e_3} 5^{e_5} 7^{e_7} \cdots.$$

Define $p = e_2 + 1$ and define $q = 3^{e_3} 5^{e_5} 7^{e_7} \cdots$. Then,

$$N = 2^{p-1} \cdot q.$$

Since $N$ is even, $e_2 \geq 1$, so $p \geq 2$. Observe that $GCD(2^{p-1}, q) = 1$ since $q$ is odd. With the fact that $\sigma_1$ is multiplicative and using Equation (2.4),

$$\sigma_1(N) = \sigma_1(2^{p-1}) \sigma_1(q) = (2^p - 1) \sigma_1(q).$$

On the other hand, since $N$ is perfect, $\sigma_1(N) = 2N = 2^p \cdot q$. Hence

$$(2^p - 1) \sigma_1(q) = 2^p \cdot q.$$

Hence

$$\sigma_1(q) = \left( \frac{2^p}{2^p - 1} \right) q = \left( \frac{2^p - 1}{2^p - 1} \right) q + \left( \frac{1}{2^p - 1} \right) q = q + \frac{q}{2^p - 1}.$$

The quantity $q/(2^p - 1)$ is a positive integer, since it equals $\sigma_1(q) - q$. It is a *proper* divisor of $q$, since $p \geq 2$ implies $2^p - 1 > 1$.

But now, notice that $\sigma_1(q)$ – by definition the sum of *all* divisors of $q$ – equals $q$ plus just one other divisor of $q$. Thus the sum of all divisors of $\sigma_1(q)$ equals the sum of only two divisors of $q$, so $q$ must have only two divisors! Hence $q$ is prime, and the two divisors are $q$ and 1. Hence $q/(2^p - 1) = 1$, and so $q = 2^p - 1$ is a prime number. By Proposition 2.36, $p$ is prime too. ∎

ALL EVEN PERFECT NUMBERS come from Euclid's construction, thanks to Euler's clever proof. But what about odd perfect numbers? Nobody has ever seen one! If an odd perfect number exists it would have to be an enormous number in many ways. Any odd perfect number would have more than 1500 digits,[29] and at least 9 distinct prime factors. Many suspect that odd perfect numbers do not exist.

[29] See "Odd perfect numbers are greater than $10^{1500}$" by P. Ochem and M. Rao, in *Mathematics of Computation* **81** (2012), pp. 1869–1877.

## Historical Notes

At the beginning of Book VII of the *Elements*, Euclid (c. 300BCE) defines the terms unit (Μονάς), number (Ἀριθμὸς), prime (Πρῶτος), composite (Σύνθετος), coprime (Πρῶτοι πρὸς ἀλλήλους), and perfect (Τέλειος). Later in Book VII, Euclid proves results which are used *today* to prove existence and uniqueness of prime factorization. The existence of factorization rests on Proposition VII.32:

> Every number is either prime or is measured by some prime number.[30]

In Proposition VII.30, Euclid states:

> If two numbers make some (number by) multiplying one another, and some prime number measures the number (so) created from them, then it will also measure one of the original (numbers).[31]

This is equivalent to our Lemma 2.12: if $p$ is prime and $p \mid bc$ then $p \mid b$ or $p \mid c$. It is the crux of our proof that factorization leads to a unique result. Euclid's proof in VII.30 is problematic, but the problems are not easily identified without looking at different notions of proportion in earlier parts of Euclid.[32]

But these proofs by Euclid should not be interpreted to mean that the "atomic metaphor" was part of Greek mathematics at the time. There is no talk of numbers being built from prime atoms at the time of Euclid. Euclid did not prove unique factorization of positive integers into primes. Factorization and its uniqueness may have been "in the air" during the Renaissance, but the first clear proof is in the *Disquisitiones Arithmeticae* (1801CE) of Gauss[33].

Euclid proves the infinitude of primes in Proposition IX.20:

> The (set of all) prime numbers is more numerous than any assigned multitude of prime numbers.[34]

Books VIII and IX are devoted to "continually proportional numbers." Geometric series are considered in Proposition IX.35:

> If there is any multitude whatsoever of continually proportional numbers, and (numbers) equal to the first are subtracted from (both) the second and the last, then as the excess of the second (number is) to the first, so the excess of the last will be to (the sum of) all those (numbers) before it.[35]

Today, we think of "continually proportional numbers" as sequences of the form $a, ax, ax^2, ax^3, \ldots, ax^e$. The last sentence above, in modern terms, gives an equality of fractions:[36]

$$\frac{\text{"the excess of the second"}}{\text{"the first"}} = \frac{\text{"the excess of the last"}}{\text{"all those before it"}}.$$

Algebraically, this is equivalent to Proposition 2.33.

$$\frac{ax - a}{a} = \frac{ax^e - a}{a + ax + \cdots + ax^{e-1}}.$$

---

The following passages of Euclid are taken from "Euclid's Elements of Geometry," the Greek text of J.L. Heiberg, edited with modern English translation by Richard Fitzpatrick.

[30] Ἅπας ἀριθμὸς ἤτοι πρῶτός ἐστιν ἢ ὑπὸ πρώτου τινὸς ἀριθμοῦ μετρεῖται.

[31] Ἐὰν δύο ἀριθμοὶ πολλαπλασιάσαντες ἀλλήλους ποιῶσί τινα, τὸν δὲ γενόμενον ἐξ αὐτῶν μετρῇ τις πρῶτος ἀριθμός, καὶ ἕνα τῶν ἐξ ἀρχῆς μετρήσει.

[32] See "Did Euclid need the Euclidean algorithm to prove unique factorization?" by D. Pengelley and F. Richman, in *The American Mathematical Monthly*, vol. 113, 3, pp. 196–205 (March 2006).

[33] See Art. 16 of the *Disquisitiones* in "Disquisitiones Arithmeticae," by Carl Friedrich Gauss, English Edition, translated by A.A. Clarke, Springer (1986).

[34] Οἱ πρῶτοι ἀριθμοὶ πλείους εἰσὶ παντὸς τοῦ προτεθέντος πλήθους πρώτων ἀριθμῶν.

[35] Ἐὰν ὦσιν ὁσοιδηποτοῦν ἀριθμοὶ ἑξῆς ἀνάλογον, ἀφαιρεθῶσι δὲ ἀπό τε τοῦ δευτέρου καὶ τοῦ ἐσχάτου ἴσοι τῷ πρώτῳ, ἔσται ὡς ἡ τοῦ δευτέρου ὑπεροχὴ πρὸς τὸν πρῶτον, οὕτως ἡ τοῦ ἐσχάτου ὑπεροχὴ πρὸς τοὺς πρὸ ἑαυτοῦ πάντας.

[36] It is generally dangerous to interpret Euclid's statements on proportion as statements about fractions. Here it seems safe enough.

PERFECT NUMBERS are defined in Book VII of Euclid's *Elements*, and his construction of perfect numbers via geometric series occurs at the end of Book IX.[37] Later, Nicomachus of Gerasa (c. 100CE) follows the Pythagorean tradition and mixes mathematics with a hefty dose of symbolism. In Chapter 13 of his Arithmetic, Nicomachus attributes to Eratosthenes the word "sieve" to describe the process of filtering out prime numbers. By Chapters 14 and 15, Nicomachus discusses **abundant** and **deficient** numbers[38] – he compares deficient numbers to "some animal... short of the natural number of limbs or parts".[39] Chapter 16 studies the perfect numbers, with a thick layer of metaphor and praise, and includes a matter-of-fact statement that perfect numbers are always even.

By 1849, Gauss (1777–1855) had assembled and analyzed a table of prime numbers up to 3 million. In a letter to Johann Encke from December of that year, Gauss describes some of his findings; in particular, he states that there are 78501 primes between 1 and 1 000 000 (he miscounted by 3), and he compares this to 78627.5, his very precise approximation to the logarithmic integral[40]

$$\mathrm{Li}(1\,000\,000) = \int_2^{1000000} \frac{dx}{\log(x)} \approx 78627.549159.$$

More generally, he states that

> behind all of its fluctuations, this frequency [of primes] is on the average inversely proportional to the logarithm, so that the number of primes below a given bound $n$ is approximately equal to $\int \frac{dn}{\log(n)}$ where the logarithm is understood to be hyperbolic.[41]

It was not until 1896 that Hadamard (1865–1963) and de la Vallée Poussin (1866–1962) independently proved the **prime number theorem**: that the estimate used by Gauss and others is accurate in the sense that

$$\lim_{n \to \infty} \frac{\pi(n)}{\mathrm{Li}(n)} = 1.$$

Hadamard and de la Vallée Poussin combined earlier insights of Riemann with contemporary breakthroughs in complex analysis to give proofs. A full discussion of the prime number theorem and its history is impossible in these notes, but we recommend Zagier's wonderful survey[42] "The first 50 million prime numbers" to start.

It is more difficult to study the statistics of prime gaps. Yitang Zhang's breakthrough[43] was the first to give a constant $C$ such that the sentence "there are infinitely many consecutive primes $(p,q)$ for which $q - p \leq C$" is true. Zhang's constant $C$ was 70 000 000. The method was refined[44] by a grand collaboration – the PolyMath Project 8 – and new insights of James Maynard, to reduce the constant $C$ to 246.

---

[37] To consider how and why Euclid might have considered this construction of perfect numbers, see "Perfect Numbers, A Mathematical Pun" by C.M. Taisbak, in *Centaurus*, vol. 20, 4, pp. 269–275 (Dec. 1976). Taisbak demonstrates how Euclid's construction might appear natural in the context of Egyptian multiplication tables (familiar in Greek arithmetic).

[38] A number $n$ is abundant if $\sigma_1(n) > 2n$, and deficient if $\sigma_1(n) < 2n$; it is perfect when $\sigma_1(n) = 2n$.

[39] This translation and others are taken from "Introduction to Arithmetic," by Nicomachus of Gerasa, translated into English by Martin Luther D'Ooge, London: MacMillan and Company, 1926.

[40] The integral from 2 to $n$ is used here in the definition of $\mathrm{Li}(n)$, whereas in the chapter, the integral from 0 to $n$ is used to define $\mathrm{li}(n)$. Since $\mathrm{li}(n) - \mathrm{Li}(n) = \log(2)$, this makes little difference. Technically though, Li is simpler since $\mathrm{li}(x) = \int_0^x \frac{dt}{\log(t)}$ traverses a singularity at $t = 1$.

[41] For this translation of Gauss, and more on his tables and computations, see "About the cover: on the distribution of primes – Gauss' tables," by Yuri Tschinkel, in the *Bulletin of the American Mathematical Society*, vol. 43, **1** (2006), pp. 89–91.

[42] "The first 50 million prime numbers," by D. Zagier, in *The Mathematical Intelligencer* vol. 1, supp. 1 (1977), pp. 7–19.

[43] "Bounded gaps between primes," by Yitang Zhang, in *Annals of Mathematics* **179** no. 3 (2014).

[44] Read "Together and alone, closing the prime gap," by Erica Klarreich, in *Quanta magazine*, November 2013.

## Exercises

1. Write the prime decompositions of 240, 1000, and 111111, and 10! (10 factorial).

2. Recall that, when $n$ is a positive integer, $n!$ (read "$n$ **factorial**") means the product $1 \times 2 \times 3 \times \cdots \times n$. Prove that if $n \geq 2$, then among the numbers

$$n! + 2, n! + 3, \ldots, n! + n,$$

   none are prime.[45]

   [45] This exercise demonstrates that there is no bound on the length of gaps between primes.

3. Let $T = \{1, 4, 7, 10, 13, 16, 19, \ldots\}$. An element of $T$ is called irreducible if it is not 1 and its only factors within $T$ are 1 and itself.

   (a) Suppose that $a \in T$ and $b \in T$, and $c$ is a positive integer. Prove that if $a = bc$, then $c \in T$.

   (b) Demonstrate that every element of $T$ can be factored as a product of irreducible elements of $T$.

   (c) Perform a sieve to find all irreducible elements of $T$ between 1 and 100.

   (d) Find three examples of elements of $T$ with *nonunique* factorizations into irreducibles.

4. Let us divide the odd positive integers into two arithmetic progressions; the *red* numbers are 1, 5, 9, 13, 17, 21, .... The *blue* numbers are 3, 7, 11, 15, 19, 23, 27, ....

   (a) Prove that the product of two red numbers is red, and the product of two blue numbers is red.

   (b) Prove that every blue number has a blue prime factor.

   (c) Prove that there are infinitely many blue prime numbers.[46]
   Hint: follow Euclid's proof, but multiply a list together, multiply the result by four, then subtract one.

   [46] There are also infinitely many red prime numbers, but this is more difficult to prove.

5. Which properties of divisibility, GCD, and LCM arise from the following properties of inequality, min, and max?

   (a) $\min(e, f) \leq \max(e, f)$.
   (b) $\min(e, f) = \max(e, f)$ if and only if $e = f$.
   (c) $\min(e, \min(f, g)) = \min(\min(e, f), g)$
   (d) $\min(2e, 2f) = 2\min(e, f)$ and $\max(2e, 2f) = 2\max(e, f)$.
   (e) $\min(g + e, g + f) = g + \min(e, f)$.

6. Prove that if $a$, $b$, and $n$ are positive integers and $a^n \mid b^n$, then $a \mid b$. Hint: use prime factorizations of $a$ and $b$.

7. Compute $\sigma_0(1000)$, $\sigma_0(750)$, and $\sigma_0(1024)$.

8. Compute $\sigma_1(100)$, $\sigma_2(750)$, and $\sigma_3(1024)$.

9. Find a short formula for $\sigma_0(p) + \sigma_1(p) + \sigma_2(p) + \cdots + \sigma_k(p)$, whenever $p$ is a prime number.

10. What is the smallest positive integer with precisely 60 positive divisors?[47]

    [47] This question was asked by Mersenne in 1644.

11. Prove that if $n$ is a positive integer, and $\sigma_0(n)$ is prime then $n$ is a power of a prime number.

12. If $n$ is a positive integer, let $\sigma_{-1}(n)$ denote the sum of the reciprocals of the positive divisors of $n$. Prove that $n$ is perfect if and only if $\sigma_{-1}(n) = 2$.

13. Look at Figure 2.8. For what sorts of numbers $n$ is it true that $\sigma_1(n) - n \approx 0$? For what sorts of numbers $n$ is it true that $\sigma_1(n) - n \approx n/2$? Why do you think the points in the figure seem to cluster just above lines of slope $1/5$, $1/2$, $3/4$, etc.?

14. Prove that if $n$ is a square number, then $n$ is not a perfect number.

15. Prove that if $n$ is a perfect number, then $2n$ is an abundant number, i.e., $\sigma_1(2n) > 4n$. Hint: use two cases – $n$ is even or $n$ is odd.

16. Describe all circumstances under which $\sigma_1(n)$ is odd.

17. Let $a$ and $b$ be positive integers. We say $a$ and $b$ are **amicable** numbers if $\sigma_1(a) = \sigma_1(b) = a + b$.[48] Prove the following result of Thabit ibn Kurrah (836–901): if $p = 3 \times 2^{n-1} - 1$ is prime and $q = 3 \times 2^n - 1$ is prime, and $r = 9 \times 2^{2n-1} - 1$ is prime, then
$$a = 2^n \times p \times q, \quad b = 2^n \times r$$
are amicable numbers. Find a pair of amicable numbers by finding such prime numbers $p$, $q$, and $r$.

    [48] Thus $a$ is perfect if $a$ is amicable with itself. Instead of saying
    $$\sigma_1(a) = \sigma_1(b) = a+b,$$
    one would usually say that the sum of the proper divisors of $a$ equals $b$, and the sum of the proper divisors of $b$ equals $a$.

18. This exercise relates to the prime decomposition of factorials.

    (a) Find the prime decomposition of $20!$.

    (b) How many zeros would you find at the end of $100!$, if you expanded it out in base ten? (Consider the twos and fives in the prime decomposition).

    (c) Let $e_p$ be the exponent of $p$ in the prime decomposition of $n!$. Prove that
    $$e_p \leq \frac{n}{p-1}.$$
    Hint: Use the geometric series
    $$p^{-1} + p^{-2} + p^{-3} + \cdots = p^{-1}(1 + p^{-1} + p^{-2} + \cdots) = \frac{1}{p-1}.$$

| | | | | |
|---|---|---|---|---|
| 20 → | 2 | 2 | | |
| 19 → | | | | |
| 18 → | 2 | | | |
| 17 → | | | | |
| 16 → | 2 | 2 | 2 | 2 |
| 15 → | | | | |
| 14 → | 2 | | | |
| 13 → | | | | |
| 12 → | 2 | 2 | | |
| 11 → | | | | |
| 10 → | 2 | | | |
| 9 → | | | | |
| 8 → | 2 | 2 | 2 | |
| 7 → | | | | |
| 6 → | 2 | | | |
| 5 → | | | | |
| 4 → | 2 | 2 | | |
| 3 → | | | | |
| 2 → | 2 | | | |
| 1 → | | | | |

Figure 2.9: Pulling all the 2s out of $20!$. The exponent of 2 in the prime decomposition of $20!$ is the number of 2s. Count them directly, or count them in columns using the floor function.

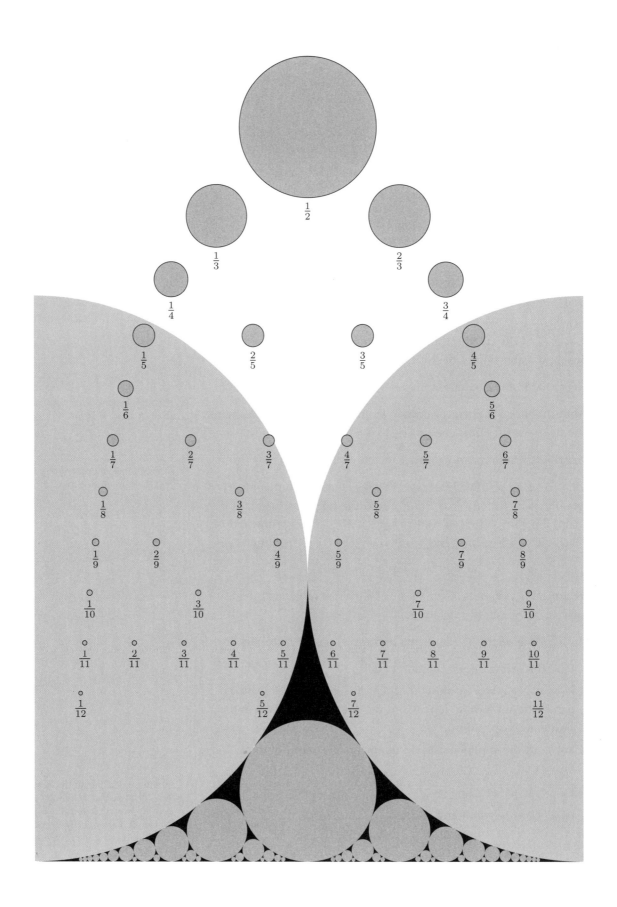

# 3
# Rational and Constructible Numbers

THE NATURAL NUMBERS are 0, 1, 2, 3, 4, .... The integers include natural numbers and their negatives: ..., −3, −2, −1, 0, 1, 2, 3, .... Within the integers we can always add, subtract, and multiply.

**The rational numbers** are those numbers[1] which are quotients of integers (without dividing by zero). Every integer is a rational number: If $n$ is an integer, then $n = n/1$ is a rational number. Within the rational numbers, we can always add, subtract, and multiply, and moreover we can divide, except by zero.

It is a bit difficult to place rational numbers. Find 2/5 on the number line. Now locate 3/7. Which one is to the right and which one is to the left?

[1] The reader might notice that throughout the book we are assuming a large set of numbers, with familiar operations and properties. This book does not contain an axiomatic construction of numbers. Instead, we take for granted the real numbers, all properties of arithmetic operations therein, equality and order, and the intermediate value theorem for polynomials (for the sake of square roots).

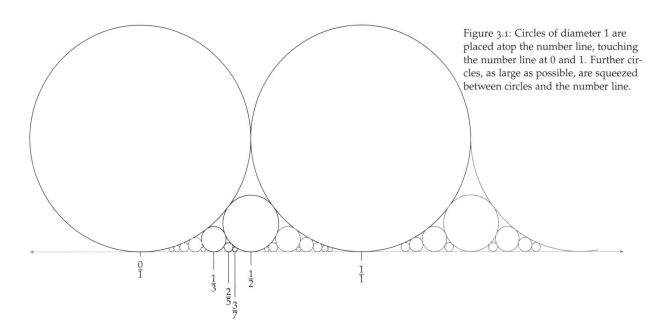

Figure 3.1: Circles of diameter 1 are placed atop the number line, touching the number line at 0 and 1. Further circles, as large as possible, are squeezed between circles and the number line.

EVERY RATIONAL NUMBER can be expressed as $a/b$ for some integers $a$ and $b$ (with $b$ nonzero). An expression $a/b$, with $a$ and $b$ integers, is called a **fraction**. The relationship[2] between fractions and rational numbers is the relationship between a word and what it means. A *fraction* is a way of writing a *rational number*. For what comes later, it is convenient to allow fractions like $1/0$ and $-17/0$, even though they do not represent rational numbers.

For example, $2/3$ and $4/6$ are *different* fractions, but they represent the *same* rational number. The *fraction* $1/0$ does not represent any rational number; only fractions with nonzero denominators represent rational numbers.

We learn in school that the different fractions $2/3$ and $4/6$ are *equal* as rational numbers. More generally, equality of rational numbers is defined by the rule

$$\frac{a}{b} = \frac{c}{d}, \text{ if } ad = bc.$$

Using this rule, we find that for nonzero integers $n$,

$$\frac{a}{b} = \frac{n \cdot a}{n \cdot b}, \text{ since } a(nb) = b(na).$$

We call a fraction $a/b$ **reduced** if $\text{GCD}(a,b) = 1$ and $b > 0$. It is also sometimes convenient[3] to call the fraction $1/0$ reduced, though it represents no rational number.

**Theorem 3.1 (Reduction of fractions)** *If $a/b$ is a fraction with $b \neq 0$, then there exists a unique reduced fraction $c/d$ such that $a/b = c/d$. Moreover, there exists a nonzero integer $n$ such that $a = nc$ and $b = nd$.*

PROOF: Since $b \neq 0$, $\text{GCD}(a,b) \neq 0$. If $b < 0$, let $n = -\text{GCD}(a,b)$; if $b > 0$, let $n = \text{GCD}(a,b)$. In either case, we have $n \mid a$ and $n \mid b$. Let $c = a/n$ and $d = b/n$. By construction,

$a = nc$ and $b = nd$ and $a/b = c/d$ and $d > 0$ and $\text{GCD}(c,d) = 1$.

The fact that $\text{GCD}(c,d) = 1$ follows from Corollary 1.29.

For uniqueness, suppose that $a/b = c/d = e/f$ for another reduced fraction $e/f$. Then

$cf = de$ and $\text{GCD}(c,d) = \text{GCD}(e,f) = 1$ and $d > 0$ and $f > 0$.

Since $c \mid cf = de$, Euclid's Lemma (Lemma 2.12) implies $c \mid e$. Similarly, $e \mid de = cf$, and by Euclid's Lemma, $e \mid c$. Likewise, we find $d \mid f$ and $f \mid d$. Therefore $c = \pm e$ and $d = \pm f$. Since $d > 0$ and $f > 0$, we have $d = f$. Since $cf = de$ and $d = f$, it follows quickly that $c = e$ as well. ∎

---

[2] **Semiotics** is the study of signs and what they signify; it is crucial to have different language to describe expressions – arrangements of symbols on a page – and their numerical meaning.

Since a fraction is just an ordered pair of integers – just written $a/b$ – one associate to a fraction a grid-point in the plane.

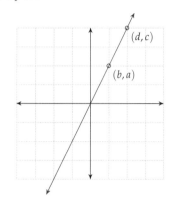

The fractions $a/b$ and $c/d$ are thought of as grid-points $(b, a)$ and $(d, c)$. The equality of rational numbers $a/b = c/d$ reflects the fact that both lie on a line of the same slope through the origin. The order, from $a/b$ to $(b, a)$, reverses since slope is defined as "rise over run," so $a$ must correspond to the rise and $b$ to the run. The rational number *is* the slope of the line.

[3] We say that a fraction of the form $a/0$ represents $\infty$ ("infinity"), if $a \neq 0$. We think of $1/0$ as the reduced fraction representation of $\infty$. Note that both $1/0$ from $-1/0$ represent the same $\infty$; there is no "negative infinity" here.

Prime decomposition can be extended from positive integers to all nonzero rational numbers. The extension to nonzero integers is ad hoc; every nonzero integer $a$ has a unique expression of the form

$$a = \pm 2^{e_2} 3^{e_3} 5^{e_5} 7^{e_7} \cdots,$$

in which the exponents $e_2, e_3, e_5, e_7, \ldots$ are natural numbers. Positive integers have sign $+$, and negative integers have sign $-$ in the above decomposition.[4]

The prime decomposition of rational numbers permits not just natural exponents but integer exponents.

**Proposition 3.2** *If $x$ is a nonzero rational number, then there exist unique integers $e_2, e_3, e_5, e_7$, etc., such that all but finitely many are zero and*

$$x = \pm 2^{e_2} 3^{e_3} 5^{e_5} 7^{e_7} \cdots.$$

PROOF: Every nonzero rational number can be uniquely expressed as a reduced fraction $x = a/b$. Dividing the prime decomposition of the nonzero integer $a$ by that of the positive integer $b$, prime by prime, leads to a decomposition of the form above.[5]

For uniqueness, consider a prime decomposition

$$x = \pm 2^{e_2} 3^{e_3} 5^{e_5} 7^{e_7} \cdots.$$

Consider the set $\{p_1, \ldots, p_s\}$ of primes with positive exponent and the set $\{q_1, \ldots, q_t\}$ of primes with negative exponent, in the prime decomposition above. Define integers

$$c = \pm p_1^{e_{p_1}} \cdots p_s^{e_{p_s}} \text{ and } d = q_1^{-e_{q_1}} \cdots q_t^{-e_{q_t}}.$$

Since the sets of primes are disjoint, $\text{GCD}(c, d) = 1$. Moreover, $d > 0$ and $x = c/d$. Since $x$ has a unique expression as a reduced fraction, $c = a$ and $d = b$. Theorem 2.15 implies that the decompositions of $c$ and $d$ above are the unique prime decompositions of $a$ and $b$, respectively. Hence the only prime decomposition of $x = a/b$ is that which arises from the unique prime decompositions of $a$ and $b$. ∎

In practice, the prime decomposition of rational numbers is no harder than the prime decomposition of numerator and denominator.

$$\frac{3}{4} = \frac{3^1}{2^2} = 2^{-2} 3^1,$$
$$\frac{11}{6} = \frac{11}{2^1 3^1} = 2^{-1} 3^{-1} 5^0 7^0 11^1.$$

As for positive integers, prime decomposition transforms multiplication of rational numbers into addition of exponents.

$$\frac{3}{4} \cdot \frac{11}{6} = 2^{-2-1} \cdot 3^{1-1} \cdot 5^0 \cdot 7^0 \cdot 11^1 = 2^{-3} 3^0 5^0 7^0 11^1.$$

[4] One might even consider $(-1)$ to be another prime; this perspective has been encouraged by John H. Conway, but adoption has not been widespread.

[5] Division yields subtraction of exponents:
$$\frac{p^{f_p}}{p^{g_p}} = p^{f_p - g_p}.$$
Thus the exponent of $p$ in the decomposition of $a/b$ is the difference of exponents in the decompositions of $a$ and $b$.

René Descartes began his *Géométrie*[6] by describing the geometry of arithmetic operations – addition, subtraction, multiplication, division, and extraction of roots. To understand questions of rationality and irrationality, we present some geometric constructions in the spirit of Descartes (1596–1650).

In Propositions I.2 and I.3, of the *Elements*, Euclid constructs the sum and difference of two line segments $x = \overline{OA}$ and $y = \overline{BC}$.

[6] See "The Geometry of René Descartes," translated from the French and Latin by D.E. Smith and M.L. Latham, Dover Publications, 1954.

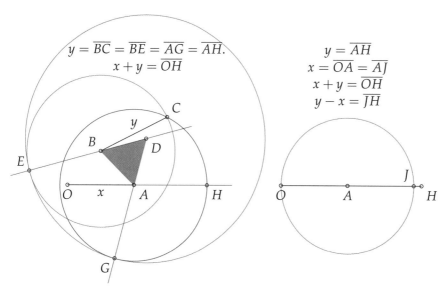

Figure 3.2: Euclid's steps are, in order: connect $A$ to $B$ by a line segment, and construct an equilateral triangle $ABD$ on that line segment (see the figure above for the method). Extend the lines $BD$ and $AD$ indefinitely. Construct the circle at center $B$, with radius $BC$. Let $E$ be the intersection of that circle with the line extending $BD$. Now construct the circle at center $D$ with radius $DE$. Let $G$ be the intersection of this circle with the line extending $AD$. Finally, construct the circle with center $A$ and radius $AG$. Let $H$ be the intersection of this circle with the line extending $OA$. Then $AH$ has length $y$, and $OH$ has length $x + y$.

To find the difference $y - x$, see the figure on the right. Draw a circle of radius $x$ with center $A$. Its intersection with $AH$ is a point $J$. Since $AH$ has length $y$, and $AJ$ has length $x$, we find that $JH$ has length $y - x$.

Descartes uses similar triangles to construct the product and quotient of two lengths $x$ and $y$, relative to a fixed unit of measure.

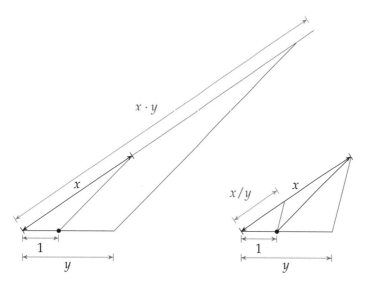

Figure 3.3: Here we arrange the segments of lengths 1, $x$, and $y$, using Euclidean constructions as above. Parallel lines (here, in blue) are constructed, yielding pairs of similar triangles. Similarity demonstrates that the red line segments have lengths $x \cdot y$ and $x/y$ as displayed.

Following Euclid,[7] Descartes also gives a geometric construction of square roots. It is based directly on Euclid's construction of the mean proportional, and the figures below demonstrate Euclid's ideas.

[7] Compare to the *Elements*, Propositions VI.13 and III.31.

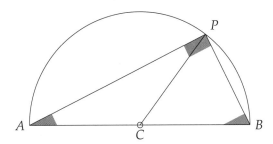
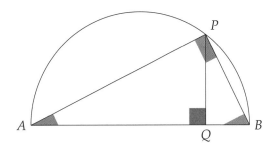

On the left, $C$ is the center of the circle and all radii are equal, so triangles $ACP$ and $BCP$ are isosceles; it follows that the two red angles equal each other, and the two blue angles equal each other. The sum of two red angles and two blue angles equals $180°$, the total of the angles of triangle $ABP$. Hence one red angle and one blue angle sum to $90°$, a right angle. The angle $APB$ at the vertex $P$ is a right angle.

On the right, drop a perpendicular from $P$ to the line $AB$. The red angles in this figure are equal, since they are both complements[8] of the same blue angle $APQ$. Similarly, the blue angles in this figure are equal, since they are both complements of the same red angle $QPB$. Therefore the right triangles $PQA$ and $BQP$ are similar triangles.

[8] **Complementary angles** are those that add to $90°$.

By similarity, we find a proportion

$$\frac{\text{length}(BQ)}{\text{length}(PQ)} = \frac{\text{length}(PQ)}{\text{length}(QA)}.$$

If we have lengths $1$ and $x$ already constructed, this proportion gives a construction of $\sqrt{x}$. Produce a line segment of length $1 + x$, and draw a semicircle whose diameter is the line segment. The altitude will be[9] $\sqrt{x}$ as in the figure on the right.

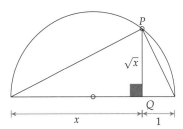

[9] Indeed, the proportion gives

$$\frac{1}{\text{length}(PQ)} = \frac{\text{length}(PQ)}{x},$$

which implies

$$\text{length}(PQ)^2 = x,$$

so the altitude $PQ$ has length $\sqrt{x}$.

**Constructible numbers** are those lengths which can be obtained, beginning with a unit line segment, and carrying out only straightedge and compass drawings; one may extend lines through any two points, draw circles whose radius is an existing segment, and that is all. The constructions here demonstrate that constructible numbers include all those which can be obtained, beginning with only 1, by addition, subtraction ($y - x$ when $y > x$), multiplication, division (not by zero, of course), and square roots. Constructible numbers include, therefore, all positive rational numbers. In addition, constructible numbers include $\sqrt{2}$ and more exotic numbers like $\sqrt{8 - \sqrt{7}}$.

## Theorem 3.3 (Algebraic characterization of constructible numbers)

*Addition, subtraction, multiplication, division, and square roots suffice[10] to describe every constructible number.*

PROOF: When $ax + by = e$ and $cx + dy = f$ describe two lines, their intersection point[11] is given by the formula below.

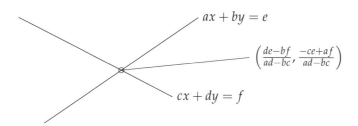

When $ax + by = e$ describes a line, and $x^2 + y^2 = r^2$ describes a circle, the intersection point(s)[12] are described by the formula below.

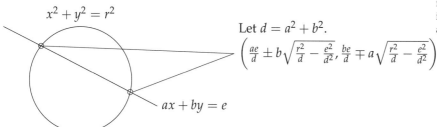

When $x^2 + y^2 = 1$ and $(x-a)^2 + (y-b)^2 = r^2$ describe two circles, the line through their intersection points (if they intersect[13]) is given by the formula below:

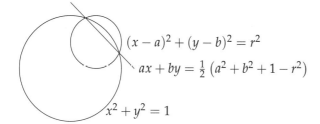

Points that arise from straightedge-compass constructions arise from these three kinds of intersections, together with shifting and scaling.[14] Only addition, subtraction, multiplication, division, and square roots are ever needed to express coordinates of constructible points. By the Pythagorean theorem, only these operations are needed to express distances between constructible points.

[10] The previous page demonstrated that numbers obtained by these operations were constructible; this page demonstrates that all constructible numbers can be given by these operations.

[11] The lines are parallel if and only if $ad - bc = 0$, in which case the formula for their intersection point does not make sense.

[12] If the quantity under the square root is positive, then there are two intersection points; if zero, there is one intersection point; if negative, the line and circle do not intersect.

[13] If the circles do not intersect, the line still makes sense but passes between the circles rather than through them. When the two circles intersect, their intersection points may be found as the intersection of the line with either circle, using the formula from the previous figure.

[14] Shifting corresponds to addition and subtraction of coordinates, and scaling corresponds to multiplication of coordinates.

Our algebraic formulae for intersections of lines and circles yields a complete description of the constructible numbers. Intersecting lines and circles also yields an old solution to an older problem.

**Theorem 3.4 (Infinitude of Pythagorean triples)** *There are infinitely many primitive*[15] ***Pythagorean triples.*** *In other words, there are infinitely many triples $(a, b, c)$ of integers, for which $\mathrm{GCD}(a, b, c) = 1$ and $a^2 + b^2 = c^2$.*

PROOF: Consider the unit circle centered at the origin, and a line through $(0, 1)$ of nonzero rational slope $m$.

The line intersects the circle at two points; $(0, 1)$ and $(u, v)$ with $u \neq 0$. As $u^2 + v^2 = 1$ and $v = 1 + mu$, substituting yields
$$u^2 + (1 + mu)^2 = 1.$$
Expanding and simplifying yields $u \cdot ((1 + m^2)u + 2m) = 0$. As $u \neq 0$, we find
$$u = -\frac{2m}{1 + m^2}, \quad v = \frac{1 - m^2}{1 + m^2}.$$

In this way, every nonzero rational number $m$ gives a pair $(u, v)$ of rational numbers such that $u^2 + v^2 = 1$. Let $a/c$ be the reduced expression for $u$ and $b/d$ the reduced expression for $v$. Then
$$\left(\frac{a}{c}\right)^2 + \left(\frac{b}{d}\right)^2 = 1, \quad \mathrm{GCD}(a, c) = \mathrm{GCD}(b, d) = 1, \quad c > 0, d > 0.$$
Expanding and multiplying through by $cd$ yields
$$a^2 d^2 + b^2 c^2 = c^2 d^2.$$

The two out of three principle implies $c^2 \mid a^2 d^2$ and $d^2 \mid b^2 c^2$. By Euclid's Lemma (Lemma 2.12), $c^2 \mid d^2$ and $d^2 \mid c^2$, and so $c = d$ (as both are positive). We may now write the equality $x^2 + y^2 = 1$ instead in the form
$$\left(\frac{a}{c}\right)^2 + \left(\frac{b}{c}\right)^2 = 1, \quad \mathrm{GCD}(a, c) = \mathrm{GCD}(b, c) = 1, \quad c > 0.$$
Multiplying through by $c^2$ yields a primitive Pythagorean triple,
$$a^2 + b^2 = c^2.$$

The steps in this process may be reversed easily enough,[16] and this sets up a one to one correspondence between two sets:

- nonzero rational numbers (the slopes $m$);
- primitive Pythagorean triples $(a, b, c)$ with $c > 0$ and $a \neq 0$.

Since the first set is infinite, so too is the second. ∎

[15] Without the primitive condition, $\mathrm{GCD}(a, b, c) = 1$, the theorem is not so interesting. Beginning with a Pythagorean triple like $(3, 4, 5)$, one can scale to form infinitely many nonprimitive triples, like $(6, 8, 10)$ and $(9, 12, 15)$ and $(12, 16, 20)$, etc.

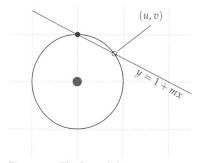

Figure 3.4: The line of slope
$$m = -1/2$$
intersects the circle at the point
$$(u, v) = \left(\frac{4}{5}, \frac{3}{5}\right).$$
As these fractions are reduced, they directly yield the primitive Pythagorean triple
$$(a, b, c) = (4, 3, 5).$$
In this two-step process, every nonzero rational slope yields a primitive Pythagorean triple.

[16] Begin with a primitive Pythagorean triple $(a, b, c)$ and let $u = a/b$ and let $v = b/c$. Then draw a line from $(0, 1)$ to $(u, v)$; the slope is certainly rational.

WHICH constructible numbers are rational? From Euclid's perspective,[17] this is a question of commensurability. Two lengths $x$ and $y$ are said to be **commensurable** if there exists a length $z$ and integers $m$ and $n$, such that $x = mz$ and $y = nz$. In other words, two lengths are commensurable if they can be both be measured precisely by the same measuring rod.

The figure in the margin can be used to demonstrate that 1 and $\sqrt{2}$ are incommensurable. Or geometrically, the leg and the hypotenuse of a 45-45-90 triangle are incommensurable. Indeed, any measuring rod that can measure both leg $\ell$ and hypotenuse $h$ can measure the differences $h - \ell$ and $2\ell - h$. These are the leg and hypotenuse of a smaller 45-45-90 triangle. Repeating this, the measuring rod must would be able to measure the sides of smaller and smaller triangles, ad infinitum. This is a contradiction – the triangles must eventually be smaller than the measuring rod itself!

In modern times, commensurability is usually rephrased in terms of rationality. Namely, if $x = mz$ and $y = nz$ for two positive integers $m$ and $n$, then $x/y = mz/nz = m/n$ is a rational number. Hence commensurable lengths are those with rational quotient.

Square, cube, and higher roots[18] are irrational, except when they are almost obviously rational. Reduction of fractions provides a powerful tool to prove irrationality; this has deprecated Euclid's geometric proofs of incommensurability.

**Proposition 3.5 (Irrationality of surds)** *Let $a/b$ be a reduced fraction (with $b > 0$). Then $\sqrt[n]{a/b}$ is a rational number if and only if $a$ and $b$ are $n^{th}$ powers of integers.*

PROOF: If $\sqrt[n]{a/b}$ is a rational number, then $\sqrt[n]{a/b} = c/d$ for some reduced fraction $c/d$. It follows that $a/b = (c/d)^n = c^n/d^n$. Since $\text{GCD}(c,d) = 1$, Corollary 2.25 implies $\text{GCD}(c^n, d^n) = 1$. Therefore, the fraction $c^n/d^n$ is reduced. By the uniqueness of reduced representatives, the equality $a/b = c^n/d^n$ implies that $a = c^n$ and $b = d^n$. Hence $a$ and $b$ are $n^{th}$ powers of integers.

Conversely, if $a$ and $b$ are $n^{th}$ powers of integers, then $a = c^n$ and $b = d^n$ for some integers $c$ and $d$. Hence $a/b = c^n/d^n = (c/d)^n$, so $\sqrt[n]{a/b} = c/d$ is a rational number. ∎

**Problem 3.6** Demonstrate that $\sqrt{17}$ is irrational.

SOLUTION: Since $17 = 17/1$, a reduced fraction, and 17 is not a square number (16 and 25 are squares of 4 and 5, and no square number lies between 16 and 25), Proposition 3.5 implies that $\sqrt{17/1}$ is irrational. ✓

[17] Book X of the *Elements*, the longest of the 13 books, studies questions of commensurability.

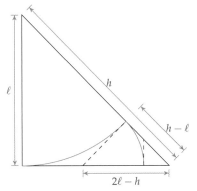

Figure 3.5: A 45-45-90 triangle. The measurement of the length $2\ell - h$ can be derived from the Pythagorean theorem: If $h^2 = 2\ell^2$ then $(2\ell - h)^2 = 2(h - \ell)^2$. The incommensurability of a leg with the hypotenuse is equivalent to the irrationality of the quotient $\sqrt{2}/1 = \sqrt{2}$.

[18] We follow the convention that $\sqrt[n]{x}$ means the *positive* $n^{th}$ root of $x$, when $n$ is even and $x$ is positive. When $n$ is odd, every real number $x$ has a unique real $n^{th}$ root. When $n$ is even and $x$ is negative, the number $x$ has no real $n^{th}$ root.

**Problem 3.7** Are the numbers $\sqrt[3]{2}$ and $3\sqrt[4]{3}$ commensurable?

SOLUTION: Two numbers are commensurable if their quotient is rational. The quotient is $\sqrt[3]{2}/3\sqrt[4]{3}$, and

$$\frac{\sqrt[3]{2}}{3\sqrt[4]{3}} = \frac{1}{3}\frac{\sqrt[12]{16}}{\sqrt[12]{27}} = \frac{1}{3}\sqrt[12]{\frac{16}{27}}.$$

Since 16/27 is a reduced fraction, and neither 16 nor 27 is a perfect twelfth power, we find that $\sqrt[12]{16/27}$ is irrational. Hence $\sqrt[3]{2}$ and $3\sqrt[4]{3}$ are not commensurable. ✓

What about more complicated constructible numbers? Numbers like $\sqrt{8-\sqrt{7}}$, for example? For **algebraic**[19] numbers, the easiest irrationality proofs use the rational root theorem.

**Theorem 3.8 (Rational Root Theorem)** *Let $a/b$ be a reduced fraction, representing a rational number $x$. Let $c_0, \ldots, c_d$ be positive integers. Then*

$$c_0 + c_1 x + \cdots + c_{d-1} x^{d-1} + c_d x^d = 0$$

*implies $a \mid c_0$ and $b \mid c_d$.*

PROOF: Noting that $x = a/b$, substitute to find

$$c_0 + c_1 \frac{a}{b} + \cdots + c_{d-1}\frac{a^{d-1}}{b^{d-1}} + c_d \frac{a^d}{b^d} = 0.$$

Multiply through by $b^d$ to obtain

$$\boxed{c_0 b^d + \boxed{c_1 a b^{d-1} + \cdots + c_{d-1} a^{d-1} b} + c_d a^d} = 0.$$

Observe that $a$ divides all terms in the red box; hence[20] $a$ divides the remaining term $c_0 b^d$. Similarly, $b$ divides all terms in the blue box; hence $b$ divides the remaining term $c_d a^d$. We have found

$$a \mid c_0 b^d \text{ and } b \mid c_d a^d.$$

Since $\mathrm{GCD}(a,b) = 1$, Corollary 2.25 implies $\mathrm{GCD}(b, a^d) = 1$ and $\mathrm{GCD}(a, b^d) = 1$, and Euclid's Lemma (Lemma 2.12) implies

$$a \mid c_0 \text{ and } b \mid c_d. \blacksquare$$

From this theorem, one may hunt for rational roots to a polynomial systematically. There are an infinite number of rational numbers, but one needs only to search the divisors of the coefficients $c_0$ and $c_d$ in order to find the numerators and denominators of rational roots.

---

[19] An algebraic number is a number $x$ which is a root of a nonzero polynomial with integer coefficients. For example, $\sqrt{2}$ is algebraic, since $x = \sqrt{2}$ satisfies $x^2 - 2 = 0$. All rational numbers are algebraic, since if $x = a/b$ is a reduced fraction, then $bx - a = 0$. All constructible numbers are algebraic, but there are many more algebraic numbers. For example, the real number $\sqrt[5]{13}$ is algebraic (since $x = \sqrt[5]{13}$ satisfies $x^5 - 13 = 0$) but not constructible. Some algebraic numbers are hard to describe without giving a polynomial; for example, if $x$ is the unique real number satisfying $x^5 - 2x + 2 = 0$, then there is no better way to describe the algebraic number $x$. Non-algebraic numbers are called **transcendental**. Famous transcendental numbers include $e$, $\pi$, and $\log(2)$.

[20] Here we use the two-out-of-three principle, when one term is zero, to give a one-out-of-two principle: if $x + y = 0$ and $a \mid y$ then $a \mid x$.

We demonstrate how to use the rational root theorem to prove the irrationality of many numbers. Let us first reinterpret the rational root theorem a bit. As stated in Theorem 3.8, if $x$ is a rational number, $x = a/b$ as a reduced fraction, and

$$c_0 + c_1 x + c_2 x^2 + \cdots + c_d x^d = 0, \tag{3.1}$$

then $a \mid c_0$ and $b \mid c_d$. If no such rational number exists, satisfying Equation (3.1) and these divisibility conditions, then any number satisfying Equation (3.1) must be irrational.

**Problem 3.9** Prove that $\sqrt{2} + \sqrt{3}$ is irrational.

SOLUTION: Let $x = \sqrt{2} + \sqrt{3}$. The square of $x$ may be computed

$$x^2 = (\sqrt{2} + \sqrt{3})^2 = 2 + 2\sqrt{2}\sqrt{3} + 3 = 5 + 2\sqrt{6}.$$

The fourth power of $x$ may be computed:

$$x^4 = x^2 \cdot x^2 = (5 + 2\sqrt{6})^2 = 25 + 20\sqrt{6} + 24 = 49 + 20\sqrt{6}.$$

Now observe that one can "cancel" the $\sqrt{6}$ terms by subtracting suitable multiples:

$$x^4 - 10x^2 = 49 + 20\sqrt{6} - 10(5 + 2\sqrt{6}) = 49 + 20\sqrt{6} - 50 - 20\sqrt{6} = -1.$$

We have found that if $x = \sqrt{2} + \sqrt{3}$ then

$$x^4 - 10x^2 + 1 = 0.$$

This allows us to apply the rational root theorem.

For if $x = a/b$ were a rational number, with $a/b$ a reduced fraction, then $a \mid 1$ and $b \mid 1$. This would imply that $x = 1$ or $x = -1$. But neither of these numbers satisfies the polynomial equation $x^4 - 10x^2 + 1 = 0$. Hence $x$ cannot be a rational number. ✓

This problem might make the reader think something like "if $x$ doesn't look like a rational number, then $x$ isn't a rational number." But not all rational ducks quack like a rational duck. Here is a very unusual-looking number, mentioned first by Daniel Shanks:[21]

$$\sqrt{5} + \sqrt{22 + 2\sqrt{5}} - \sqrt{11 + 2\sqrt{29}} - \sqrt{16 - 2\sqrt{29} + 2\sqrt{55 - 10\sqrt{29}}}.$$

This beast of a constructible number certainly looks irrational. But it is a most rational number; it equals zero.

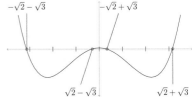

Figure 3.6: The graph of $y = x^4 - 10x^2 + 1$ intersects the x-axis at four points. The rightmost point is $\sqrt{2} + \sqrt{3}$. The rational root theorem shows that all four intersection points $\pm\sqrt{2} \pm \sqrt{3}$ are irrational, since they do not equal $\pm 1$.

[21] See "Incredible Identities," by D. Shanks, in the *Fibonacci Quarterly*, **12**, pp. 271–281 (1974).

Before returning to rational numbers, we give one more connection between straightedge-compass constructions and number theory.

Proposition I.1, of the *Elements* constructs an equilateral triangle. But for another approach, begin with a unit length 1 and some other length $x$, less than 1. Then try to construct a polygon, by tracing chords inside a unit circle, at distance $x$ from the center:

Figure 3.7: Euclid's construction of an equilateral triangle on the segment $AB$.

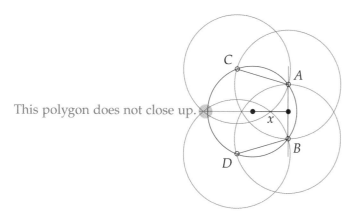

Figure 3.8: Begin with a unit circle, and draw a chord at distance $x$ from the circle's center; this chord will intersect the circle at two points, $A$ and $B$. Now draw circles with centers $A$ and $B$, sharing the same radius $AB$. These circles will intersect the unit circle at new points $C$ and $D$. Connect $AC$ and $BD$. Carry on in this manner, inscribing a polygon in the unit circle. If you get lucky with your choice of initial length $x$, the polygon will close up perfectly.

If you begin with $x = \cos(\pi/n)$, then this construction "closes up" to give a regular[22] polygon with $n$ sides. It follows that

**Proposition 3.10** *If $n$ is an integer, $n \geq 3$, and $\cos(\pi/n)$ is a constructible number, then one can inscribe a regular $n$-gon in a unit circle with only straightedge and compass.*

Not only did Euclid construct a regular triangle in his *Elements*. In Proposition IV.6, Euclid inscribes a square in a given circle, and by Proposition IV.11, Euclid inscribes a regular pentagon. In light of the proposition above, the inscriptions of regular triangle, quadrilateral, and pentagon are possible because $\cos(\pi/3)$ and $\cos(\pi/4)$ and $\cos(\pi/5)$ are constructible numbers:

$$\cos\left(\frac{\pi}{3}\right) = \frac{1}{2}, \quad \cos\left(\frac{\pi}{4}\right) = \frac{\sqrt{2}}{2}, \quad \cos\left(\frac{\pi}{5}\right) = \frac{1+\sqrt{5}}{4}.$$

By March 30, 1796, the almost-19-year-old Gauss wrote in his journal of a construction of the 17-gon. The relevant trigonometric identity appears in his *Disquisitiones*. The 17-gon is constructible, because $\cos(\pi/17)$ equals an irrational but constructible number:

[22] Here **regular** means equilateral (all sides have the same length) and equiangular (all angles have the same measure). For triangles, equilateral is equivalent to equiangular. But for quadrilaterals, note that all rectangles are equiangular, and all rhombi are equilateral. Only the squares are regular quadrilaterals.

A **Fermat prime** is a prime number $p$ of the form $2^{2^n} + 1$. The first few Fermat primes are

$$2^{2^0} + 1 = 3, \quad 2^{2^1} + 1 = 5, \quad 2^{2^2} + 1 = 17.$$

Gauss's method generalizes to prove constructibility of a $p$-gon, whenever $p$ is a Fermat prime. Unfortunately, there are only five known Fermat primes:

$$3, 5, 17, 257, 65537.$$

$$\frac{1}{8}\sqrt{30 + 2\sqrt{17} + 2\sqrt{2}\left(\sqrt{34 + 6\sqrt{17} + \sqrt{2}(\sqrt{17} - 1)\sqrt{17 - \sqrt{17}} - 8\sqrt{2}\sqrt{17 + \sqrt{17}}} + \sqrt{17 - \sqrt{17}}\right)}.$$

The reader may be happy to return immediately to rational numbers.

THE WRONG WAY to add fractions is to add the numerators and add the denominators. So we use a new symbol for this operation:
$$\frac{a}{b} \vee \frac{c}{d} = \frac{a+c}{b+d}.$$
Critical is the fact that this is an operation on *fractions* and not on *rational numbers*; the effect of this operation depends on whether the fractions are reduced or not. For example,
$$\frac{2}{4} \vee \frac{1}{3} = \frac{3}{7} \text{ and } \frac{1}{2} \vee \frac{1}{3} = \frac{2}{5}.$$
Though $2/4 = 1/2$, the results $3/7$ and $2/5$ are different rational numbers! This is just one sign that $\vee$ cannot be addition.

This operation is called the **mediant**. We say that $2/5$ is the mediant fraction of $1/2$ and $1/3$. A geometric interpretation explains the name.

**Proposition 3.11 (Mediant fractions lie between)** *Let $a/b$ and $c/d$ be fractions with $b > 0$ and $d > 0$. Then the mediant $(a/b) \vee (c/d)$ represents a rational number that lies between $a/b$ and $c/d$.*

PROOF: Since $b > 0$ and $d > 0$, $b + d > 0$. It follows that $(a/b) \vee (c/d) = (a+c)/(b+d)$ has positive denominator; it represents a rational number. To see that the mediant lies between $a/b$ and $c/d$, we view the rational numbers as slopes in the right half[23] of the plane.

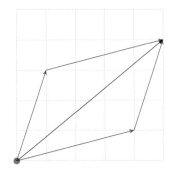

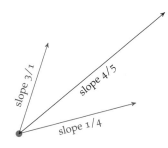

[23] It is crucial that we work in the right half of the plane for this geometric reasoning. "Betweenness" of vectors in the right half of the plane reflects betweenness of slopes. If, instead, we worked in the top half of the plane, this would no longer be true. Consider the figure below:

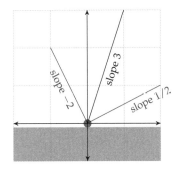

Certainly, the black line is between the red and blue. But 3 is not between $1/2$ and $-2$. The slope of a line does not vary continuously, as a line rotates past vertical.

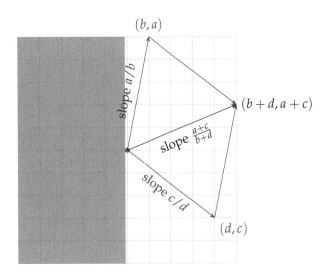

The diagonal black line lies between the blue and red edges in the right half-plane; hence the slope of the black line is between the slopes of the blue and red edges. Thus the rational number $(a/b) \vee (c/d)$ is between $(a/b)$ and $(c/d)$. ∎

**Problem 3.12** Which is greater, 2/5 or 3/7?

SOLUTION: Observe that
$$\frac{3}{7} = \frac{2}{5} \vee \frac{1}{2}.$$
Hence 3/7 is between 2/5 and 1/2. Since 2/5 < 1/2, we find that
$$\frac{2}{5} < \frac{3}{7} < \frac{1}{2}.$$  ✓

Figure 3.9: The deviations between slopes of 1/2, 2/5, and 3/7 are almost indiscernable; but still, 3/7 is between 1/2 and 2/5.

This method is quite powerful, even with larger fractions.

**Problem 3.13** Which is greater, 11/17 or 17/30?

SOLUTION: Observe the following mediant identity:
$$\frac{17}{30} = \frac{11}{17} \vee \frac{6}{13}.$$
Since 6/13 is less than a half, and 11/17 is more than a half, we find that
$$\frac{6}{13} < \frac{17}{30} < \frac{11}{17}.$$  ✓

The mediant fraction lies between two given reduced fractions. But *where* between them? The mediant is rarely the mean; it is not simply the average of two rational numbers.[24] To locate the mediant, we introduce **kissing fractions**.

**Definition 3.14** Let $a/b$ and $c/d$ be fractions. We say $a/b$ kisses $c/d$ if $ad - bc = \pm 1$. For notation, we denote kissing[25] with a heart:
$$\frac{a}{b} \heartsuit \frac{c}{d} \text{ means that } ad - bc = 1 \text{ or } ad - bc = -1.$$

Each fraction has its place on the number line. To see the fractions kiss, you must draw their Ford circles. Lying atop the number line at $a/b$, the **Ford circle** is the circle of diameter $1/b^2$.

[24] It could not be an honest operation on rationals like the mean, since it is not an operation on rational numbers at all!

[25] Note that if $\frac{a}{b} \heartsuit \frac{c}{d}$ then $\frac{c}{d} \heartsuit \frac{a}{b}$. The equality $ad - bc = \pm 1$ is equivalent to the equality $bc - ad = \mp 1$. The heart symbol is symmetric, just like the kissing relation.

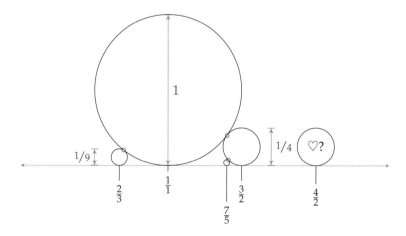

Figure 3.10: The Ford circles on top of 1/1, 2/3, 7/5, 3/2, and 7/3. Their diameters are 1, 1/9, 1/25, 1/4, and 1/9, respectively. The circles *osculate*, or *kiss*, at their point of tangency. Observe three kisses:
$$\frac{2}{3} \heartsuit \frac{1}{1}.$$
$$\frac{1}{1} \heartsuit \frac{3}{2}.$$
$$\frac{3}{2} \heartsuit \frac{7}{5}.$$

FRACTIONS KISS precisely when their Ford circles are tangent.

**Theorem 3.15 (Kissing fractions have tangent Ford circles)** *Let $a/b$ and $c/d$ be fractions with nonzero denominators. Then $a/b$ kisses $c/d$ if and only if the Ford circle atop $a/b$ is tangent to the Ford circle atop $c/d$.*

PROOF: Let $\Delta = ad - bc$; thus $\Delta = \pm 1$ if and only if $a/b$ kisses $c/d$. The number $\Delta$ occurs naturally when subtracting fractions:

$$\frac{a}{b} - \frac{c}{d} = \frac{ad - bc}{bd} = \frac{\Delta}{bd}.$$

Draw the Ford circles atop $a/b$ and $c/d$.

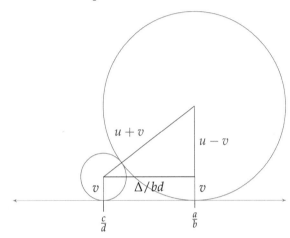

Let $u = 1/2b^2$ and $v = 1/2d^2$, the radii of the two circles. The distance between their centers, by the Pythagorean theorem, equals

$$\sqrt{\left(\frac{\Delta}{bd}\right)^2 + (u-v)^2}.$$

The circles are tangent if and only if the distance between their centers is the sum of the radii $(u+v)$, as pictured. Hence the circles are tangent if and only if

$$\left(\frac{\Delta}{bd}\right)^2 + (u-v)^2 = (u+v)^2.$$

This is equivalent to

$$\frac{\Delta^2}{b^2 d^2} = (u+v)^2 - (u-v)^2 = 4uv.$$

But notice that $4uv = 1/b^2 d^2$. So the circles are tangent if and only if

$$\frac{\Delta^2}{b^2 d^2} = \frac{1}{b^2 d^2}.$$

Hence the circles are tangent if and only if $\Delta^2 = 1$, if and only if $\Delta = \pm 1$, if and only if the fractions kiss. ∎

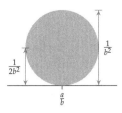

Figure 3.11: The Ford circle atop $a/b$ has diameter $1/b^2$ and radius $1/2b^2$.

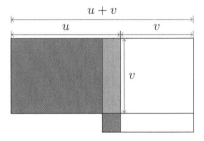

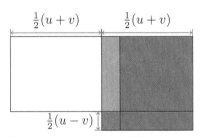

Figure 3.12: Based on Euclid, Proposition II.5, and related figures (unpublished) by W. Casselman. This figure demonstrates the identity

$$\left(\frac{u+v}{2}\right)^2 - \left(\frac{u-v}{2}\right)^2 = uv.$$

The left side corresponds to all colored regions except for the blue, in the bottom figure. The right side comes from arranging the red and green regions as in the top figure. Multiplying through by 4 gives the identity

$$(u+v)^2 - (u-v)^2 = 4uv$$

used in this proof.

KISSING FRACTIONS lead inevitably to more kissing. The following result demonstrates that when two fractions kiss, the mediant fraction gets in on the action.

**Proposition 3.16** *Let $a/b$ and $c/d$ be two kissing fractions. Then the mediant $(a/b) \vee (c/d)$ kisses both $a/b$ and $c/d$.*

PROOF: The proof is algebraic, beginning with the kissing assumption: $ad - bc = \pm 1$. A fortuitous cancellation yields

$$(a+c)b - (b+d)a = ab + cb - ba - da = cb - da = -(ad - bc) = \mp 1.$$

The mediant is $(a/b) \vee (c/d) = (a+c)/(b+d)$, and the above computation implies

$$\frac{a+c}{b+d} \heartsuit \frac{a}{b}.$$

The same computation, reversing the roles of $a/b$ and $c/d$, implies that

$$\frac{a+c}{b+d} \heartsuit \frac{c}{d}. \blacksquare$$

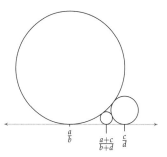

Figure 3.13: Given $(a/b)\heartsuit(c/d)$, we find that
$$\frac{a}{b} \heartsuit \frac{a+c}{b+d} \heartsuit \frac{c}{d}.$$

The geometric consequence is the following: if one begins with two kissing fractions, and continues production of fractions by mediants, then the resulting Ford circles are squeezed tight.

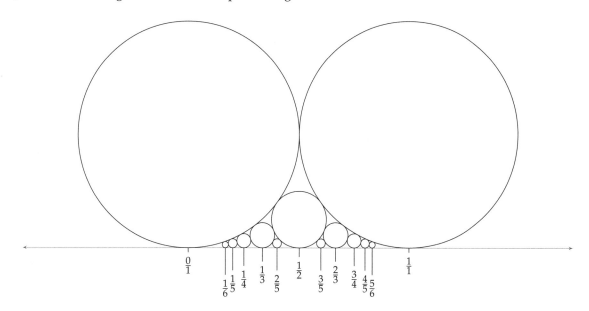

Figure 3.14: Mediants, kissing fractions and tangent Ford circles.

Begin with the kissing fractions $0/1$ and $1/1$. Their mediant, $1/2$ kisses both, and therefore the Ford circle on $1/2$ squeezes perfectly between the two large circles. The mediant of $0/1$ and $1/2$ is $1/3$. The Ford circle on $1/3$ squeezes perfectly between the circle at $1/2$ and the circle at $0/1$.

Which fractions kiss? The kissing equation $ad - bc = \pm 1$ requires $\text{GCD}(a, b) = 1$ and $\text{GCD}(c, d) = 1$. Indeed, if $g \mid a$ and $g \mid b$, then $g \mid ad - bc$, so $g \mid \pm 1$, so $g = \pm 1$. Hence $\text{GCD}(a, b) = 1$, and the same argument yields $\text{GCD}(c, d) = 1$. Nonreduced fractions cannot kiss.

**Proposition 3.17** *If* $\text{GCD}(a, b) \neq 1$, *then* $a/b$ *kisses no other fractions.*

In contrast, reduced fractions find many mates.

**Proposition 3.18** *Let* $a/b$ *be a reduced fraction. Then* $a/b$ *kisses infinitely many fractions. If* $b > 1$ *then among the reduced fractions kissing* $a/b$, *there are exactly two with denominator smaller than* $b$. *These two fractions kiss each other and have mediant* $a/b$.

PROOF: Since $\text{GCD}(a, b) = 1$, there exists a solution $(x_0, y_0)$ to the linear Diophantine equation

$$ay - bx = 1.$$

Moreover, by Theorem 1.21, there are infinitely many solutions,[26] one for every integer $n$:

$$y = y_0 + bn, \quad x = x_0 + an.$$

[26] The signs here are slightly different from Theorem 1.21, but the result follows directly if one keeps track of signs carefully.

Each such solution $(y, x)$ gives a fraction $x/y$ which kisses $a/b$.

If $b > 1$, then among the numbers $y_0 + bn$ (as $n$ varies over all integers), exactly two are smaller than $b$ in absolute value.

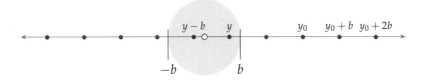

Figure 3.15: The numbers $y_0 + bn$ are highlighted with dots on the number line, spaced at intervals of length $b$. Zero is not among the highlighted numbers since $y = 0$ and $ay - bx = 1$ implies $bx = 1$, but $b > 1$, a contradiction. Thus exactly two dots can fit into the circle of radius $b$.

Among the two, call the positive number $y$, so its negative neighbor is $y - b$. These give two reduced fractions

$$\frac{x}{y} \heartsuit \frac{a}{b}, \quad \frac{a-x}{b-y} \heartsuit \frac{a}{b}.$$

The fractions $(x/y)$ and $(a - x)/(b - y)$ kiss (a direct algebraic computation suffices), and their mediant is $a/b$. ∎

This Theorem has a geometric interpretation using Ford circles.

**Corollary 3.19** *Let* $a/b$ *be a reduced fraction. Then the Ford circle atop* $a/b$ *is tangent to infinitely many other Ford circles; if* $b > 1$, *then two of these have larger diameters and kiss each other.*

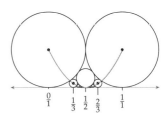

Figure 3.16: Infinitely many fractions kiss the reduced fraction $1/2$, among which two are bigger. Centers of their Ford circles lie on the parabola

$$y = 2 \cdot (x - 1/2)^2.$$

As the proof of the Theorem uses linear Diophantine equations, we can find kissing fractions by using the Euclidean algorithm.

**Problem 3.20** Find a fraction which kisses 71/83.

SOLUTION: We look for integers $x, y$ such that $(x/y) \heartsuit (71/83)$; this is equivalent to the statement $71y - 83x = \pm 1$. To solve this Diophantine equation, we use the Euclidean algorithm.

$$83 = 1(71) + 12$$
$$71 = 5(12) + 11$$
$$12 = 1(11) + 1.$$

Solving for 1 (review Problem 1.15) yields

$$1 = 12 - 1(11)$$
$$= 12 - 1(71 - 5(12)) = 6(12) - 1(71)$$
$$= 6(83 - 1(71)) - 1(71)$$
$$= 6(83) - 7(71).$$

We have found that $71(-7) - 83(-6) = 1$, so $71(7) - 83(6) = -1$, and

$$\frac{71}{83} \heartsuit \frac{6}{7}. \quad \checkmark$$

By finding kissing fractions, of smaller and smaller denominators, we can work backwards until the denominator equals 1.

**Theorem 3.21 (Integers generate all reduced fractions via mediants)** *Begin with the integer fractions $n/1$ for every integer $n$. Whenever fractions kiss, take their mediants, and continue this process indefinitely. Only reduced fractions will occur, and all reduced fractions will eventually occur.*

PROOF: Note first that only fractions with positive denominators will occur; the mediant of two fractions with positive denominator is another fraction with positive denominator. Moreover, since mediant fractions get in on the action, and kissing fractions (with positive denominators) are always reduced, this process *only* yields reduced fractions.

To see that *all* reduced fractions eventually occur, we begin with a reduced fraction $a/b$. If $b = 1$, then $a/b = a/1$ occurred at the beginning of the process. Otherwise, the previous Theorem implies that $a/b$ is the mediant of two reduced fractions $x/y$ and $(a-x)/(b-y)$ with smaller denominator. Repeat the process on $x/y$ and $(a-x)/(b-y)$ and so on, until all denominators equal 1. This traces the ancestry of $a/b$ back to fractions of denominator 1. ∎

In this way, Ford circles and kissing fractions resolve the problem of locating rational numbers on the number line. Each rational number lands in its place when its Ford circle is carried on its back.

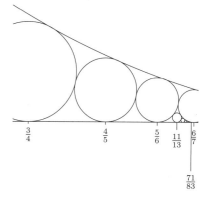

Why does 71/83 occur? First, note that it is a mediant:

$$\frac{71}{83} = \frac{6}{7} \vee \frac{65}{76}.$$

These are both mediants of fractions with smaller denominators:

$$\frac{6}{7} = \frac{1}{1} \vee \frac{5}{6}, \quad \frac{65}{76} = \frac{6}{7} \vee \frac{59}{69}.$$
$$\frac{5}{6} = \frac{1}{1} \vee \frac{4}{5}, \quad \frac{59}{69} = \frac{6}{7} \vee \frac{53}{62}.$$
$$\frac{4}{5} = \frac{1}{1} \vee \frac{3}{4}, \quad \frac{53}{62} = \frac{6}{7} \vee \frac{47}{55}.$$
$$\frac{3}{4} = \frac{1}{1} \vee \frac{2}{3}, \quad \frac{47}{55} = \frac{6}{7} \vee \frac{41}{48}.$$
$$\frac{2}{3} = \frac{1}{1} \vee \frac{1}{2}, \quad \frac{41}{48} = \frac{6}{7} \vee \frac{35}{41}.$$
$$\frac{1}{2} = \frac{1}{1} \vee \frac{0}{1}, \quad \frac{35}{45} = \frac{6}{7} \vee \frac{29}{34}.$$
$$\frac{29}{34} = \frac{6}{7} \vee \frac{23}{27}, \quad \frac{23}{27} = \frac{6}{7} \vee \frac{17}{20}.$$
$$\frac{17}{20} = \frac{6}{7} \vee \frac{11}{13}, \quad \frac{11}{13} = \frac{6}{7} \vee \frac{5}{6}.$$

These facts can be used as directions, to begin with Ford circles on 0/1 and 1/1, and nest down to the Ford circle on top of 71/83.

WE APPROXIMATE real numbers by rational numbers every time we write a decimal expansion. When we write $\pi \approx 3.14$, we are saying that $\pi$ is approximately equal to the rational number $314/100$. But this rational approximation is not particularly good; the approximation $22/7 = 3.\overline{142857}$ is better, and for a much smaller *price*.

The **price**[27] of a fraction is the size of its denominator. So the fraction $314/100$ costs 100 dollars; even reducing it to $107/50$, it still costs 50 dollars. For 50 dollars, we can approximate $\pi$ within an error of about 0.0016. But for only 7 dollars, we may purchase the fraction $22/7$, which approximates $\pi$ within an error of about 0.0013 – a better approximation for lower price!

**Diophantine approximation** studies the question: given a real number $x$, what is the relationship between the *accuracy* of a rational approximation and its *cost*.

The first Diophantine approximation theorem follows from the geometry of the number line.

[27] Edward Burger taught me to think of the denominator of a fraction as a price or cost, in the context of Diophantine approximation. See "Making Transcendence Transparent", by E. Burger and R. Tubbs, Springer (2004) for much more.

**Proposition 3.22** *Let $x$ be a real number. Let $b$ be a positive integer. Then there exists a rational number $a/b$ such that*

$$\left| x - \frac{a}{b} \right| \leq \frac{1}{2b}.$$

*In other words, given $b$ dollars, one can approximate every real number $x$ within a distance of $1/2b$.*

PROOF: Plot $x$ on the number line, and mark every fraction with denominator $b$. Surround each fraction with denominator $b$ by a circle[28] of diameter $1/b$ to completely cover the number line by circles.

[28] Intervals suffice but circles look better.

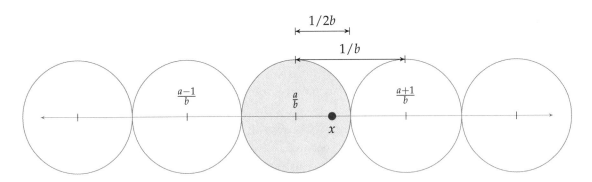

If $a/b$ is the fraction whose circle contains $x$, then the distance between $x$ and $a/b$ is bounded by the radius of the circle.

$$\left| x - \frac{a}{b} \right| \leq \frac{1}{2b}.$$

∎

The previous theorem guarantees that $b$ dollars buys an approximation within $1/2b$. Ford circles can give closer approximations, often for a better price.

Suppose that $x$ is a real number which lies in the shadow of a Ford circle. If $a/b$ is the rational number touching the Ford circle, then as the diameter of the Ford circle is $1/b^2$, we observe that

$$\left| x - \frac{a}{b} \right| < \frac{1}{2b^2}.$$

**Theorem 3.23 (Dirichlet approximation theorem)** *Let $x$ be an irrational real number. Then there exist infinitely many reduced fractions $a/b$ such that*

$$\left| x - \frac{a}{b} \right| < \frac{1}{2b^2}.$$

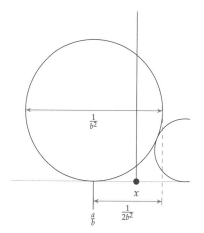

PROOF: To demonstrate the theorem, it suffices to prove that $x$ lies in the shadows of infinitely many Ford circles. To begin, $x$ lies in the shadow of one Ford circle: either the Ford circle atop $\lfloor x \rfloor / 1$ or the Ford circle atop $\lceil x \rceil / 1$ overshadows $x$.

Now consider a Ford circle atop $a/b$ whose shadow contains $x$, and all of the Ford circles tangent to it.

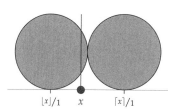

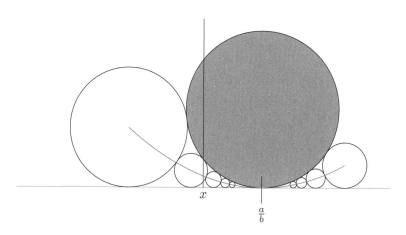

These Ford circles kissing $a/b$ form an unbroken[29] chain, covering all points in the shadow, except $a/b$ itself. As $x$ is irrational, $x \neq a/b$, and the vertical line above $x$ must pass through the chain.

Hence when $x$ lies in the shadow of one Ford circle, it lies in the shadow of a smaller Ford circle; this continues indefinitely, to demonstrate that $x$ lies in the shadows of infinitely many Ford circles. ∎

[29] The chain is unbroken since each Ford circle is obtained from an adjacent Ford circle by taking its mediant with $a/b$. The x-coordinates of the centers along the chain approach $a/b$, since

$$\lim_{n \to \pm\infty} \frac{x_0 + an}{y_0 + bn} = \frac{a}{b}.$$

Hence all points, arbitrarily close to $a/b$, are covered by the chain.

## Historical notes

This chapter owes much to *La Geometrie* (published in 1637) of René Descartes (1596 – 1650CE), which brought algebraic techniques to bear results on classical Greek geometry.[30] Our perspective differs from Descartes, as we place arithmetic in the center and geometry in its service. In Book 1 of his *Geometrie*, Descartes writes[31]

> And if it can be solved by ordinary geometry, that is, by the use of straight lines and circles traced on a plane surface, when the last equation shall have been entirely solved there will remain at most only the square of an unknown quantity, equal to the product of its root by some known quantity, increased or decreased by some other quantity also known.

Descartes states that, whatever Euclidean constructions are carried out, the resulting equations can be solved with nothing more the quadratic formula. In effect, our Theorem 3.3 is due to Descartes.

PYTHAGOREAN TRIPLES occur whenever right triangles arise, e.g. in the construction of altars in Vedic India. Many authors[32] have suggested that a Babylonian tablet, Plimpton 322, is a table of Pythagorean triples. But Eleanor Robson has offered an alternative hypothesis,[33] based on contemporary mathematics of Mesopotamia such as reciprocal pairs and completing the square.

In Lemma 1 before Proposition X.29 of the *Elements*, Euclid gives a recipe to produce Pythagorean triples. Euclid's recipe can be modernized into an algebraic recipe: Begin with positive integers $x$ and $y$, both even or both odd, with $x > y$; thus the difference $x - y$ and sum $x + y$ are both even. There is an equality of integers

$$xy + \left(\frac{x-y}{2}\right)^2 = \left(\frac{x+y}{2}\right)^2.$$

If $xy$ is a square number, then this gives a Pythagorean triple.[34]

Diophantus, in Book II, Problem 8 of his *Arithmetica*, demonstrates a similar recipe. Pierre de Fermat (c. 1601–1665) owned Bachet's Latin edition of the *Arithmetica*, and that is where Fermat wrote his famous marginal note:

> Cubum autem in duos cubos, aut quadratoquadratum in duos quadratoquadatos & generaliter nullam in infinitum ultra quadratum potestatem in duos eius dem nominis fas est dividere cuius rei demonstrationem mirabilem sane detexi. Hanc marginis exiguitas non caperet.[35]

Here Fermat states that, to express a cube as two cubes or a fourth-power as two fourth-powers, and so on, is impossible. Fermat claims a wonderful proof of this impossibility, too small to fit in the margin.

---

[30] See "The Geometry of René Descartes," translated from the French and Latin by David Eugene Smith and Marcia L. Latham, Dover Publications, Inc., New York, NY 1954.

[31] The French from 1637 reads "Et que si elle peut estre resolue par la Geometrie ordinaire, c'est a dire, en ne se seruant que de lignes droites & circulaires tracées fur une superficie plate, lorsque la derniere Equation aura esté entierement démeslée il n'y restera tout au plus qu'un quarré inconnu, esgal a ce qui se produist de l'Addition, ou soustraction de sa racine multipliée par quelque quantité connue, & de quelque autre quantité aussy connue."

[32] See "Sherlock Holmes in Babylon," by R. C. Buck, in the *American Mathematical Monthly* **87** (1980) pp. 335–345, which refers back to "The Exact Sciences in Antiquity," by O. Neugebauer (original, 1951) for two examples.

[33] See "Neither Sherlock Holmes nor Babylon: a reassessment of Plimpton 322," by E. Robson, in *Historia Mathematica* **28**, pp. 167–202 (2001).

[34] For example, choose $x = 9$ and $y = 1$, so that $xy = 3^2$ is a square as required. Then substitution yields $(x-y)/2 = 4$ and $(x+y)/2 = 5$, and the Pythagorean triple $3^2 + 4^2 = 5^2$. If one chooses $x = 32$ and $y = 2$, then $xy = 64 = 8^2$; substitution yields $(x-y)/2 = 15$ and $(x+y)/2 = 17$, and the Pythagorean triple $8^2 + 15^2 = 17^2$.

[35] Pierre de Fermat's marginal note was included in a 1670 edition of the *Arithmetica*, published by Fermat's son. Read "Fermat's Enigma," by Simon Singh, Anchor Books ed., New York (1998) for a good popular account of the history of the Last Theorem.

**Fermat's Last Theorem** refers to this marginal suggestion of Fermat. Here it is, stated in modern terms.

**Theorem 3.24 (Fermat's Last Theorem)** *If $n$ is a positive integer, and*[36] *$n > 2$, then the equation $x^n + y^n = z^n$ has no solutions for which $x$, $y$, and $z$ are all positive integers.*

Over 300 years, efforts to prove Fermat's conjecture led to outstanding discoveries in number theory, even as progress on the conjecture itself was incremental.[37]

Following suggestions of G. Frey (1986), and proving the $\epsilon$-conjecture of J.P. Serre, K. Ribet (1990) demonstrated that a 1957 conjecture of Shimura and Taniyama implied Fermat's Last Theorem. By proving enough cases of Shimura and Taniyama's conjecture, Andew Wiles with Richard Taylor proved Fermat's Last Theorem in 1995.[38]

FORD CIRCLES are named for L.R. Ford, who first revealed[39] the connections between the circles and Diophantine approximation.

We have given only two results in Diophantine approximation of an irrational number $x$; the more powerful result guarantees the existence of infinitely many rational numbers $a/b$ for which $|x - a/b| \leq 1/2b^2$.

There are two ways one might try to improve this: by increasing the constant 2 or by increasing the exponent 2. Ford improved the constant from 2 to the best possible $\sqrt{5}$ (note that $\sqrt{5} \approx 2.236 > 2$):[40]

$$\left| x - \frac{a}{b} \right| \leq \frac{1}{\sqrt{5}b^2}, \text{ for infinitely many rational numbers } \frac{a}{b}.$$

Improving the exponent in this estimate is more difficult. A series of results, beginning with Liouville (1844), through results of Thue and Siegel, culminated with Roth's proof[41] that the exponent cannot be improved for *algebraic* numbers:

**Theorem 3.25 (Thue-Siegel-Roth Theorem)** *Let $x$ be an algebraic irrational number. Then for any positive number $\epsilon$ and constant $C$, the inequality*

$$\left| x - \frac{a}{b} \right| \leq \frac{C}{b^{2+\epsilon}}$$

*is only satisfied for finitely many fractions $a/b$.*

In other words, one cannot improve the exponent, even slightly, from 2 to $2 + \epsilon$, at least not for *algebraic* numbers. This gives an effective way of proving transcendence of some numbers – if it is possible to approximate an irrational number extremely well by infinitely many rational numbers, then it must be transcendental!

---

[36] If $n = 2$, then the equation $x^2 + y^2 = z^2$ has infinitely many solutions in positive integers. They are the Pythagorean triples.

[37] Fermat proved the Theorem when $n = 4$. Euler proved the Theorem when $n = 3$, or at least his arguments can be fixed to make a proof in this case. Legendre and Dirichlet proved the Theorem when $n = 5$. Lamé and Lebesgue (*not* of Lebesgue measure fame) proved the Theorem when $n = 7$. The Theorem was divided into two cases later, and Germain proved the "first case" for odd primes up to 100.

[38] "Modular elliptic curves and Fermat's Last Theorem," by A. Wiles and "Ring theoretic properties of certain Hecke algebras," by R. Taylor and A. Wiles, in *Annals of Mathematics*, vol. 141 **3** (1995).

[39] "Fractions," by L. R. Ford, in *The American Mathematical Monthly*, vol. 45, **9** (Nov.,1938), pp. 586–601.

[40] The best-possible constant $\sqrt{5}$ was found earlier, by different methods, by A. Hurwitz, in "Über die angenäherte Darstellung der Irrationalzahlen durch rationale Brüche," in *Mathematische Annalen*, vol. 39 (1891) pp. 279–284.

[41] "Rational approximations to algebraic numbers," by K. F. Roth, in *Mathematika*, vol. 2, **1** (1955), pp. 1–20. The story does not completely end with Roth. Instead of increasing the exponent from $b^2$ to $b^{2+\epsilon}$, on can try a slighter change from $b^2$ to $b^2 \log(b)^{1+\epsilon}$. This gives the conjecture of S. Lang (1927–2005)

## Exercises

1. Beginning with a line segment of length 1, construct line segments of length 3, 1/3, 4/3, and $2\sqrt{3}/3$.

2. Prove that there are infinitely many positive integer triples $(x, y, z)$ such that $x^2 + 2y^2 = 3z^2$. Hint: Follow the proof for Pythagorean triples, replacing the point $(0, 1)$ by the point $(1, 1)$.

3. Demonstrate that $\sqrt{5/7}$, $\sqrt{3} + \sqrt{5}$, and $\sqrt{8} - \sqrt{7}$ are irrational.

4. (Challenge) Prove that if $p$ and $q$ are distinct prime numbers, then $\sqrt{p} + \sqrt{q}$ is irrational.

5. The number $e$ is defined as the sum of the reciprocals of the factorials,[42] If $e$ were rational, let $n$ be its denominator when represented as a fraction, let $x$ be the sum of the terms up to $1/n!$ and let $y$ be the sum of the rest of the terms. Demonstrate in this case that $n! \cdot x$ is an integer and $n! \cdot e$ is an integer, and that $0 < n! \cdot y < 1$. Use this to achieve a contradiction, and fill in the steps to prove that $e$ is irrational.

    [42] The convergent series defining $e$ is
    $$e = \frac{1}{1} + \frac{1}{1} + \frac{1}{2} + \frac{1}{6} + \frac{1}{24} + \frac{1}{120} + \cdots.$$
    Hint for proving $n! \cdot y < 1$: the best way to bound a series is to compare it with a geometric series.

6. (a) Use the half-angle identity to prove that if an $n$-gon is constructible, then a $2n$-gon is constructible.

   (b) Prove that if $GCD(a, b) = 1$, and an $a$-gon is constructible and a $b$-gon is constructible, then an $(ab)$-gon is constructible. Hint: the equation $ax + by = 1$ implies $x \cdot \pi/a + y \cdot \pi/b = \pi/ab$. Use sum-difference identities for the cosine..

   (c) Let $e_2$ be a natural number, and let $p_1, \ldots, p_s$ be distinct Fermat primes. Let $N = 2^{e_2} \cdot p_1 \cdots p_n$. Use the fact that $p$-gons are constructible for every Fermat prime $p$, to demonstrate that the $N$-gon is constructible.[43]

   [43] More difficult is the following result: if $p$ is a prime number, and $p \neq 2$ and $p$ is *not* a Fermat prime, then the $p$-gon is *not* constructible. For example, the 7-gon is not constructible. The proof is a famous application of Galois theory.

7. Use mediants to compare 13/21 and 17/27.

8. Consider the sequence of Fibonacci numbers, defined recursively by $F_0 = 0$, $F_1 = 1$, and $F_n = F_{n-1} + F_{n-2}$ for all $n \geq 2$. Let $r_n$ be the sequence of fractions obtained from consecutive Fibonacci numbers: $r_n = F_n/F_{n-1}$ for all $n \geq 1$. Thus $r_n$ is the sequence of fractions:
$$\frac{1}{0}, \frac{1}{1}, \frac{2}{1}, \frac{3}{2}, \frac{5}{3}, \frac{8}{5}, \ldots.$$

   (a) Prove that each fraction in the sequence $r_n$ is the mediant of the previous two fractions, and kisses the previous two fractions.

   (b) Prove that every fraction in the sequence $r_n$ is reduced.

   (c) Draw a diagram of the Ford circles above the fractions in the sequence $r_n$. Prove any observations you can make.

(d) Prove that the sequence $r_n$ converges to some real number.

(e) Let $\phi = 1/2 \cdot (1 + \sqrt{5})$. This is called the **golden ratio**. Prove that $\lim_{n \to \infty} r_n = \phi$. Hint: squeeze $\phi$ between consecutive terms in the sequence.

9. Find fractions which kiss 7/13, 77/133, and 101/37.

10. Write a formula which describes all fractions which kiss 7/11.

11. If $P$ is a point in the plane, and $\ell$ is a line that does not pass through $P$, then the set of points equidistant from $P$ and $\ell$ forms a parabola. Prove that, given a Ford circle $C$, all centers of Ford circles adjacent to $C$ lie on a parabola.

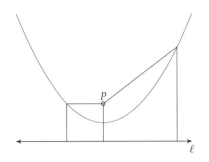

12. Prove that if Fermat's Last Theorem is true for all prime exponents $p$, and for the exponent 4, then Fermat's Last Theorem is true for all exponents $n \geq 3$.

13. Let $x$ be a real number, and let $n$ be a positive integer.[44]

    (a) For each integer $b$ between 1 and $n$, let $r_b = bx - \lfloor bx \rfloor$. The set $S = \{0, r_1, r_2, \ldots, r_n, 1\}$ contains $n + 2$ real numbers between 0 and 1. Prove that there are elements $s, t \in S$ such that
    $$0 < t - s \leq \frac{1}{n+1}.$$

    [44] This exercise leads the reader through a proof of Dirichlet's Approximation Theorem (proven by P.G.L. Dirichlet, c.1840CE), slightly weaker than our result from Ford circles, but stronger than the naïve bound.

    (b) Prove that there exist integers $a, b$ such that $1 \leq b \leq n$ and
    $$|bx - a| \leq \frac{1}{n+1}.$$
    Hint: consider the previous part. Three cases are needed: $s = 0$, $t = r_b$, or $s = r_b$, $t = 1$, or $s = r_b$, $t = r_c$.

    (c) Prove that if $|bx - a| \leq \frac{1}{n+1}$ and $b \leq n$, then
    $$\left|x - \frac{a}{b}\right| < \frac{1}{b^2}.$$
    Conclude that if $x$ is a real number, there exist infinitely many fractions $a/b$ satifying the above inequality.

14. Use the Thue-Siegel-Roth theorem (Theorem 3.25) to prove that there are only finitely many integer solutions $(x, y)$ to the equation
    $$x^3 - 5y^3 = 100.$$
    Hint: Prove[45] that $|\sqrt[3]{5+x} - \sqrt[3]{5}| < 1/3 \cdot 5^{-2/3} x$ when $x > 0$. Use this to show that if $(a, b)$ is any solution to the equation, then $a/b$ is very close to $\sqrt[3]{5}$.

    [45] The graph of $y = \sqrt[3]{x}$ near $x = 5$ lies beneath its tangent lines; use the derivative at $x = 5$ to prove the estimate.

15. Use the Thue-Siegel-Roth theorem to give an example of a decimal expansion which represents a transcendental number.

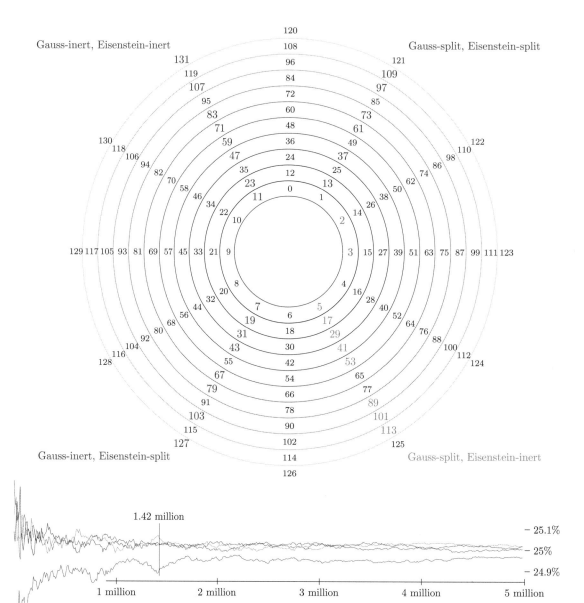

Figure 3.17: Among the primes up to 1.42 million, about 24.89% are at 1 o'clock (red) and about 25.066% are at 5 o'clock (orange).

# 4
# Gaussian and Eisenstein Integers

NUMBER THEORY flows deductively from division-with-remainder, through the Euclidean algorithm, to prime decomposition. When other number systems[1] admit division-with-remainder, a form of prime decomposition follows. We discuss two number systems in this chapter, the Gaussian and Eisenstein integers, which admit division-with-remainder and its consequences.

These number systems are interesting not only for *parallels to* the ordinary integers; they also *shed light on* the ordinary integers. In this chapter, and later in this book, we find that studying larger and stranger number systems bears fruit for the original, natural numbers. We generalize with purpose.

GAUSSIAN AND EISENSTEIN integers are named for Johann Carl Friedrich Gauss (1777-1855) and Ferdinand Gotthold Max Eisenstein (1823 – 1852). In 1844, Eisenstein wrote,

> The basic results in the theory of those complex numbers of the form $a + b\rho$, where $\rho$ is an imaginary **cube** root of of unity, have still not been recorded. But we believe, because of the great analogy between these complex numbers and the usually so-called complex numbers of the form $a + b\sqrt{-1}$, although they are similar in regards to the divisibility of numbers, decomposition into simple factors, the theory of complex primes, etc., our propositions may not be known. Compare, along the way, to the second chapter of Gauss on the biquadratic residues... [2]

As Gauss developed the arithmetic of numbers of the form $a + bi$ (where $a$ and $b$ are integers and $i^2 = -1$), Eisenstein developed the arithmetic of numbers of the form $a + b\omega$ (where $a$ and $b$ are integers and $\omega^3 = 1$ and $\omega \neq 1$). Both were driven by the study of reciprocity laws, a topic that occupies a later chapter of this book.

[1] The term "number system" is informal here. A more precise term is "ring" – a set endowed with elements called 0 and 1, operations called addition and multiplication, and satisfying axioms of identity and additive inverse, commutativity, associativity, and distributivity. A **Euclidean domain** is an integral domain which admits a form of division-with-remainder. The integers form a Euclidean domain.

[2] Loosely translated from "Beweis des Reciprocitätssatzes für die cubischen Reste in der Theorie der aus dritten Wurzeln der Einheit zusammengesetzten complexen Zahlen," by F. Eisenstein, in *Journal für die reine und angewandte Mathematik*, **27** (1844), pp. 289–310.

Natural numbers, integers, rational numbers, and constructible numbers are placed on the *line* of real numbers; we do not axiomatically study the real numbers, but we exploit their arithmetic and geometry. The commutative, associative, and distributive properties of addition and multiplication hold for real numbers. The real numbers are ordered, from left to right on the number line, and order is preserve by addition and positive scaling.

Gaussian and Eisenstein integers are placed on the *plane* of complex numbers; two dimensions are required. To understand the arithmetic of Gaussian and Eisenstein integers, we first review the arithmetic and geometry of complex numbers.

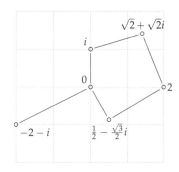

Figure 4.1: Complex numbers, plotted in the plane. They have been joined by lines to form a constellation, to compare with the opposite page.

COMPLEX numbers are points on the plane. The point with coordinates $(x, y)$, when thought of as a complex number, is called $x + yi$. We leave out zeros when convenient, so that $1 = 1 + 0i$ corresponds to the point $(1, 0)$ and $i = 0 + 1i$ corresponds to the point $(0, 1)$. When we talk about a complex number, we do not talk about its x-coordinate and y-coordinate; we talk instead about its **real part** and **imaginary part**. The real part of $2 + 3i$ is 2; the imaginary part of $2 + 3i$ is 3.

Addition of complex numbers is just addition of vectors.[3] To add vectors, add x-coordinates and y-coordinates separately:

$$(x, y) + (u, v) = (x + u, y + v).$$

To add complex numbers, add real and imaginary parts separately:

$$(x + yi) + (u + vi) = (x + u) + (y + v)i.$$

The difference between complex numbers and vectors is just notation, when only addition is involved.

Multiplication of complex numbers requires familiar commutative, associative, and distributive properties and one new fact:

$$i \cdot i = -1.$$

[3] Similarly, subtraction of complex numbers is just subtraction of vectors. To subtract vectors, subtract x-coordinates and y-coordinates separately:

$$(x, y) - (u, v) = (x - u, y - v).$$

To subtract complex numbers, subtract real and imaginary parts separately:

$$(x + yi) - (u + vi) = (x - u) + (y - v)i.$$

**Problem 4.1** Multiply $1 + i$ and $3 - i$.

SOLUTION: We compute the product

$$\begin{aligned}
(1 + i) \cdot (3 - i) &= 1 \cdot (3 - i) + i \cdot (3 - i) &&\text{by the distributive law;} \\
&= 3 - i + (i \cdot 3) + (i \cdot (-i)) &&\text{by the distributive law;} \\
&= 3 - i + 3i - (i \cdot i) &&\text{by the commutative law;} \\
&= 3 - i + 3i - (-1) &&\text{since } i \cdot i = -1; \\
&= 4 + 2i &&\text{collecting terms.} \quad \checkmark
\end{aligned}$$

To understand multiplication geometrically, one should understand complex numbers in polar form. In polar form, one writes[4] $r \cdot \mathbf{e}(\theta)$ for the point at distance $r$ from the origin, and at angle $\theta$ (measured in radians, counterclockwise from the east horizon).

[4] Some texts use the notation $\mathbf{cis}(\theta)$ instead of $\mathbf{e}(\theta)$, since

$$\mathbf{e}(\theta) = \cos(\theta) + i \cdot \sin(\theta).$$

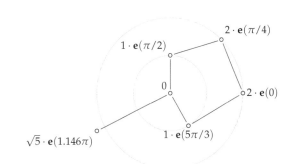

Figure 4.2: The same complex numbers are marked: on the left, they are described by real and imaginary parts; on the right they are described in polar form.

**Problem 4.2** Express the complex number $1 + i$ in polar form.

The point $1 + i$ has xy-coordinates $(1,1)$. Its distance from the origin is $\sqrt{1^2 + 1^2} = \sqrt{2}$, by the Pythagorean theorem. Its angle, measured from the east horizon, is $45°$, or $\pi/4$ radians. Hence

$$1 + i = \sqrt{2} \cdot \mathbf{e}(\pi/4).$$

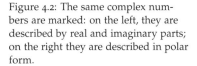

Polar form is optimal for multiplication of complex numbers.

**Proposition 4.3 (Polar multiplication of complex numbers)** *Let $r \cdot \mathbf{e}(\theta)$ and $s \cdot \mathbf{e}(\phi)$ be two complex numbers in polar form. Then*

$$(r \cdot \mathbf{e}(\theta)) \cdot (s \cdot \mathbf{e}(\phi)) = rs \cdot \mathbf{e}(\theta + \phi).$$

*In other words, the radius of the product is the product of the radii, and the angle of the product is the sum of the angles.*

Note that the number 0 has multiple representations in polar form. Most often, we just write 0, but we could also write $0 \cdot \mathbf{e}(0)$.

PROOF: The proof follows[5] from addition and subtraction formulae for sine and cosine, beginning with the observation

$$r \cdot \mathbf{e}(\theta) = r\cos(\theta) + i \cdot r\sin(\theta).$$

or more swiftly from laws of exponents and Euler's famous formula[6]

$$\mathbf{e}(\theta) = e^{i\theta}.$$

∎

[5] See the exercises for a hint.

[6] In 1748, Euler published his "Introductio in analysin infinitorum," (E101 in the Euler Archive); in Chapter 8, §132, we find a trigonometric argument that implies this theorem. In §138, we find Euler's formula, written

$$e^{+v\sqrt{-1}} = \cos .v + \sqrt{-1} \cdot \sin .v.$$

The most famous consequence of Euler's formula arises from setting $\theta = \pi$. Since $\mathbf{e}(\pi) = -1$, Euler's formula implies

$$e^{i\pi} + 1 = 0.$$

MULTIPLICATION rotates and scales the complex[7] plane. Fix a complex number $z = r \cdot \mathbf{e}(\theta)$. Multiplication of other complex numbers by $z$ moves them closer or further from the origin, by a factor of $r$, and rotates them counterclockwise around the origin, by an angle of $\theta$.

**Problem 4.4** Describe the geometric effect of multiplication by $1 + i$.

We may express the number $1 + i$ in polar form as $\sqrt{2} \cdot \mathbf{e}(\pi/4)$. Therefore, multiplication by $1 + i$ has the effect of moving points further from the origin, by a factor of $\sqrt{2}$, then[8] rotating them counterclockwise by an angle of $45°$. This is displayed below.

[7] Multiplication reflects and scales the real number line; multiplication by a positive number $r$ stretches or contracts the line by a factor of $r$. Multiplication by $-1$ reflects the real number line across the origin.

[8] One may scale then rotate, or rotate then scale. The geometric operations commute, just as multiplication commutes.

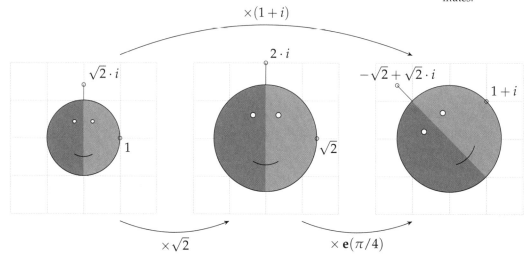

Figure 4.3: Multiplication by $1 + i$ is broken into two steps: multiplication by $\sqrt{2}$ having a scaling effect, and multiplication by $\mathbf{e}(\pi/4)$ having a rotation effect.

Rotation changes neither length nor area. However, scaling by a factor of $r$ changes lengths by a factor of $r$ and areas by a factor of $r^2$. In the figure above, the blue circle on the left has radius 1, and area $\pi$. Multiplying by $1 + i = \sqrt{2} \cdot \mathbf{e}(\pi/4)$ scales lengths by $\sqrt{2}$ and areas by 2. The two larger blue circles have radius $\sqrt{2}$ and area $2\pi$.

If $z$ is a complex number, the **absolute value** of $z$ is the distance from $z$ to 0. This can be expressed algebraically in two ways. If $z$ is given as $x + yi$ with real numbers $x$ and $y$, then by the Pythagorean theorem,

$$|x + yi| = \sqrt{x^2 + y^2}.$$

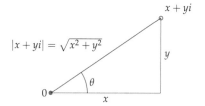

On the other hand, if $z$ is given in polar form, $z = r \cdot \mathbf{e}(\theta)$, then $r$ captures the absolute value by definition:

$$|r \cdot \mathbf{e}(\theta)| = r.$$

From Proposition 4.3, it follows that the absolute value is multiplicative:

$$|z \cdot w| = |z| \cdot |w|.$$

COMPLEX CONJUGATION is an operation on complex numbers that has no analogue among real numbers. If $x + yi$ is a complex number, with $x$ and $y$ real numbers, then its **complex conjugate** is defined by

$$\overline{x + yi} = x - yi.$$

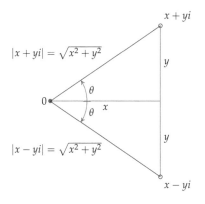

The real part stays the same, and the imaginary part is negated. In polar form, the complex conjugate of a complex number $r \cdot \mathbf{e}(\theta)$ is

$$\overline{r \cdot \mathbf{e}(\theta)} = r \cdot \mathbf{e}(-\theta).$$

The absolute value stays the same, and the angle is negated[9].

Complex conjugation reflects points across the real axis. Reflecting twice returns any point to itself; for every complex number $z$, $\overline{\overline{z}} = z$.

[9] Negating the angle has the effect of switching counterclockwise rotation to clockwise rotation.

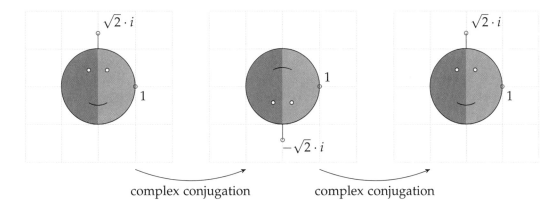

complex conjugation   complex conjugation

More surprising are the following facts, which are succinctly stated in one sentence: complex conjugation is a "field automorphism" of the complex numbers. A proof fits in the margin.

**Proposition 4.5 (Conjugation is a field automorphism)** *For all complex numbers $z$ and $w$,*

$$\overline{z} + \overline{w} = \overline{z + w};$$
$$\overline{z} - \overline{w} = \overline{z - w};$$
$$\overline{z} \cdot \overline{w} = \overline{zw}.$$

The product of a complex number and its conjugate is the square of the absolute value.

$$|z|^2 = z \cdot \overline{z}.$$

Indeed, if $z = x + yi$, then $\overline{z} = x - yi$, and so

$$z \cdot \overline{z} = (x + yi)(x - yi) = x^2 - xyi + xyi - y^2(i \cdot i) = x^2 + y^2 = |z|^2.$$

Here are brief algebraic proofs. Let $z = x + yi$ and $w = u + vi$, for real numbers $x, y, u, v$. Then

$$\overline{z} \pm \overline{w} = (x - yi) \pm (u - vi)$$
$$= (x \pm u) - (y \pm v)i$$
$$= \overline{(x \pm u) + (y \pm v)i}$$
$$= \overline{z \pm w}.$$

For multiplication, put $z$ and $w$ in polar form, with $z = r\,\mathbf{e}(\theta)$ and $w = s\,\mathbf{e}(\phi)$. Then

$$\overline{z} \cdot \overline{w} = (r\,\mathbf{e}(-\theta)) \cdot (s\,\mathbf{e}(-\phi))$$
$$= rs \cdot \mathbf{e}(-\theta - \phi)$$
$$= \overline{rs \cdot \mathbf{e}(\theta + \phi)}$$
$$= \overline{(r\,\mathbf{e}(\theta)) \cdot (s\,\mathbf{e}(\phi))}$$
$$= \overline{z \cdot w}.$$

**Gaussian integers** are complex numbers $x + yi$ in which $x$ and $y$ are *integers*. Examples of Gaussian integers are $0$, $1 + 17i$, $3i$, $-2 + 5i$, and $13$. All integers are Gaussian integers. Numbers like $1/2$ and $\sqrt{2} + \pi i$ are complex numbers, but not Gaussian integers.

Sums, differences, conjugates, and products of Gaussian integers are again Gaussian integers. Indeed, these operations on complex numbers involve only addition, subtraction, and multiplication[10] among the real and imaginary parts. So since sums, differences, and products of integers are integers, so too it holds for Gaussian integers. Zero and one are Gaussian integers.

We say that a Gaussian integer $z$ divides a Gaussian integer $w$, if there exists a Gaussian integer $m$ such that $w = z \cdot m$. In this case, we write $z \mid w$; we might also say that $z$ goes into $w$, or that $w$ is a (Gaussian) multiple of $z$. Since nonzero Gaussian integers have minimum absolute value 1, and absolute values are multiplicative, we have a familiar connection between divisibility and size of numbers.

**Proposition 4.6** *If $z \mid w$ and $w \neq 0$, then $|z| \leq |w|$.*

If $z = x + yi$ is a Gaussian integer, then $|z|^2 = x^2 + y^2$ is a natural number. The **Gaussian units** are the numbers $1$, $-1$, $i$, and $-i$. They are the only Gaussian integers whose absolute value is one. They are also the only Gaussian integers that divide 1. Indeed, Gaussian divisors of 1 must be nonzero and must have absolute value less than or equal to 1; the only possibilities are the four units $1$, $-1$, $i$, and $-i$.

Gaussian integers lie on a grid of unit-side squares. If $z$ is a Gaussian integer, then multiples of $z$ also lie on a grid of squares.

**Problem 4.7** Draw the multiples of the Gaussian integer $1 + 2i$.

SOLUTION: In polar form, $1 + 2i = \sqrt{5} \cdot \mathbf{e}(\theta)$, where $\theta \approx 1.107 \approx 63.43°$. Multiples of $1 + 2i$ are obtained by scaling the Gaussian integers by $\sqrt{5}$ and rotating them counterclockwise by 63.43 degrees. ✓

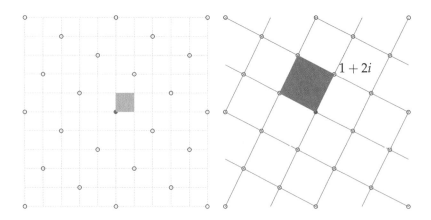

[10] To multiply complex numbers,
$$(x+yi)(u+vi) = (xu - yv) + (xv + yu)i.$$
If $x$, $y$, $u$, and $v$ are integers, then so too are $(xu - yv)$ and $(xv + yu)$ integers.

Proof of Proposition 4.6: If $z \mid w$ and $w \neq 0$, then there exists $m$ such that $w = zm$. Since $w \neq 0$, $m \neq 0$. Hence $1 \leq |m|$. Therefore
$$|z| = |z| \cdot 1 \leq |z| \cdot |m| = |w|.$$

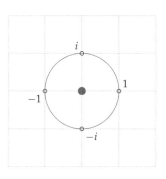

Figure 4.4: The Gaussian units, highlighted in red. They are the only Gaussian integers (grid points) whose absolute value is one (which lie on the unit circle). No Gaussian integers except zero lie inside the circle.

Figure 4.5: Gaussian integers lie at the green grid points. Multiples of $1 + 2i$ lie on the red grid points. The red square is obtained by multiplying the green square by $1 + 2i$; the area of the green square is 1, and the area of the red square is 5.

**Eisenstein integers** are complex numbers $x + y\omega$, in which $x$ and $y$ are integers, and $\omega$ is the very special complex number

$$\omega = \mathbf{e}(2\pi/3).$$

The best way to perform arithmetic on Eisenstein integers is to use the following two properties of $\omega$; the second follows from the first.

$$\omega^2 = -1 - \omega, \text{ and } \omega^3 = 1.$$

Sums, differences, conjugates, and products[11] of Eisenstein integers are Eisenstein integers. If $z = x + y\omega$ is an Eisenstein integer, then

$$|z|^2 = z\bar{z} = (x + y\omega)(x + y\omega^2);$$
$$= x^2 + xy\omega + xy\omega^2 + y^2\omega^3;$$
$$= x^2 - xy + y^2.$$

As $x$ and $y$ are integers, $|z|^2 = x^2 - xy + y^2$ is a natural number.

Divisibility of Eisenstein integers is defined analogously to divisibility of Gaussian integers. Since the nonzero Eisenstein integers have minimum absolute value 1, Proposition 4.6 holds for Eisenstein integers. The **Eisenstein units** are the numbers $1, -1, \omega, -\omega, \omega^2, -\omega^2$; these are the only Eisenstein integers which are divisors of 1.

Eisenstein integers lie on a grid of unit-side triangles. If $z$ is an Eisenstein integer, then multiples of $z$ also lie on a grid of triangles.

**Problem 4.8** Draw the multiples of the Eisenstein integer $1 - 2\omega$.

SOLUTION: Above we found that $|x + y\omega|^2 = x^2 - xy + y^2$. Thus we compute:

$$|1 - 2\omega| = \sqrt{1^2 - (1)(-2) + 2^2} = \sqrt{7}.$$

Multiplication by $1 - 2\omega$ rotates (clockwise by about $40.85°$) and scales by a factor of $\sqrt{7}$. After plotting a few multiples by hand, we can find the triangular grid below. ✓

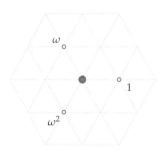

Figure 4.6: Observe $1 + \omega + \omega^2 = 0$ (the vectors add to zero) and $\bar{\omega} = \omega^2$ (reflect $\omega$ across the x-axis).

[11] If $x + y\omega$ and $u + v\omega$ are Eisenstein integers, then we compute

$$(x + y\omega) \cdot (u + v\omega)$$
$$= xu + xv\omega + yu\omega + yv\omega^2$$
$$= (xu - yv) + (xv + yu - yv)\omega.$$

If $x, y, u,$ and $v$ are integers, then so too are $(xu - yv)$ and $(xv + yu - yv)$ integers.

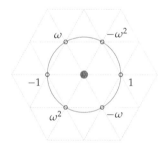

Figure 4.7: The Eisenstein units, highlighted in red. They are the only Eisenstein integers – grid points – whose absolute value is one. No Eisenstein integers except zero lie inside the circle.

 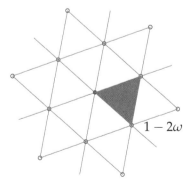

Figure 4.8: Eisenstein integers lie at the green grid points. Multiples of $1 - 2\omega$ lie on the red grid points. The area of the green triangle is $\sqrt{3}/4$. The area of the red triangle is $7\sqrt{3}/4$.

How many Gaussian integers lie within a circular area around zero? Draw a circle of radius $r$; for each Gaussian integer $z$ within the circle, draw a unit square centered $z$.

Figure 4.9: The square centered at $z$. Its area equals 1. The half-diagonal (circumradius) has length $\sqrt{2}/2$.

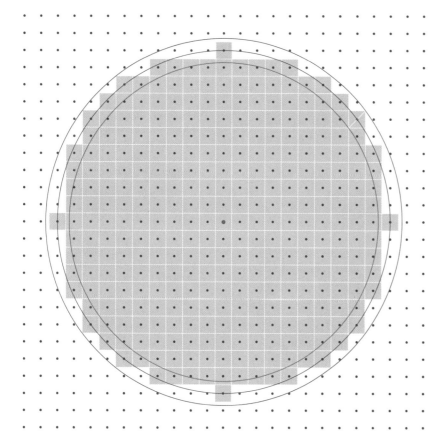

Figure 4.10: Circles of radius $r - \sqrt{2}/2$, $r$, and $r + \sqrt{2}/2$, in green, red, and blue respectively. Red dots are Gaussian integers $z$ within the red circle, i.e. those that satisfy $|z| \leq r$. Note that the green circle is completely within the area covered by squares, and the area covered by squares is completely within the blue circle.

Gauss demonstrated that the area of the "blocky" red shaded area is approximately equal to the area of the red circle, with error bounded by a linear function of the radius.

Let $G(r)$ be the number of Gaussian integers within[12] the circle of radius $r$. As the green circle is completely within the area covered by squares, which in turn is completely within the blue circle, we may compare their areas.[13] A bit of algebra yields the estimate below.

$$\pi(r - \sqrt{2}/2)^2 \leq G(r) \leq \pi(r + \sqrt{2}/2)^2;$$
$$\pi r^2 - \pi\sqrt{2}r + \pi/2 \leq G(r) \leq \pi r^2 + \pi\sqrt{2}r + \pi/2;$$
$$-\pi\sqrt{2}r + \pi/2 \leq G(r) - \pi r^2 \leq \pi\sqrt{2}r + \pi/2.$$

This bounds the difference between $G(r)$ and $\pi r^2$:

$$\left| G(r) - \pi r^2 \right| \leq \pi\sqrt{2} \cdot r + \frac{\pi}{2}.$$

But a much stronger bound is expected.

**Conjecture 4.9 (Gauss's Circle Problem)** *For every positive real number $\epsilon$, there exists a positive constant $K_\epsilon$ such that*

$$\left| G(r) - \pi r^2 \right| \leq K_\epsilon r^{\frac{1}{2} + \epsilon} \text{ for all positive } r.$$

[12] Counting points on the circle itself

[13] As the area of one square is 1 and $G(r)$ is the number of Gaussian integers within a circle of radius $r$, the square-covered area is the product $G(r) \cdot 1 = G(r)$.

In other words, we can estimate $G(r)$ by the quadratic function $\pi r^2$, with error bounded by a linear function of $r$. But it is expected that the error can be bounded by a function which grows almost as slowly as $\sqrt{r}$. In 1906, Sierpinski demonstrated that the error can be bounded by a constant times $r^{2/3}$.

How many Eisenstein integers are within a circular area around zero? Draw a circle of radius $r$; for each Eisenstein integer $z$ within the circle, draw a hexagon centered at $z$.

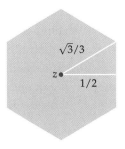

Figure 4.11: The hexagon centered at $z$. Its area equals $\sqrt{3}/2$. The circumradius equals $\sqrt{3}/3$.

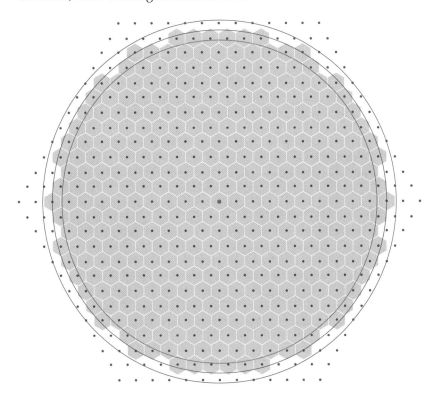

Figure 4.12: Circles of radius $r - \sqrt{3}/3$, $r$, and $r + \sqrt{3}/3$, in green, red, and blue respectively. Red dots are Eisenstein integers $z$ within the red circle, i.e. those that satisfy $|z| \leq r$. Note that the green circle is completely within the area covered by hexagons, and the area covered by hexagons is completely within the blue circle.

Let $E(r)$ be the number of Eisenstein integers within[14] the circle of radius $r$. As the blue circle encloses the area covered by hexagons, which contains the green circle, we may compare their areas.[15]

$$\pi(r - \tfrac{\sqrt{3}}{3})^2 \leq E(r)\tfrac{\sqrt{3}}{2} \leq \pi(r + \tfrac{\sqrt{3}}{3})^2 ;$$

$$\pi r^2 - 2\pi\tfrac{\sqrt{3}}{3}r + \pi/3 \leq E(r)\tfrac{\sqrt{3}}{2} \leq \pi r^2 + 2\pi\tfrac{\sqrt{3}}{3}r + \pi/3 ;$$

$$-2\pi\tfrac{\sqrt{3}}{3}r + \pi/3 \leq E(r)\tfrac{\sqrt{3}}{2} - \pi r^2 \leq 2\pi\tfrac{\sqrt{3}}{3}r + \pi/3 .$$

Multiplying through by $\tfrac{2}{\sqrt{3}}$ demonstrates that

$$\left| E(r) - \frac{2\pi}{\sqrt{3}} r^2 \right| \leq \frac{4\pi}{3} r + \frac{2\pi}{3\sqrt{3}}.$$

As in the Gaussian case, a much stronger bound is expected.

**Conjecture 4.10** *For every positive real number $\epsilon$, there exists a positive constant $K_\epsilon$ such that*

$$\left| E(r) - \frac{2\pi}{\sqrt{3}} r^2 \right| \leq K_\epsilon r^{\frac{1}{2}+\epsilon} \text{ for all positive } r.$$

[14] Counting points on the circle itself

[15] As the area of one hexagon is $\sqrt{3}/2$ and $E(r)$ is the number of Eisenstein integers within a circle of radius $r$, the hexagon-covered area is the product $E(r) \cdot \sqrt{3}/2$.

In other words, we can estimate $E(r)$ by the quadratic function $2\pi r^2/\sqrt{3}$, with error bounded by a linear function of $r$. But it is expected that the error can be bounded by a function which grows almost as slowly as $\sqrt{r}$.

DIVISION WITH REMAINDER comes from the geometry of multiples. Why can we divide $a$ by $b$, to obtain an expression $a = q(b) + r$, with $r$ smaller than $b$? It is because the multiples of a nonzero $b$ – the possible expressions $q(b)$ – slice the number line into intervals of length $b$. The quotient $q$ tells us *in which* interval $a$ lies, and the remainder $r$ tells us *where in* the interval $a$ lies.

Similarly, when $b$ is a nonzero Gaussian integer, the Gaussian multiples of $b$ slice the plane into squares. And similarly, when $b$ is a nonzero Eisenstein integer, the Eisenstein multiples of $b$ slice the plane into triangles. One may deduce a division-with-remainder, $a = q(b) + r$ in the Gaussian and Eisenstein contexts, in which $q$ determines *in which* square or triangle $a$ lies, and $r$ determines *where* in the square or triangle $a$ lies.

**Theorem 4.11 (Gaussian/Eisenstein division with remainder)**
*Let both $a$ and $b$ be Gaussian integers or both be Eisenstein integers, with $b \neq 0$. Then there exist Gaussian or Eisenstein integers $q$ and $r$, such that $a = q(b) + r$, and*

$$|r| \leq |b|\sqrt{2}/2 \quad \text{or} \quad |r| \leq |b|\sqrt{3}/3,$$

*in the Gaussian or Eisenstein cases, respectively.*

PROOF: The multiples of $b$ slice the plane into squares of side-length $|b|$ in the Gaussian context, or triangles of side-length $|b|$ in the Eisenstein context. Let $q(b)$ be a multiple of $b$ which is as close as possible to $a$; in particular, $a$ lies in a square or triangle, or side-length $|b|$, with vertex at $q(b)$. How far could $a$ be from this vertex $q(b)$?

In the square case, the farthest possible distance between $a$ and $q(b)$ is half the diagonal of the square, which is $|b|\sqrt{2}/2$. In the triangular case, the farthest possible distance between $a$ and $q(b)$ is two-thirds of the altitude of the triangle, which is $|b|\sqrt{3}/3$.

Now, let $r = a - q(b)$; we find that $a = q(b) + r$, and $|r|$ is the distance between $a$ and $q(b)$, so

$$|r| \leq |b|\sqrt{2}/2 \quad \text{or} \quad |r| \leq |b|\sqrt{3}/3,$$

in the Gaussian or Eisenstein case, respectively. ∎

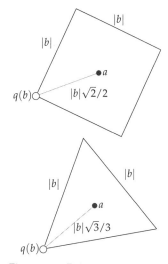

Figure 4.13: Points $a$ in a square or triangle, as far as possible from the vertices. Observe also that

$$\sqrt{2}/2 \approx 0.707 < 1,$$
$$\sqrt{3}/3 \approx 0.577 < 1.$$

Since $\sqrt{2}/2 < 1$ and $\sqrt{3}/3 < 1$, we find that after division with remainder $a = q(b) + r$, the remainder may be taken to be smaller, as measured by absolute value, than the divisor $b$.

**Problem 4.12** Divide 10 by $4 + 3i$, with remainder.

SOLUTION: We wish to find Gaussian integers $q$ and $r$, with $10 = q(4 + 3i) + r$, and with $|r|$ as small as possible. Begin by plotting the multiples of $4 + 3i$.

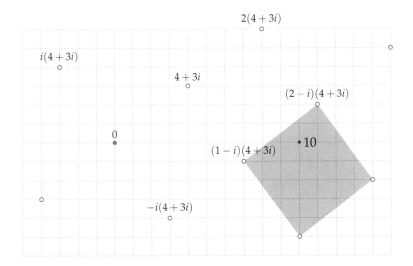

Figure 4.14: Gaussian integers lie at the grid points. Multiples of $4 + 3i$ are highlighted. The integer 10 is contained in the pale red square, whose closest vertex is $(2 - i)(4 + 3i)$. The red square has side-length 5, since $|4 + 3i| = 5$ using the Pythagorean triple $3, 4, 5$.

Near 10, the closest multiple of $4 + 3i$ is $(2 - i)(4 + 3i)$. The diagram illustrates the equality

$$10 = (2 - i)(4 + 3i) + (-1 - 2i).$$

Thus, if $q = 2 - i$, and $r = -1 - 2i$, then $10 = q(4 + 3i) + r$. Furthermore we have

$$|r| = \sqrt{5} \approx 2.236, \text{ and } |b|\sqrt{2}/2 = 5\sqrt{2}/2 \approx 3.535.$$

The remainder is smaller than the divisor, as in Theorem 4.11. ✓

THE EUCLIDEAN ALGORITHM follows directly from division with remainder. By continuing to divide with remainder, we could go further than the previous problem to compute the Gaussian greatest common divisor of 10 and $4 + 3i$. Each step below is a Gaussian division with remainder.

$$10 = (2 - i) \cdot (4 + 3i) + (-1 - 2i)$$
$$4 + 3i = (-2 - i)(-1 - 2i) + 0.$$

The first nonzero remainder is the last nonzero remainder, and

$$\text{GCD}(10, 4 + 3i) = (-1 - 2i).$$

In other words, $(-1 - 2i)$ is a common divisor of 10 and $4 + 3i$, and any common divisor is a divisor of $(-1 - 2i)$.

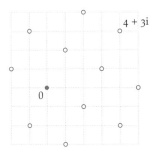

Figure 4.15: Multiples of $(-1 - 2i)$ are highlighted in blue. Note that $4 + 3i$ lies on a blue dot – it is a multiple of $(-1 - 2i)$.

A PRIME NUMBER is a positive integer $p$, whose positive divisors are 1 and $p$. For purposes of generalization, it is convenient to consider $-p$ just as prime as $p$. The numbers 1 and $-1$ are called the unit integers – they are the integer divisors of 1. A **prime integer** is an integer $p$ satisfying two conditions:

1. $p$ is not a unit.

2. If $x$ and $y$ are integers, and $xy = p$, then $x$ or $y$ is a unit.

If $p$ is a prime number, then $p$ and $-p$ are prime integers; all prime integers arise from prime numbers in this way.[16]

This definition adapts to the Gaussian and Eisenstein contexts. The Gaussian units are $\pm 1$ and $\pm i$. The Eisenstein units are $\pm 1$, $\pm \omega$, and $\pm \omega^2$. Hereafter, we write "G/E" to mean "Gaussian or Eisenstein" – a sentence about G/E integers, primes, etc. will stand for two sentences: one about Gaussian integers, primes, etc., and another about Eisenstein integers, primes, etc..

A G/E integer $q$ is called **prime** if it satisfies two conditions:

1. $q$ is not a unit.

2. If $z$ and $w$ are G/E integers, and $zw = q$, then $z$ or $w$ is a unit.

By the same sequence of logical steps described in Chapters 1 and 2, outlined on the opposite page[17], one can deduce the following existence and uniqueness of prime decomposition.

**Theorem 4.13 (Prime decomposition for Gaussian/Eisenstein integers)**
*If $z$ is a nonzero G/E integer, then $z$ can be expressed as a product of G/E primes. This prime decomposition is unique, in the sense that any two such decompositions differ only by rearrangement and units.*[18]

**Problem 4.14** Factor 10 into Gaussian primes.

SOLUTION: First, factor 10 as an integer to find $10 = 2 \times 5$. Although 2 and 5 are prime integers, they might not be prime *Gaussian* integers. To look for factors of 2, we must check for Gaussian integers whose absolute value is between 1 and $\sqrt{2}$; indeed, we find that $(1+i)$ is a prime factor of 2, and so is $1-i$.[19]

$$2 = (1+i) \cdot (1-i).$$

To look for factors of 5, we must check Gaussian integers whose absolute value is between 1 and $\sqrt{5}$; indeed, we find that $(2+i) \mid 5$:

$$5 = (2+i) \cdot (2-i).$$

The numbers $2 \pm i$ are Gaussian primes. Hence

$$10 = 2 \times 5 = (1+i) \cdot (1-i) \cdot (2+i) \cdot (2-i).$$

This is a factorization of 10 into Gaussian primes. ✓

[16] If $p$ is a prime number, $x$ and $y$ are integers, and $xy \mid \pm p$, then $|x| \cdot |y| \mid p$. But the only positive divisors of $p$ are 1 and $p$; hence either $|x| = 1$ or $|y| = 1$. Hence $x$ or $y$ is a unit. Therefore, when $p$ is a prime number, $\pm p$ are both prime integers.

[17] The opposite page applies to the integer, Gaussian integer, and Eisenstein integer contexts.

[18] We find such differences also for ordinary integers. For example,
$$100 = 2 \times 5 \times 2 \times 5$$
$$= (-2) \times 2 \times 5 \times (-5).$$
Both are decompositions of 100 into prime integers; they differ only by rearrangement and units.

[19] In fact,
$$1 - i = (-i) \cdot (1+i),$$
and so the Gaussian primes $(1-i)$ and $(1+i)$ are no different than the prime integers 7 and $-7$. We say that $1+i$ and $1-i$ are associate primes, but that is not important for solving this problem.

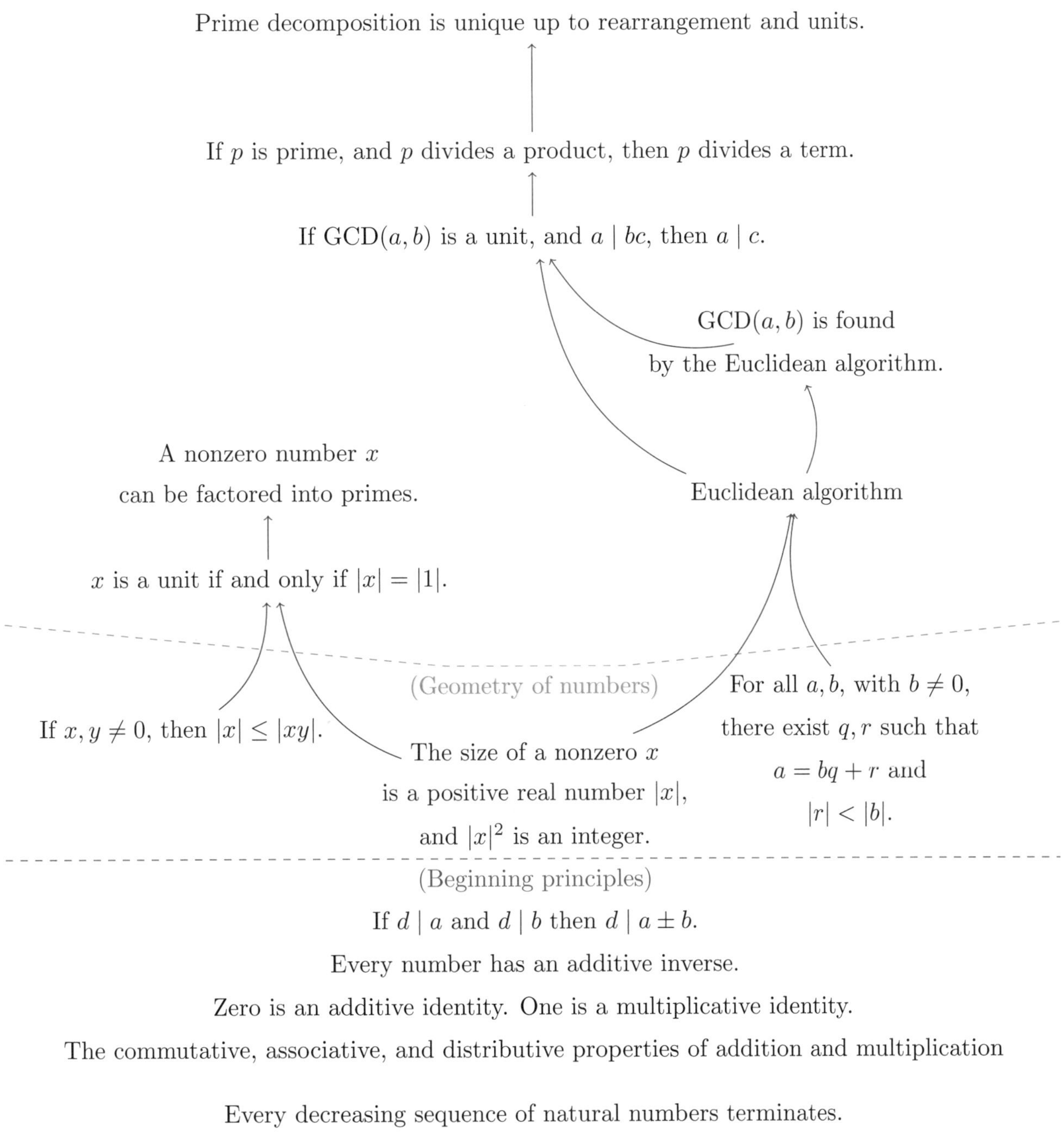

Prime decomposition is unique up to rearrangement and units.
↑
|
If $p$ is prime, and $p$ divides a product, then $p$ divides a term.
↑
|
If $\mathrm{GCD}(a,b)$ is a unit, and $a \mid bc$, then $a \mid c$.

$\mathrm{GCD}(a,b)$ is found
by the Euclidean algorithm.

A nonzero number $x$
can be factored into primes.

Euclidean algorithm

$x$ is a unit if and only if $|x| = |1|$.

- - - - - - - - - - - - - - - - - - - - - - - - - - - - - - - - - - - - - - - - - - - - - - - - - - - - - - - - - - - -

(Geometry of numbers)

For all $a, b$, with $b \neq 0$,

If $x, y \neq 0$, then $|x| \leq |xy|$.

there exist $q, r$ such that

The size of a nonzero $x$

$a = bq + r$ and

is a positive real number $|x|$,

$|r| < |b|$.

and $|x|^2$ is an integer.

- - - - - - - - - - - - - - - - - - - - - - - - - - - - - - - - - - - - - - - - - - - - - - - - - - - - - - - - - - - -

(Beginning principles)

If $d \mid a$ and $d \mid b$ then $d \mid a \pm b$.

Every number has an additive inverse.

Zero is an additive identity. One is a multiplicative identity.

The commutative, associative, and distributive properties of addition and multiplication

Every decreasing sequence of natural numbers terminates.

Figure 4.16: The deductive progression leading to existence and uniqueness of prime decomposition.

GAUSSIAN AND EISENSTEIN PRIMES occur in constellations – groupings of 4, 6, 8, or 12 arranged around a circle.

**Proposition 4.15** *If q is a G/E prime, and u is a unit, then uq is a G/E prime.*

PROOF: If $q$ is a G/E prime then $q$ is not a unit, and thus $uq$ is not a unit. If $uq = zw$ then $q = (u^{-1}z)w$. Hence $u^{-1}z$ or $w$ must be a unit. If $u^{-1}z$ is a unit, then $z$ is a unit. Hence $z$ is a unit or $w$ is a unit. Therefore $uq$ is prime. ∎

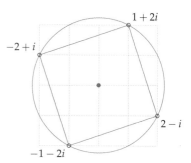

Figure 4.17: The Gaussian integer $1 + 2i$ is prime. Its primagon contains four Gaussian primes: $-2+i, -1+2i, -2-i,$ and $1-2i$. The primagon is inscribed in the circle of radius $|1+2i| = \sqrt{5}$.

Since there are four Gaussian units, and six Eisenstein units, we find that G/E primes come in groups of four or six. We call such a group a **primagon**; Gaussian primagons are arranged in a square, and Eisenstein primagons are arranged in a regular hexagon. All points on a primagon have the same absolute value.

**Proposition 4.16** *If q is a G/E prime then $\bar{q}$ is a G/E prime.*

PROOF: If $q$ is a G/E prime then $q$ is not a unit. Hence $\bar{q}$ is not a unit. If $\bar{q} = zw$, then $q = \bar{\bar{q}} = \bar{z}\bar{w}$. Since $q$ is a G/E prime, $\bar{z}$ or $\bar{w}$ must be a G/E unit; hence $z$ or $w$ must be a G/E unit. Thus $\bar{q}$ is prime. ∎

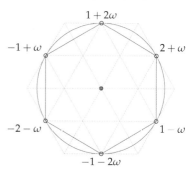

Figure 4.18: The Eisenstein integer $1 - \omega$ is prime. Its primagon contains six Eisenstein primes: $1-\omega, 2+\omega, 1+2\omega, -1+\omega, -2-\omega, -1-2\omega$. The primagon is inscribed in the circle of radius $|1 - \omega| = \sqrt{3}$.

When working with three number systems – ordinary integers, Gaussian integers, and Eisenstein integers – one must be extremely careful about context, especially when the context changes. The meanings of the words "multiple" and "prime" and "factor" and "unit" depend on the number system. We highlight the contextual words for emphasis, as the context changes within a single sentence.

Observe that if $z$ is a G/E integer, and $z$ is a real number, then $z$ must be an ordinary integer. For example, if $z = a + bi$ is a Gaussian integer, and $z$ is a real number, then $b = 0$; thus $z = a$ is an ordinary integer.

**Proposition 4.17** *Suppose that $z, w$ are integers. Then $z \mid w$ as G/E integers if and only if $z \mid w$ as integers.*

PROOF: If $z, w$ are integers and $z \mid w$ as G/E integers, then $w = zm$ for some G/E integer $m$. Taking conjugates yields

$$w = \bar{w} = \overline{zm} = \bar{z}\bar{m} = z\bar{m}, \text{ since } w = \bar{w} \text{ and } z = \bar{z}.$$

Figure 4.19: The only G/E integers on the real number line are the ordinary integers.

Since $w = zm = z\bar{m}$, we find that $z = w = 0$ or else $m = \bar{m}$. When $z = w = 0$, we find that $z \mid w$ as integers. When $m = \bar{m}$, $m$ is an ordinary integer, so again $z \mid w$ as integers.

Conversely, if $z \mid w$ as integers, then $w = zm$ for some integer $m$. Since $m$ is also a G/E integer, $z \mid w$ as G/E integers. ∎

We classify G/E primes based on the orientation of their primagons. In what follows, let $q$ be a G/E prime. We say $q$ has type (I), (R), or (S) according to the following possible orientations of its primagon.

**(I)** The primagon has an integer vertex and coincides with its complex conjugate.

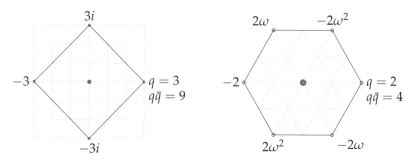

We use the letters (I), (R), and (S) as they corresopnd to the common terms **inert**, **ramified**, and **split**.

Figure 4.20: In case (I), the primagon contains a point on the x-axis; it follows that for any vertex $q$ of the primagon, $q\bar{q}$ is a square integer. Here the primagon equals its conjugate. In the Gaussian case, the primagon of type (I) has diagonal edges of slopes $\pm 1$. In the Eisenstein case, the primagon of type (I) has a pair of horizontal edges.

**(R)** The primagon coincides with its complex conjugate, but does not have an integer vertex.

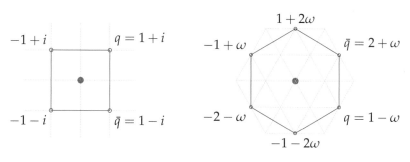

Figure 4.21: In case (R), the prime $q$ does not lie on the x-axis. In the Gaussian case, the primagon of $q$ has horizontal and vertical edges. In the Eisenstein case, the primagon of $q$ has a pair of vertical edges.

**(S)** The primagon of $q$ does not coincide with its conjugate.

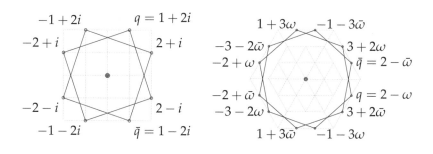

Figure 4.22: In case (S), the prime $q$ belongs to one primagon, and $\bar{q}$ belongs to a distinct mirror-image primagon. In the Gaussian case, these primagons together contain 8 primes. In the Eisenstein case, these primagons together contain 12 primes.

On the next pages, we explore the relationship between ordinary primes and G/E primes. We will find a correspondence – that every G/E primagon "lies above" an ordinary prime, and every ordinary prime "lies below" one or two G/E primagons.

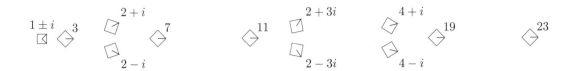

Let $p$ be a prime number and let $q$ be a G/E prime. We say that $p$ **lies below** $q$ if $q$ is a G/E factor of $p$, i.e., $q \mid p$. When $p$ lies below $q$, $p$ lies below the entire primagon of $q$. Also, since $p = \bar{p}$, we find that $q \mid p$ if and only if $\bar{q} \mid p$. When $p$ lies below $q$, $p$ lies below the primagons of $q$ and $\bar{q}$.

**Theorem 4.18 (Lifting primes)** *Let $p$ be a prime number. Then exactly one of the following two statements is true:*

1. *$p$ lies below a prime $q$ of type (R) or (S), and $p = q\bar{q}$. Moreover, $p$ lies below the primagons of $q$ and $\bar{q}$ and no others.*

2. *$p$ itself is a G/E prime of type (I). $p$ lies below its own primagon and no others.*

PROOF: Let $q$ be a G/E prime factor of $p$; there exists a G/E integer $m$ such that $p = qm$, whence $p = \bar{q}\bar{m}$ as well. Therfore $p^2 = q\bar{q}m\bar{m}$. Since $q\bar{q}$ is a positive integer divisor of $p^2$, and $q\bar{q} \neq 1$, we find that $q\bar{q} = p$ or $q\bar{q} = p^2$.

If $q\bar{q} = p^2$ then $m\bar{m} = 1$ and hence $m$ is a unit. Since $p = qm$, we find that $p$ itself is a G/E prime of type (I). The only G/E prime factors of $p$ are those in its own primagon.

If $q\bar{q} = p$ then $q$ cannot be of type (I); indeed, if $q$ were of type (I), then $q\bar{q}$ would be a square integer. But primes are not square. Thus if $q\bar{q} = p$ then $q$ must be of type (R) or (S), and the equality $p = q\bar{q}$ is a prime decomposition of $p$ as a G/E integer. Thus the only prime factors of $p$ are those in the primagon(s) of $q$ and $\bar{q}$. Going a bit further, the types (R) and (S) can be distinguished by whether the primagons of $q$ and $\bar{q}$ coincide or not. ∎

A prime number $p$ lies below one or two G/E primagons. Factoring $p$ into G/E primes is a way of looking up.

**Corollary 4.19** *If $p$ lies below a G/E prime $q$, then $p$ is a factor of $q\bar{q}$.*

Figure 4.23: The prime numbers, lying below Gaussian primagons. One vertex in each primagon is labeled; the orientation of the primagon is highlighted and scale is ignored. The only primagon of type (R) is the one containing $1 \pm i$. The primagons of type (I) are squares containing 3, 7, 11, 19, and 23.

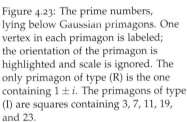

Figure 4.24: Hasse diagrams for this proof. The two diagrams on the left consist of G/E integers, and the diagram on the right consists of positive integers. The top-bottom orientation of the Hasse diagram is reversed from the "lying below" terminology.

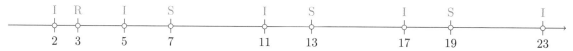

Figure 4.25: The Eisenstein primagons, lying above the Eisenstein primes. One vertex in each primagon is labeled; the orientation of the primagon is highlighted and scale is ignored. The only primagon of type (R) is the one containing $1 - \omega$ and $1 - \bar{\omega}$. The primagons of type (I) contain 2, 5, 11, 17, and 23.

Let $q$ be a G/E prime and let $p$ be a prime number. We say that $q$ **lies above** $p$ if $p \mid q\bar{q}$. When $q$ lies above $p$, every prime in the primagon of $q$ lies above $p$. Also, when $q$ lies above $p$, $\bar{q}$ lies above $p$ (since $q = \bar{\bar{q}}$). When $q$ lies above $p$, the primagon of $q$ and the primagon of $\bar{q}$ lie above $p$.

**Theorem 4.20 (Lowering primes)** *Let $q$ be a G/E prime. Then $q$ lies above exactly one prime number $p$. Exactly one of the following two statements is true about $p$ and $q$:*

1. *$q$ is a prime of type (R) or (S), and $q\bar{q} = p$.*

2. *$q$ is a G/E prime of type (I), $q\bar{q} = p^2$, and $p$ lies in the primagon of $q$.*

PROOF: Let $p$ be a (positive integer) prime factor of the positive integer $q\bar{q}$. By Theorem 4.18, we find that $p$ is a G/E prime of type (I) or else $p = r\bar{r}$ for some G/E prime $r$.

If $p$ is a G/E prime, then $p$ is a G/E factor of $q\bar{q}$, so by uniqueness of G/E prime decomposition, $p$ is a unit multiple of $q$ or of $\bar{q}$. But $p \mid q$ if and only if $p \mid \bar{q}$ since $p = \bar{p}$. Therefore, $p$ is contained in the primagon of $q$, $q$ is a prime of type (I), and $q$ lies above $p$. In this case $q\bar{q} = p^2$, and so $q$ lies above $p$ and no other prime.

If $p = r\bar{r}$ for some G/E prime $r$, then $q\bar{q}$ is a multiple of $r\bar{r}$. By uniqueness of prime decomposition, we find that $r$ is a unit multiple of $q$ or of $\bar{q}$. Hence $p = r\bar{r} = q\bar{q}$. Hence $q$ is a prime of type (R) or (S), and $p$ is the only prime factor of $q\bar{q}$. ∎

**Corollary 4.21** *If $p$ is a factor of $q\bar{q}$, then $q$ is a G/E prime factor of $p$.*

Together, the two corollaries justify the complementary terminology "lying above" and "lying below":

$q$ lies above $p$ if and only if $p$ lies below $q$.

EVERY PRIME NUMBER $p$ lies below a Gaussian prime $q$. We say that $p$ is **inert** if $q$ has type (I) and **ramified** if $q$ has type (R) and **split** if $q$ has type (S). The meaning of these words depends on the context; on this page the context is *Gaussian*.

**Proposition 4.22** *The only prime number which ramifies is 2.*

PROOF: A Gaussian primagon of type (R) is a square, whose corners lie on the diagonals pictured. But if $q$ is a corner of such a primagon, then $q$ is a multiple of the prime $1 + i$. Hence the only Gaussian primagon of type (R) is that which contains $1 + i$, which lies above 2.

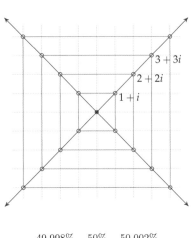

We can determine whether a prime number is inert or splits.

**Proposition 4.23** *Let $p$ be an odd prime number. Then $p$ splits if and only if $p$ can be expressed as the sum of two squares, i.e., $p = x^2 + y^2$ for some integers $x$ and $y$.*

PROOF: Since $p \neq 2$, the previous result implies that $p$ is inert or $p$ splits. If $p$ lies below a prime $q = x + yi$ of type (S), then $p = q\bar{q}$. Hence $p = (x + yi)(x - yi) = x^2 + y^2$.

Conversely, if $p = x^2 + y^2$ for some integers $x$ and $y$, then $p = (x + yi)(x - yi)$. Let $q = x + yi$, so we observe that $p = q\bar{q}$. It follows that $q$ is of type (S). ■

In a letter dated December 25, 1640, Fermat wrote the following observation to Mersenne.

**Theorem 4.24 (Fermat's Christmas Theorem)** *Let $p$ be an odd prime number. Then $p$ can be expressed as the sum of two squares if and only if $p - 1$ is a multiple of 4.*

The proof will have to wait until a later chapter – see Theorem 8.7 and Theorem 10.15. For now it gives us a rapid test to see whether a prime number splits, ramifies, or is inert. If $p$ is a prime number, then either

(R) $p = 2$ and $p$ ramifies,

(S) $p - 1$ is a multiple of 4 and $p$ splits, or

(I) $p - 3$ is a multiple of 4 and $p$ is inert.

Asymptotically, a prime number has an equal chance of being split or being inert. **Chebyshev's bias** is the term used to describe the slight bias towards being inert – among the prime numbers up to a large number $x$, the inert primes frequently outnumber the split primes.

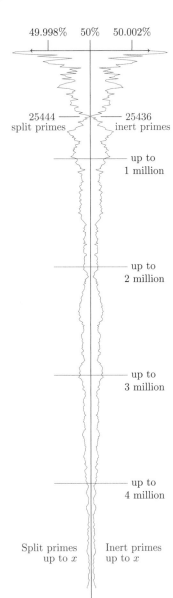

EVERY PRIME NUMBER $p$ lies below an Eisenstein prime $q$. We say that $p$ is **inert** if $q$ has type (I) and **ramified** if $q$ has type (R) and **split** if $q$ has type (S). The meaning of these words depends on the context; on this page the context is *Eisenstein*.

**Proposition 4.25** *The only prime number which ramifies is* 3.

PROOF: An Eisenstein primagon of type (R) is a hexagon with a pair of vertical edges. But if $q$ is a corner of such a primagon, then $q$ is a multiple of the prime $1 - \omega$. Hence the only Eisenstein primagon of type (R) is that which contains $1 - \omega$. ∎

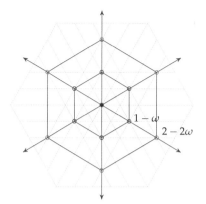

We can determine whether a prime number is inert or splits.

**Proposition 4.26** *Let $p$ be a prime number with $p \neq 3$. Then $p$ splits if and only if there exist integers $x$ and $y$ such that $p = x^2 - xy + y^2$.*

PROOF: Since $p \neq 3$, the previous result implies that $p$ is inert or $p$ splits. If $p$ lies below a prime $q = x + y\omega$ of type (S), then $p = q\bar{q}$. Hence $p = (x + y\omega)(x + y\bar{\omega}) = x^2 - xy + y^2$.

Conversely, if $p = x^2 - xy + y^2$ for some integers $x$ and $y$, then $p = (x + y\omega)(x + y\bar{\omega})$. Let $q = x + yi$, so we observe that $p = q\bar{q}$. It follows that $q$ is of type (S). ∎

In a letter to Pascal, dated September 25, 1654, Fermat observed that if $p$ is a prime number, exceeding by one a multiple of 3, then $p$ can be expressed as a square plus three-times-a-square. This comes close to the following theorem.

**Theorem 4.27** *Let $p$ be a prime number with $p \neq 3$. Then $p$ can be expressed as $x^2 - xy + y^2$ if and only if $p - 1$ is a multiple of 3.*

The proof will wait until a later chapter – see Theorem 10.16. For now it gives us a rapid test to see whether a prime number splits, ramifies, or is inert. If $p$ is a prime number, then either

(R) $p = 3$ and $p$ ramifies,

(S) $p - 1$ is a multiple of 3 and $p$ splits, or

(I) $p - 2$ is a multiple of 3 and $p$ is inert.

In the Eisenstein case as in the Gaussian case, the apparent even likelihood with which primes are split or inert belies Chebyshev's bias towards inert primes. Between 1 and 4 000 000, about 49.9547% of primes are Eisenstein-split and 50.0453% are Eisenstein-inert. Chebyshev's bias is subtle but **persistent**.

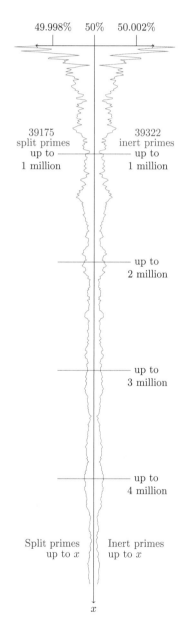

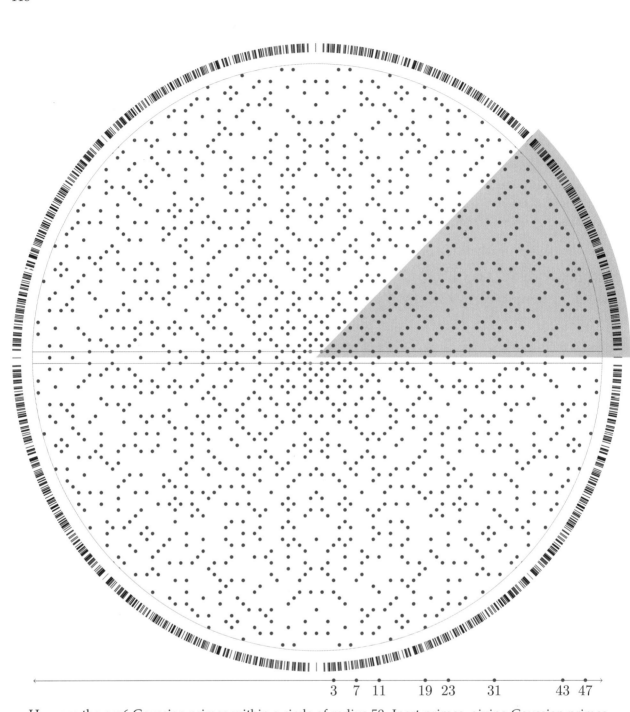

Here are the 1476 Gaussian primes within a circle of radius 50. Inert primes, giving Gaussian primes of type (I), are red dots. The ramified prime 2 yields four Gaussian primes $\pm 1 \pm i$, $\pm 1 \mp i$, in green. Other primes, in blue, come in sets of eight composed of two square primagons. Gaussian primes along the parallel horizontal lines arise from ordinary primes of the form $x^2 + 1$.

The highlighted area is a "fundamental domain" – all primes can be obtained from those in the highlighted area, by reflection and rotation. The black ticks around the circumference record the angles of the primes. Hecke (1918,1920) proved these angles are equidistributed: primes are as likely to be found in any one sector as any other sector of equal angular measure, as the circle's radius approaches infinity.

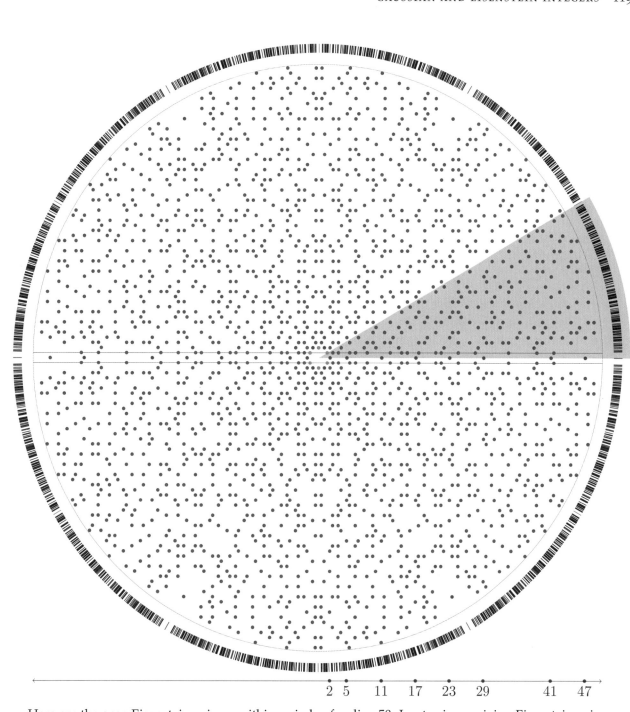

Here are the 2190 Eisenstein primes within a circle of radius 50. Inert primes, giving Eisenstein primes of type (I), are red dots. The ramified prime 3 yields six Eisenstein primes $\pm 1 \mp \omega, \pm \omega \mp \omega^2, \pm \omega^2 \mp 1$, in green. Other primes, in blue, come in sets of twelve composed of two hexagonal primagons. Eisenstein primes along the parallel horizontal lines arise from ordinary primes of the form $x^2 - x + 1$.

The highlighted area is a "fundamental domain" – all Eisenstein primes can be obtained from those in the highlighted area, by reflection and rotation. The snowflake-like symmetry comes from the six Eisenstein units and complex conjugation.

## Historical notes

In his 1832 commentary[20] on biquadratic reciprocity – a topic for much later in this book – Gauss introduced what we call the Gaussian integers. In §30, Gauss writes:

> Already in 1805, when we had begun to dedicate our thoughts, we verified that the real source of a general theory would be found in the extension of the field of arithmetic, that we indicated in Article 1...
>
> ...when the field of arithmetic is extended to *imaginary* quantities, those that constitute, without restriction, numbers of the form $a + bi$, with $i$ standing for the customary quantity $\sqrt{-1}$, and the indeterminates $a, b$ real integers between $-\infty$ and $+\infty$. We will call these numbers *complex integer numbers*...[21]

The *numeros integros complexos* of Gauss are now called Gaussian integers. By §34, Gauss accomplishes the classification "numerorum primorum complexorum" (of Gaussian primes) into three sorts corresponding to our types (R), (I), and (S). By §37, Gauss demonstrates the unique decomposition of Gaussian integers into Gaussian primes.

In a footnote to §30, Gauss writes,

> The theory of cubic residues is built in a similar manner, by considering numbers of the form $a + bh$, where $h$ is the imaginary root of the equation $h^3 - 1 = 0$, for example $h = -1/2 + \sqrt{3/4} \cdot i$...[22]

Eisenstein carried out this program of Gauss, developing the theory of Eisenstein integers and proving a theorem of cubic reciprocity. His proof was published in the *Journal für die reine und angewandte Mathematik* (also known as Crelle's journal) in 1844 – a volume containing no less than 16 works of the 21-year old Eisenstein.[23]

The term "primagon" is nonstandard – a compromise between an awkward expression like "a prime together with all of its associates" and the more general notion of "(principal) ideal". In generalizing the notion of primality to the Gaussian context, Gauss realized the importance of working with not only a single prime number, but an entire group of 4 or 8 primes. Later, in more general contexts (e.g. if one works with a fifth root of unity instead of a cube or fourth root), it became clear that a better notion is that of an "ideal" number. In this perspective, one works not with a prime number, but with a "prime ideal". In this way, one works simultaneously with all multiples of numbers, and a version of prime decomposition holds in great generality.

The Gauss circle problem arises in "De Nexu inter Multitudinem Classium, in quas Formae Binariae Secundi Gradus Distribuuntur, Earumque Det Erminantem," two commentaries from 1834 and 1837.[24] There Gauss considers a more general question: how many

---

[20] The following quotes are from C.F. Gauss, "Theoria Residuorum Biquadraticorum: commentatio secunda," in *Werke*, vol. 2, Göttingen (1876).

[21] Translated from the Latin: "Cui rei quum inde ab anno 1805 meditationes nostras dicare coepisseums, mox certiores facti sumus, fontem genuinum theoriae generalis in campo arithmeticae promoto quaerendum esse, uti iam in art. 1 addigitavimus...
...quando campus arithmeticae ad quantitates *imaginarias* extenditur, ita ut absque restrictione ipsius obiectum constituant numeri formae $a + bi$, denotantibus $i$ pro more quantitatem imaginariam $\sqrt{-1}$, atque $a, b$ indefinite omnes numeros reales integros inter $-\infty$ et $+\infty$. Tales numeros vocabimus *numeros integros complexos*..."

[22] Translated from the Latin: "Theoria residuorum cubicorum simili modo superstruenda eset considerationi numerorum formae $a + bh$, ubi $h$ est radix imaginaria aequationis $h^3 - 1 = 0$, puta $h = -1/2 + \sqrt{3/4} \cdot i$..."

[23] For a short biography of Eisenstein, and most notably an English translation of Eisenstein's own Curriculum Vita, see M. Schmitz, "The life of Gotthold Ferdinand Eisenstein," in *Res. Lett. Inf. Math. Sci.*, vol. 6 (2004).

[24] See *Werke*, vol. 2, pp.269–291.

points in a rectangular lattice lie within an expanding family of ellipses. Let $E(t)$ be an ellipse with vertical minor radius $rt$ and horizontal major radius $st$. Its area is given by the formula

$$A(t) = \pi r s t^2.$$

Consider a rectangular lattice of grid-points, aligned with horizontal and vertical edges, in which each rectangle has area $a$. Let $R(t)$ be the number of grid-points within the ellipse $E(t)$. Then,

$$\lim_{t \to \infty} \frac{A(t)}{R(t)} = a.$$

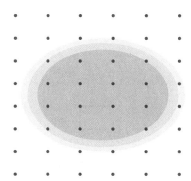

Figure 4.26: The number of red dots within the ellipse, $R(t)$ should be approximately the area of the ellipse $A(t)$ divided by the area of a single rectangle $a$. The approximation $R(t) \approx A(t)/a$ becomes $A(t)/R(t) \approx a$, which becomes equality in the limit as $t$ tends to infinity.

[25] M. Rubinstein and P. Sarnak, "Chebyshev's Bias," in *Experiment. Math.*, vol. 3, 3 (1994).

CHEBYSHEV'S BIAS is a term coined in the eponymous article by Sarnak and Rubinstein.[25] Their attribution comes from a letter, dated March 10, 1853, from Chebyshev to Fuss, in which Chebyshev writes,

> In studying the limiting behavior of the functions which describe the quantity of prime numbers of the form $4n + 1$ and those of the form $4n + 3$, up to a very large limit, I came to realize that... the value, for the numbers $4n + 3$, is larger than that for the numbers $4n + 1$; thus, if from the quantity of the prime numbers of the form $4n + 3$, one subtracts that of the prime numbers of the form $4n + 1$, and one divides the result of the difference by the quantity $\frac{\sqrt{x}}{\log x}$, one finds many values of $x$ for which, this quotient will approach 1 as closely as we please.[26]

Assuming two conjectures – the Grand Riemann Hypothesis (a grander version of the Riemann Hypothesis from Chapter 2) and its more refined Grand Simplicity Hypothesis – Sarnak and Rubinstein are able to prove a bias as observed by Chebyshev. They demonstrate that prime numbers of the form $4n + 3$ outnumber those of the form $4n + 1$ over 99% of the "time" (where the meaning of "time" is a bit subtle).

The study of the distribution of Gaussian and Eisenstein primes can be broken into three parts. First is the distribution of ordinary primes themselves, discussed in Chapter 2. Second is the determination of which primes split and which are inert – the statistical answer is half and half, asymptotically, but Chebyshev's bias gives a light preference towards inert primes. Third is the distribution of the *angles* of the split primes. The equidistribution of the prime angles is a direct consequence of E. Hecke's 1920 paper "Eine neue Art von Zetafunktionen und ihre Beziehungen zur Verteilung der Primzahlen. II", in *Math Z.* **6**. Subtler phenomena may arise in how these angles tend towards equidistribution as larger discs of primes are considered.

[26] Translated from the French: "En cherchant l'expression limitative des fonctions qui déterminent la totalité des nombres premiers de la forme $4n + 1$ et de ceux de la forme $4n + 3$, pris au-dessous d'une limite très grande, je suis parvenu à reconnaître que... la valeur, pour les nombres $4n + 3$, est plus grande que celle pour les nombres $4n + 1$; ainsi, si de la totalité des nombres premiers de la forme $4n + 3$, on retranche celle des nombres premiers de la forme $4n + 1$, et que l'on divise ensuite cette différence par la quantité $\sqrt{x}/\log x$, on trouvera plusieurs valeurs de $x$ telles, que ce quotient s'approchera de l'unité aussi près qu'on le voudra." This was published in *Bull. Classe Phys. Acad. Imp. Sci. St. Petersburg* **11** (1853).

## Exercises

1. (Gaussian arithmetic): Express your answers in the form $x + yi$.

   $1 + 3i + 2 - 5i =$ _____ ; $\quad (1 + 3i) \cdot (2 - 5i) =$ _____ ;

   $(1 + i)^{10} =$ _____ ; $\quad 1 + 2i + 3 + 4i + \cdots + 99 + 100i =$ _____ .

2. (Eisenstein arithmetic): Express your answers in the form $x + y\omega$.

   $2 + 5\omega - 1 - 7\omega =$ _____ ; $\quad (2 + 3\omega) \cdot (5 - \omega) =$ _____ ;

   $(1 - \omega)^5 =$ _____ ; $\quad 0 + 1\omega + 2\omega^2 + 3\omega^3 + 4\omega^4 + \cdots 99\omega^{99} =$ _____ .

3. Factor 85 into Gaussian primes. Factor 85 into Eisenstein primes.

4. Plot the Gaussian multiples of $3 + i$ on graph paper. Divide 11 by $3 + i$ with remainder, as Gaussian integers.

5. Let $z$ be a complex number. Prove that if $z$ is a Gaussian integer *and* an Eisenstein integer, then $z$ is an ordinary integer. Hint: Use the irrationality of $\sqrt{3}$.

6. Use the identity $\mathbf{e}(\theta) = \cos(\theta) + i \sin(\theta)$ to prove that $\mathbf{e}(\theta + \phi) = \mathbf{e}(\theta) \cdot \mathbf{e}(\phi)$ for any angles $\theta$ and $\phi$. Hint: recall the formulae for $\cos(\theta + \phi)$ and $\sin(\theta + \phi)$.

7. Compute $\cos(72°)$ by following these steps.

   (a) Let $\zeta = \mathbf{e}(2\pi/5)$. Express $\zeta$ in real and imaginary parts using trigonometric functions.

   (b) Demonstrate that $1 + \zeta + \zeta^2 + \zeta^3 + \zeta^4 = 0$.

   (c) Let $z = \zeta + \bar{\zeta}$. Use the previous identity to demonstrate that $z$ is a root of a nonzero quadratic polynomial with integer coefficients.

   (d) Use the quadratic polynomial to compute $z$, and use this to compute $\cos(72°)$.

   (e) Demonstrate that $\cos(2\pi/7)$ is a root of a nonzero cubic polynomial with integer coefficients.

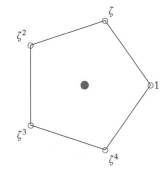

8. Let $n$ be a positive integer. Let $\zeta = \mathbf{e}(2\pi/n)$. Prove that
   $$1 + \zeta + \zeta^2 + \cdots + \zeta^{n-1} = 0.$$
   Hint: show that the sum equals itself times $\zeta$.

9. Suppose that $q$ is a Gaussian or Eisenstein prime number, and let $p$ be the prime number lying below $q$. Let $S$ be the set of G/E multiples of $q$. Prove that $S \cap \mathbb{Z}$ is the set of integer multiples of $p$.

10. Let $p$ be a prime number. Prove that if $p$ can be expressed as $a^2 + 3b^2$ for two integers $a, b$, then $p$ can be expressed as $x^2 - xy + y^2$ for two integers $x, y$. Hint: A linear substitution suffices.

11. Prove that there are infinitely many prime numbers in the arithmetic progression $1, 5, 9, 13, 17, 21, 25, 29, \ldots$. Hint: If not, multiply them together and add $i$; the result has a Gaussian prime factor $q$ of type (S). Demonstrate that $q\bar{q}$ is a "new" prime.

12. Prove that there are infinitely many prime numbers in the arithmetic progression $1, 4, 7, 10, 13, 16, 19, \ldots$. Follow the idea of the previous problem, but use Eisenstein integers.

13. Let $x$ be an integer. Prove that $x + i$ is a Gaussian prime number if and only if $x^2 + 1$ is an ordinary prime number.[27] Prove that $x + \omega$ is an Eisenstein prime number if and only if $x^2 - x + 1$ is an ordinary prime number.

[27] It is expected that there are infinitely many prime numbers of the form $x^2 + 1$; nobody knows how to prove this, but an equivalent conjecture is that there are infinitely many Gaussian prime numbers of the form $x + i$.

14. The following exercises answer the question: how many ways can one decompose a natural number into a sum of two squares?

    (a) Use the uniqueness of Gaussian prime decomposition to prove the following: If $p$ is a prime number and $p - 1$ is a multiple of 4, then there are exactly four pairs of integers $(x, y)$ satisfying $x^2 + y^2 = p$.

    (b) Prove that if $m = x^2 + y^2$ and $n = u^2 + v^2$, then $mn$ can also be expressed as the sum of two squares. (Hint: $z\bar{z}w\bar{w} = zw\bar{z}\bar{w}$.)

    (c) Let $n$ be a positive integer, with prime decomposition $n = 2^{e_2} 3^{e_3} 5^{e_5} \cdots$. Prove that $n$ can be expressed as the sum of two squares if and only if $e_p$ is even whenever $p$ is a prime which is three more than a multiple of 4.

    (d) (Challenge) Use the prime decomposition of $n$ to determine how many ways one can express $n$ as the sum of two squares.

15. A Gaussian rational number is a complex number of the form $z/w$ in which $w \neq 0$ and $z, w$ are Gaussian integers. The Ford sphere on $z/w$ is the sphere of diameter $1/(w\bar{w})$, lying atop the complex plane at $z/w$. Demonstrate that the Ford spheres on $z/w$ and $u/v$ are tangent precisely when $zv - wu$ is a Gaussian unit, i.e., when the Gaussian rational numbers kiss.

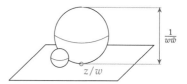

Figure 4.27: Kissing Ford spheres. For a more complete treatment, see L.R. Ford, "Fractions," in *The American Mathematical Monthly*, vol.45, **9** (Nov., 1938).

16. Consider the set of complex numbers of the form $a + b\sigma$, where $a$ and $b$ are integers and $\sigma = \sqrt{-2}$. Draw a grid of parallelograms whose grid-points consist of these numbers. Demonstrate that division with remainder is possible in this context; in other words, demonstrate that if $x, y$ are such numbers, with $y \neq 0$, then there exist numbers $q, r$ such that $x = qy + r$, and $|r| < |y|$.

# Part II

# Modular Arithmetic

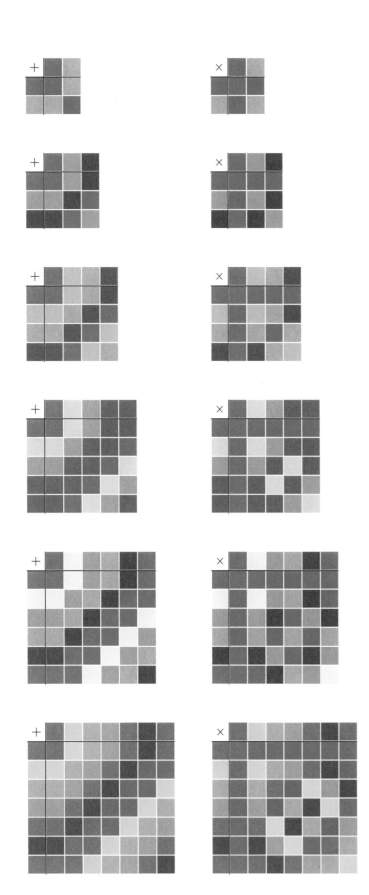

# 5
# The Modular Worlds

MODULO[1] 7, seven is the same as zero. In this strange "modular" world – the world of integers mod 7 – we add and multiply integers with familiar rules, but equality must be radically broadened. This broadened equality is called **congruence**, written $7 \equiv 0 \mod 7$, and spoken "7 is congruent to 0 modulo 7."

Insisting that $7 \equiv 0$, and that properties of $=$ are shared by $\equiv$, we must also admit that $14 = 7 + 7 \equiv 0 + 0 = 0$. And as $14 \equiv 0$, and $7 \equiv 0$, we must have $21 = 14 + 7 \equiv 0 + 0 \equiv 0$. In this way we find $0 \equiv 7 \equiv 14 \equiv 21 \equiv 28 \equiv \cdots$.

Negative integers are also affected; since $7 \equiv 0$ and $-7 + 7 = 0$, we find that $-7 + 0 \equiv 0$, and so $-7 \equiv 0$. Proceeding as for positive integers, we find that

$$\cdots \equiv -28 \equiv -21 \equiv -14 \equiv -7 \equiv 0 \equiv 7 \equiv 14 \equiv 21 \equiv 28 \equiv \cdots$$

All multiples of 7 must be congruent to zero, as soon as $7 \equiv 0$.

Since $7 \equiv 0$, $8 = 1 + 7 \equiv 1 + 0 = 1$ and $9 = 2 + 7 \equiv 2 + 0 = 2$, etc.. Every integer belongs to exactly one row in the figure below.

$$\cdots \quad -21 \equiv -14 \equiv -7 \equiv 0 \equiv 7 \equiv 14 \equiv 21 \equiv \cdots$$
$$\cdots \quad -20 \equiv -13 \equiv -6 \equiv 1 \equiv 8 \equiv 15 \equiv 22 \equiv \cdots$$
$$\cdots \quad -19 \equiv -12 \equiv -5 \equiv 2 \equiv 9 \equiv 16 \equiv 23 \equiv \cdots$$
$$\cdots \quad -18 \equiv -11 \equiv -4 \equiv 3 \equiv 10 \equiv 17 \equiv 24 \equiv \cdots$$
$$\cdots \quad -17 \equiv -10 \equiv -3 \equiv 4 \equiv 11 \equiv 18 \equiv 25 \equiv \cdots$$
$$\cdots \quad -16 \equiv -9 \equiv -2 \equiv 5 \equiv 12 \equiv 19 \equiv 26 \equiv \cdots$$
$$\cdots \quad -15 \equiv -8 \equiv -1 \equiv 6 \equiv 13 \equiv 20 \equiv 27 \equiv \cdots$$

Modulo 7, every integer is congruent to one of seven numbers:

$$0, 1, 2, 3, 4, 5, 6.$$

[1] On August 15, 2012, *The Economist* published a column on the word "modulo" titled "The award for nerdiest preposition goes to..."

Figure 5.1: The rows are called **congruence classes** mod 7. On the opposite page, each congruence class (modulo 2, 3, 4, 5, 6, 7) is provided a color, and addition and multiplication of congruence classes is tabulated.

OUR MODULAR WORLDS contain only finitely many numbers. To perform modular arithmetic, one needs only ordinary arithmetic and a **simplification** procedure. We describe two styles of simplification, based on two systems of representatives.

*The natural representatives* mod 7 are $0, 1, 2, 3, 4, 5, 6$.

*The minimal representatives* mod 7 are $-3, -2, -1, 0, 1, 2, 3$.

These are called **systems of representatives**[2] because every integer is congruent to a unique representative from each system. Using the natural representatives, we would simplify

$$11 \equiv 4 \bmod 7, \quad 23 \equiv 2 \bmod 7, \quad -100 \equiv 5 \bmod 7.$$

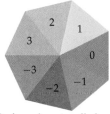

Using the minimal representatives, we would simplify

$$11 \equiv -3 \bmod 7, \quad 23 \equiv 2 \bmod 7, \quad -100 \equiv -2 \bmod 7.$$

[2] A set $S$ of numbers is called a system of representatives, mod $m$, if it has the following two properties:
1. Every number $n$ is congruent to an element of $S$, mod $m$.
2. If $a \in S$ and $b \in S$, and $a \equiv b \bmod m$, then $a = b$ (as integers).

To simplify larger numbers, one can use division with remainder. For example, $100 \div 7 = 14$ with a remainder of 2. Translating into multiplication and addition, $100 = 14(7) + 2$. Therefore,

$$100 = 14(7) + 2 \equiv 14(0) + 2 = 2 \bmod 7.$$

**Problem 5.1** Simplify the following numbers, mod 7, using natural representatives: $7, 702, -10, 6998, 1000$.

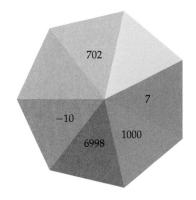

SOLUTION: The natural representatives are 0, 2, 4, 5, 6 respectively.

(a) $7 \equiv 0 \bmod 7$, by subtracting 7.

(b) $702 \equiv 2 \bmod 7$, by subtracting 700.

(c) $-10 \equiv -3 \equiv 4 \bmod 7$, by adding 7 twice.

(d) $6998 \equiv -2 \equiv 5 \bmod 7$, by subtracting 7000 then adding 7.

(e) $1000 \equiv 6 \bmod 7$ since $1000 \div 7$ equals 142 with a remainder 6. ✓

Computer programmers often think about "mod" as an operation – a computer scientist might write "10 mod 7 = 3", because 10 divided by 7 leaves a remainder 3. This can be fine for computation.

But, although it stretches the imagination, there are great rewards for those who can broaden their conception of equality to accept congruence in a modular world. The rewards are greater for those who can pass from one modular world to another, and back to the ordinary world of integers, without losing a step.

| 0 | 1 | 2 | 3 | 4 | 5 | 6 | 7 | 8 | 9 | 10 | 11 | 12 | 13 | 14 | 15 | 16 |

To ADD OR MULTIPLY, mod 7, just add and multiply as usual, but exploit the following at every opportunity:

If you see a number bigger than 7, simplify it.

Using this rule, we can assemble sums and products, mod 7, into addition and multiplication tables like those on primary school walls. First we use natural representatives.

| + | 0 | 1 | 2 | 3 | 4 | 5 | 6 |
|---|---|---|---|---|---|---|---|
| 0 | 0 | 1 | 2 | 3 | 4 | 5 | 6 |
| 1 | 1 | 2 | 3 | 4 | 5 | 6 | 0 |
| 2 | 2 | 3 | 4 | 5 | 6 | 0 | 1 |
| 3 | 3 | 4 | 5 | 6 | 0 | 1 | 2 |
| 4 | 4 | 5 | 6 | 0 | 1 | 2 | 3 |
| 5 | 5 | 6 | 0 | 1 | 2 | 3 | 4 |
| 6 | 6 | 0 | 1 | 2 | 3 | 4 | 5 |

| × | 0 | 1 | 2 | 3 | 4 | 5 | 6 |
|---|---|---|---|---|---|---|---|
| 0 | 0 | 0 | 0 | 0 | 0 | 0 | 0 |
| 1 | 0 | 1 | 2 | 3 | 4 | 5 | 6 |
| 2 | 0 | 2 | 4 | 6 | 1 | 3 | 5 |
| 3 | 0 | 3 | 6 | 2 | 5 | 1 | 4 |
| 4 | 0 | 4 | 1 | 5 | 2 | 6 | 3 |
| 5 | 0 | 5 | 3 | 1 | 6 | 4 | 2 |
| 6 | 0 | 6 | 5 | 4 | 3 | 2 | 1 |

Second, we use minimal representatives.

| + | −3 | −2 | −1 | 0 | 1 | 2 | 3 |
|---|---|---|---|---|---|---|---|
| −3 | 1 | 2 | 3 | −3 | −2 | −1 | 0 |
| −2 | 2 | 3 | −3 | −2 | −1 | 0 | 1 |
| −1 | 3 | −3 | −2 | −1 | 0 | 1 | 2 |
| 0 | −3 | −2 | −1 | 0 | 1 | 2 | 3 |
| 1 | −2 | −1 | 0 | 1 | 2 | 3 | −3 |
| 2 | −1 | 0 | 1 | 2 | 3 | −3 | −2 |
| 3 | 0 | 1 | 2 | 3 | −3 | −2 | −1 |

| × | −3 | −2 | −1 | 0 | 1 | 2 | 3 |
|---|---|---|---|---|---|---|---|
| −3 | 2 | −1 | 3 | 0 | −3 | 1 | −2 |
| −2 | −1 | −3 | 2 | 0 | −2 | 3 | 1 |
| −1 | 3 | 2 | 1 | 0 | −1 | −2 | −3 |
| 0 | 0 | 0 | 0 | 0 | 0 | 0 | 0 |
| 1 | −3 | −2 | −1 | 0 | 1 | 2 | 3 |
| 2 | 1 | 3 | −2 | 0 | 2 | −3 | −1 |
| 3 | −2 | 1 | −3 | 0 | 3 | −1 | 2 |

A few familiar properties of addition and multiplication are visible in these tables. Commutativity of addition and multiplication is visible, as the symmetry across the diagonal ∖ in all four tables. The identity property of zero ($0 + x = x$) is visible, as is the identity property of one ($1 \times x = x$). The reader may find more patterns with the minimal representatives.

The associative and distributive properties also hold in modular arithmetic, though they are not as easy to see.

To evaluate $52 \times 52$, modulo 7, one might try two routes.

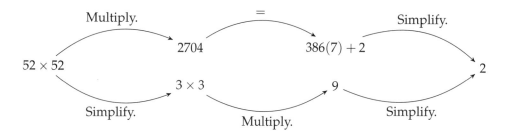

Why do both routes lead to the same solution? Our next theorem will guarantee it. Before proving the theorem, we must formalize the definition of congruence.[3]

**Definition 5.2** Let $m$ be a positive integer (called the **modulus**). We say that two integers $x$ and $y$ are **congruent** modulo $m$, and write $x \equiv y \bmod m$, if their difference $x - y$ is a multiple of $m$.

So when we write $x \equiv y \bmod 11$, we mean that $x - y$ is a multiple of 11. Equivalently, $x - y = 11t$ for some integer $t$. Equivalently, this means that $x = y + 11t$ for some integer $t$. Being "congruent to zero" has a most important interpretation: the sentence $x \equiv 0 \bmod m$ means that $x - 0$ is a multiple of $m$, i.e., that $x$ is a multiple of $m$. The ability to frame divisibility in terms of congruences will be powerful in this and later chapters.

**Proposition 5.3 (Well-definedness of arithmetic mod $m$)** *If $x \equiv x'$ and $y \equiv y' \bmod m$, then*

$$x \pm y \equiv x' \pm y' \text{ and } x \cdot y \equiv x' \cdot y' \bmod m.$$

PROOF: If $x \equiv x' \bmod m$ and $y \equiv y' \bmod m$, then there exist integers $u$ and $v$ for which $x = x' + um$ and $y = y' + vm$. Hence

$$\begin{aligned} x \pm y &= (x' + um) \pm (y' + vm) \\ &= (x' \pm y') + (um \pm vm) \\ &= x' \pm y' + (u \pm v)m. \end{aligned}$$

Hence $x \pm y \equiv x' \pm y' \bmod m$.

For multiplication, the computation is a bit more intricate.

$$\begin{aligned} x \cdot y &= (x' + um) \cdot (y' + vm) \\ &= (x' \cdot y') + x'vm + y'um + uvm^2 \\ &= (x' \cdot y') + (x'v + y'u + uvm)m. \end{aligned}$$

Hence $x \cdot y \equiv x' \cdot y' \bmod m$. ∎

Now we can see why both routes to evaluating $52 \times 52$ lead to the same result mod 7. Since $52 \equiv 3 \bmod 7$, the theorem implies that $52 \times 52 \equiv 3 \times 3 \bmod 7$.

[3] The discussion of congruence on the previous pages began with an assertion that $m \equiv 0$ (modulo $m$), and considered the effects of such an assertion. The definition given here is the innovation that began Gauss's *Disquisitiones Arithmeticae* (1801).

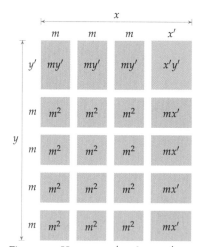

Figure 5.2: Here $x \equiv x'$ and $y \equiv y'$, mod $m$. Indeed, $x - x'$ and $y - y'$ are multiples of $m$. All rectangles except the red corner have area a multiple of $m$, and therefore are congruent to zero modulo $m$. Hence $xy \equiv x'y' \bmod m$.

**Problem 5.4** Restate the following congruences, using the language of multiples, and using an auxiliary variable.

(a) $3x \equiv 1 \bmod 11$.

(b) $5x \equiv 0 \bmod 3$.

(c) If $xy \equiv 0 \bmod p$, then $x \equiv 0 \bmod p$ or $y \equiv 0 \bmod p$.

SOLUTION: (a) $3x - 1$ is a multiple of 11. There exists an integer $m$ such that $3x - 1 = 11m$.

(b) $5x$ is a multiple of 3. There exists an integer $m$ such that $5x = 3m$.

(c) If $xy$ is a multiple of $p$, then $x$ is a multiple of $p$ or $y$ is a multiple of $p$. ✓

**Problem 5.5** Translate the following sentences into congruences.

(a) $x + y$ is a multiple of 10.

(b) If $x$ is odd then $3x + 1$ is even.

(c) If $x$ is a multiple of $y$ and $y$ is a multiple of $z$, then $x$ is a multiple of $z$.

(d) $10^{10}$ leaves a remainder of 1 when it is divided by 11.

SOLUTION: (a) $x + y \equiv 0 \bmod 10$.

(b) If $x \equiv 1 \bmod 2$ then $3x + 1 \equiv 0 \bmod 2$.

(c) If $x \equiv 0 \bmod y$ and $y \equiv 0 \bmod z$, then $x \equiv 0 \bmod z$.

(d) $10^{10} \equiv 1 \bmod 11$. ✓

**Problem 5.6** Simplify $11^3$ modulo 13.

SOLUTION: To compute $11^3 = 11 \times 11 \times 11$ modulo 13, we simplify first using minimal representatives. Since $11 \equiv -2 \bmod 13$, we find

$$11^3 = 11 \times 11 \times 11 \equiv (-2) \times (-2) \times (-2) = -8 \equiv 5 \bmod 13.$$

Thus $11^3 \equiv 5 \bmod 13$. ✓

Such computations, while they may seem mechanical, have nonobvious consequences. We can deduce that when $11^3$ is divided by 13, it leaves a remainder 5. And we deduced that fact without thinking about a single number bigger than 13!

Without modular arithmetic, one can compute $11^3 = 1331$. Dividing by 13 yields a remainder of 5.

EXTRA CARE is required when working with exponents in modular arithmetic. Exponents in modular arithmetic, as in ordinary arithmetic, abbreviate repeated multiplication (when the exponents are natural numbers). Many "rules" of exponents are really observations about repetition, and so they hold in the modular setting as well. For example, the rule $a^{b+c} = a^b a^c$ reflects the fact:

$$\underbrace{a \times \cdots \times a}_{\text{repeated } b+c \text{ times}} = \underbrace{(a \times \cdots \times a)}_{\text{repeated } b \text{ times}} \times \underbrace{(a \times \cdots \times a)}_{\text{repeated } c \text{ times}}.$$

To see the rule $a^{bc} = (a^b)^c$, the left side describes $\underbrace{a \times \cdots \times a}_{\text{repeated } bc \text{ times}}$, and the right side describes the repeated multiplication in the margin.

**Problem 5.7** Simplify $9^{10}$ mod 7.

SOLUTION: Exponentiation abbreviates repeated multiplication.

$$9^{10} = 9 \times 9 \times 9 \times 9 \times 9 \times 9 \times 9 \times 9 \times 9 \times 9.$$

Since $9 \equiv 2$ mod 7, we find

$$9^{10} \equiv 2 \times 2 \times 2 \times 2 \times 2 \times 2 \times 2 \times 2 \times 2 \times 2 \text{ mod } 7.$$

Since $2 \times 2 \times 2 \equiv 1$ mod 7, we simplify further,

$$9^{10} \equiv (2 \times 2 \times 2) \times (2 \times 2 \times 2) \times (2 \times 2 \times 2) \times 2 \equiv 1 \times 1 \times 1 \times 2.$$

Hence $9^{10} \equiv 2$ mod 7. ✓

On the other hand, suppose that we simplified the *exponent*[4] mod 7 as well. We would find that

$$9^{10} \stackrel{?}{\equiv} 9^3 \equiv 2^3 = 8 \equiv 1 \text{ mod } 7.$$

This is an error! One *cannot* naïvely simplify exponents!

**Problem 5.8** Simplify $2^{100}$, mod 11.

SOLUTION: Rather than evaluating $2^{100}$ directly, we simplify along the way.[5] To begin,

$$2^{100} = 2^{5 \times 20} = (2^5)^{20}.$$

Notice that $2^5 = 32 \equiv -1$ mod 11. Hence

$$2^{100} \equiv (32)^{20} \equiv (-1)^{20} \text{ mod } 11.$$

But now we evaluate as usual, and notice that $(-1)^{20} = 1$ (since 20 is even). Hence

$$2^{100} \equiv 1 \text{ mod } 11. \quad ✓$$

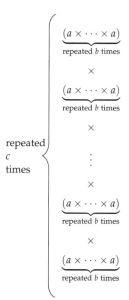

Figure 5.3: The exponent $(a^b)^c$ via repeated multiplication. When we write "repeated $b$ times," we mean that the variable $a$ occurs $b$ times.

[4] What *is* true: if $a \equiv b$ mod $m$, and $e$ is a positive integer, then

$$a^e \equiv b^e \text{ mod } m.$$

Hence, one can simplify the *bases* when performing modular arithmetic, but one cannot easily simplify the exponents.

[5] As you read this solution, convert all exponents to repeated multiplications to convince yourself that each step is justified.

As ten-fingered creatures, we typically write numbers in a base-ten system. In this system, a positive integer $N$ has a units digit $u$, a tens digit $t$, a hundreds digit $h$, a thousands digit $s$, etc., and the relationship between $N$ and its digits can be expressed algebraically:

$$N = u + 10t + 100h + 1000s + \cdots.$$

Since $10 \equiv 1 \bmod 3$ and $10 \equiv 1 \bmod 9$, our base-ten system is convenient for simplifying numbers modulo 3 and 9.

**Proposition 5.9 (Divisibility by 3 and 9)** *Let $N$ be a positive integer. Let $S$ be the sum of the base-ten digits of $N$. Then $N \equiv S \bmod 3$ and $N \equiv S \bmod 9$.*

PROOF: Since $10 \equiv 1 \pmod{3 \text{ or } 9}$, we find that $100 = 10^2 \equiv 1$ and $1000 = 10^3 \equiv 1$, etc.. If $u, t, h, s$, etc., are the digits of $N$, then

$$N = u + 10t + 100h + 1000s + \cdots \equiv u + t + h + s + \cdots = S \bmod 3 \text{ or } 9.$$

∎

**Corollary 5.10** *Let $N$ be a positive integer. Then $N$ is divisible by 3 or 9 if and only if the sum of the digits of $N$ is divisible by 3 or 9.*

PROOF: $N$ is divisible by 3 or 9 precisely when $N \equiv 0$ modulo 3 or 9. By the previous theorem, this occurs when $S \equiv 0$ modulo 3 or 9.

**Problem 5.11** Is 385 divisible by 9?

SOLUTION: Applying the previous theorem twice, we find

$$385 \equiv 3 + 8 + 5 = 16 \equiv 1 + 6 \equiv 7 \bmod 9.$$

Therefore 385 is not a multiple of 9. ✓

This method can be extended and combined with other modular tricks, to test divisibility of humongous numbers.

**Problem 5.12** Is $385^{100} + 2$ divisible by 9?

SOLUTION: We evaluate the expression, modulo 9.

$$385^{100} + 2 \equiv 7^{100} + 2 \bmod 9.$$

Now $7 \equiv -2 \pmod{9}$, and $(-2)^3 = -8 \equiv 1 \pmod{9}$. Therefore,

$$7^{100} + 2 \equiv (-2)^{100} + 2 \equiv (-2)^{3 \cdot 33 + 1} + 2 \equiv 1^{33} \cdot (-2) + 2 \equiv 0 \bmod 9.$$

Hence $385^{100} + 2$ is divisible by 9. ✓

Having mastered modular arithmetic, we turn our attention to algebra in the modular worlds. We can ask traditional algebraic questions in any modular world, as long as we stay within the realm of integers, addition, subtraction, and multiplication.

**Problem 5.13** Find all solutions to the congruence $3x \equiv 5 \bmod m$, when $m$ equals $2, 3, 4, 5, 6, 7, 8$.

SOLUTION: We look for solutions for each modulus, using a variety of methods.

*Mod 2:* Since $3 \equiv 5 \equiv 1 \bmod 2$, the congruence is equivalent to $x \equiv 1 \bmod 2$. There is one solution: $x \equiv 1 \bmod 2$.

*Mod 3:* Since $3 \equiv 0 \bmod 3$ and $5 \equiv 2 \bmod 3$, the congruence is equivalent to $0x \equiv 2 \bmod 3$. But since $0 \not\equiv 2 \bmod 3$, there are *no* solutions modulo 3.

*Mod 4:* Since $5 \equiv 1 \bmod 4$, the congruence is equivalent to $3x \equiv 1 \bmod 4$. The possible solutions are 0, 1, 2, and 3 (in this small world of four numbers). Checking each one, we find that $x \equiv 3 \bmod 4$ is the only solution.

*Mod 5:* Since $5 \equiv 0 \bmod 5$, the congruence is equivalent to $3x \equiv 0 \bmod 5$. This congruence states that $3x$ is a multiple of 5; since $GCD(3, 5) = 1$, it is equivalent to the statement that $x$ is a multiple of 5. Hence $x \equiv 0 \bmod 5$ is the only solution.

*Mod 6:* We cannot simplify the congruence $3x \equiv 5 \bmod 6$, so we start looking for solutions. Notice that as $x$ ranges over the integers, $3x \equiv 0 \bmod 6$ or $3x \equiv 3 \bmod 6$. But $0 \not\equiv 5 \bmod 6$ and $3 \not\equiv 5 \bmod 6$. Hence the congruence has *no solutions*.

*Mod 7:* We cannot simplify the congruence $3x \equiv 5 \bmod 7$, so we start looking for solutions. Trying $x \equiv 0, 1, 2, 3, 4, 5, 6 \bmod 7$, we find the only solution is $x \equiv 4 \bmod 7$. There is only this one soluion.

*Mod 8:* We look for solutions, trying[6] $x \equiv 0, 1, 2, 3, 4, -1, -2, -3$. The only solution we find is $x \equiv -1 \equiv 7 \bmod 8$. ✓

The solubility of congruences depends on the modular world in which we look for solutions. Understanding this dependence is one of the most important problems in number theory. In fact, we will demonstrate that the congruence $3x \equiv 5 \bmod m$ has a solution precisely when the modulus $m$ is not divisible by 3. For such moduli, there is in fact only one solution. This is exhibited in the table in the margin.

| Modulus ($m$) | Solution ($x$) |
|---|---|
| 1 | 0 |
| 2 | 1 |
| 3 | None |
| 4 | 3 |
| 5 | 0 |
| 6 | None |
| 7 | 4 |
| 8 | 7 |
| 9 | None |
| 10 | 5 |
| 11 | 9 |
| 12 | None |
| 13 | 6 |
| 14 | 11 |
| 15 | None |
| 16 | 7 |
| 17 | 13 |
| 18 | None |
| 19 | 8 |
| 20 | 15 |
| 21 | None |
| 22 | 9 |
| 23 | 17 |
| 24 | None |
| 25 | 10 |
| 26 | 19 |
| 27 | None |
| 28 | 11 |
| 29 | 21 |
| 30 | None |

Table 5.1: Solutions to $3x \equiv 5 \bmod m$, for various $1 \leq m \leq 30$.

[6] We use minimal representatives for a change.

A ONE-WAY STREET links the existence of solutions to a Diophantine equation to the existence of solutions to congruences in the modular worlds. For example, consider the linear Diophantine equation

$$15x + 39y = 12. \tag{E}$$

This equation (E) has a solution[7], since $GCD(15, 39) = 3$ and $3 \mid 12$. Moreover, for every positive modulus $m$, a solution to (E) will produce a solution to

[7] One solution is $x = -7, y = 3$.

$$15x + 39y \equiv 12 \bmod m, \tag{E mod $m$}$$

since equality of integers implies congruence in every modular world. We formalize this idea below.

**Proposition 5.14** *Let (E) be an equation, involving only integers, variables, addition, multiplication, and equality. If (E) has an integer solution, then the congruence (E mod m) has a solution for **every** modulus m.*

The *converse* to this theorem is not true! Consider, for example, the equation

$$x^2 + y^2 + z^2 + w^2 = -1. \tag{E}$$

This equation $E$ has no integer solutions; indeed, perfect squares cannot be negative and so sums of perfect squares cannot be negative.

On the other hand, a beautiful old theorem of Lagrange[8] states:

[8] The proof of this theorem is a bit beyond the scope of this text, though it was proven in 1770. A proof is sketched in a later exercise.

**Theorem 5.15 (Lagrange's Four-Square Theorem)** *Every natural number can be expressed as the sum of four squares.*

Hence if $m$ is a positive integer, then $m - 1$ can be expressed as $x^2 + y^2 + z^2 + w^2$ for some integers $x, y, z, w$. For such integers,

$$x^2 + y^2 + z^2 + w^2 = m - 1 \equiv -1 \bmod m.$$

Therefore, for every modulus $m$, the congruence

$$x^2 + y^2 + z^2 + w^2 \equiv -1 \bmod m \tag{E mod $m$}$$

has a solution.

In this example, the congruences (E mod $m$) have solutions for *every* modulus, but the Diophantine equation (E) has no solutions. The failure of (E) to have solutions reflects its failure to have *real* solutions. There are no real numbers $x, y, z, w$ satisfying $x^2 + y^2 + z^2 + w^2 = -1$.

A surprisingly simple example[9] is given by the equation

$$2x^2 + 7y^2 = 1. \tag{E}$$

The congruences (E mod $m$) have solutions for every modulus $m$ (this is not obvious!), and there are real solutions to (E), but there can be no integer solutions to (E) since $2x^2 + 7y^2 \geq 2$ when it is nonzero.

| $n$ | Sum of 4 squares |
|---|---|
| 0 | $0^2 + 0^2 + 0^2 + 0^2$ |
| 1 | $1^2 + 0^2 + 0^2 + 0^2$ |
| 2 | $1^2 + 1^2 + 0^2 + 0^2$ |
| 3 | $1^2 + 1^2 + 1^2 + 0^2$ |
| 4 | $2^2 + 0^2 + 0^2 + 0^2$ |
| 5 | $2^2 + 1^2 + 0^2 + 0^2$ |
| 6 | $2^2 + 1^2 + 1^2 + 0^2$ |
| 7 | $2^2 + 1^2 + 1^2 + 1^2$ |
| 8 | $2^2 + 2^2 + 0^2 + 0^2$ |
| 9 | $3^2 + 0^2 + 0^2 + 0^2$ |
| 10 | $3^2 + 1^2 + 0^2 + 0^2$ |

Table 5.2: A table of numbers $n$, and representations of $n$ as the sum of 4 squares. Multiple representations are possible but not displayed.

[9] From a *MathOverflow* post by Keith Conrad, "Diophantine equation with no integer solutions, but with solutions modulo every integer," URL (version: 2010-11-27): http://mathoverflow.net/q/47528.

EXISTENCE of solutions to a Diophantine equation implies existence of solutions to congruences. Logically then, *nonexistence* of solutions of congruences *prevents* the existence of solutions to a Diophantine equation. In this way, congruences can be used to prove negative results about very complicated Diophantine equations.

Solving the following two problems requires a strong intuition for modular arithmetic, and this will develop over the next few chapters.

**Problem 5.16** Prove that the Diophantine equation $x^4 + 16y^7 = 2011$ has no solutions.

SOLUTION: If the equation had an integer solution $(x, y)$, then $(x, y)$ would also provide a solution to the congruence[10]

$$x^4 + 16y^7 \equiv 2011 \bmod 4.$$

But observe that $16 \equiv 0 \bmod 4$ and $2011 \equiv 3 \bmod 4$. So we would find a solution to the congruence

$$x^4 \equiv 3 \bmod 4.$$

But only 0 and 1 are fourth powers, modulo 4. Hence $x^4 \equiv 3 \bmod 4$ has no solutions. Therefore the original equation cannot have any integer solutions. ✓

[10] Intuition leads one to consider the congruence modulo 4. This is a worthwhile approach, since $16 \equiv 0 \bmod 4$; considering the congruence modulo 4 effectively gets rid of a variable as it turns $16y^7$ to 0.

We directly compute
$$0^4 \equiv 0, \quad 1^4 \equiv 1,$$
$$2^4 \equiv 0, \quad 3^4 \equiv 1 \bmod 4.$$

**Problem 5.17** Prove that the Diophantine equation $x^3 + 700y = 140002$ has no solutions.

SOLUTION: Modulo 7, the coefficients simplify easily: $700 \equiv 0 \bmod 7$ and $140002 \equiv 2 \bmod 7$. An integer solution $(x, y)$ to the equation $x^3 + 700y = 140002$ would give a solution to the congruence

$$x^3 \equiv 2 \bmod 7.$$

But only 0, 1, and $-1$ are cubes, modulo 7:

$$0^3 \equiv 0, \ 1^3 \equiv 1, \ 2^3 \equiv 1, \ 3^3 \equiv -1, \ 4^3 \equiv 1, \ 5^3 \equiv -1, \ 6^3 \equiv -1 \bmod 7.$$

Since 2 is not a cube, modulo 7, there are no solutions to $x^3 \equiv 2 \bmod 7$. Hence there are no integer solutions to the original equation $x^3 + 700y = 140002$. ✓

THIS METHOD – reducing a Diophantine equation to a congruence – is quite powerful for *excluding* the possibility of solutions to a Diophantine equation. It relies entirely on an ability to exclude solutions to a congruence, and it relies also on some intuition for choosing the right modulus. But both of these osbtacles can be overcome, theoretically and computationally.

SOLVING LINEAR CONGRUENCES is equivalent to solving linear Diophantine equations. This is an exercise in translation: One must never forget the definition of congruence. For example, consider two integers $a$ and $b$, and a congruence of the form

$$ax \equiv b \bmod m.$$

An integer $x$ is a solution to this congruence if and only if

$$ax - b \text{ is a multiple of } m.$$

This is true, if and only if there exists an integer $y$ such that

$$ax - b = my.$$

Solving the congruence $ax \equiv b \bmod m$ becomes equivalent[11] to solving the linear Diophantine equation $ax - my = b$. From Theorem 1.14 we find a theoretical result.

[11] The variable $y$ in the Diophantine equation is auxiliary to the congruence. One may solve the congruence $ax \equiv b \bmod m$ by solving the linear Diophantine equation $ax - my = b$, and then "forgetting" about $y$.

**Proposition 5.18 (Solubility of linear congruences)** *The congruence $ax \equiv b \bmod m$ has a solution if and only if $\mathrm{GCD}(a, m)$ divides $b$.*

The Euclidean algorithm allows one to solve linear Diophantine equations, and thus to solve linear congruences.

**Problem 5.19** Solve the congruence $23x \equiv 1 \bmod 50$.

SOLUTION: If one could find integers $x, y$ satisfying $23x + 50y = 1$, the integer $x$ would satisfy $23x \equiv 1 \bmod 50$. So we apply the Euclidean algorithm to 50 and 23, down and up.

$$50 = 2(23) + 4 \qquad 1 = 4 - 1(3)$$
$$23 = 5(4) + 3 \qquad\quad = 4 - 1(23 - 5(4)) = 6(4) - 1(23)$$
$$4 = 1(3) + 1. \qquad\quad = 6(50 - 2(23)) - 1(23) = 6(50) - 13(23).$$

We have found that $6(50) - 13(23) = 1$ and so

$$23(-13) \equiv 1 \bmod 50.$$

Since $-13 \equiv 37 \bmod 50$, we find that $x \equiv 37 \bmod 50$ is a solution to the linear congruence $23x \equiv 1 \bmod 50$.[12]

[12] We prefer natural representatives. $x \equiv -13 \bmod 50$ is a valid solution, but we slightly prefer the equivalent solution $x \equiv 37 \bmod 50$.

Note that a solution to a congruence gives more than a single solution to an equation. The congruence $x \equiv 37 \bmod 50$ states that

$$x = \ldots \text{ or } -63 \text{ or } -13 \text{ or } 37 \text{ or } 87 \text{ or } 137 \text{ or } \ldots.$$

In this way, congruences can be a convenient way of describing the solution set to a linear Diophantine equation.

AN ADDITIVE INVERSE of a number $a$ is a partner $b$ for which $a + b = 0$. In the integers, additive inverses are given by negatives: the additive inverse of 7 is $-7$, since $7 + (-7) = 0$. This also holds in modular arithmetic; but using natural representatives, the additive inverse of $a \bmod m$ is $m - a \bmod m$ since (for every modulus $m$)

$$a + (m - a) = m = \equiv 0 \bmod m.$$

For example, 13 and 37 are additive inverses, modulo 50.

A MULTIPLICATIVE INVERSE of a number $a$ is a partner $b$ for which $ab = 1$.[13] In a modular world, the multiplicative inverse of a number $a \pmod{m}$ is a partner $b$ for which $ab \equiv 1 \bmod m$.

**Theorem 5.20 (Existence of modular inverses)** *Let $m$ be a positive integer. Let $a$ be an integer. Then $a$ has a multiplicative inverse, mod $m$, if and only if $GCD(a, m) = 1$. Moreover, any two multiplicative inverses of $a$, mod $m$, are congruent mod $m$.*

PROOF: Proposition 5.18 states that the equation $ab \equiv 1 \bmod m$ has a solution if and only if $GCD(a, m) = 1$; the resulting solution $b$ is precisely a multiplicative inverse of $a$. Next, suppose $b$ and $c$ are multiplicative inverses of $a$, mod $m$. Thus $ab \equiv ac \equiv 1 \bmod m$. Hence

$$b = b \cdot 1 \equiv b(ac) = bac = abc = (ab)c \equiv 1 \cdot c = c \bmod m.$$

Therefore $b \equiv c \bmod m$. ∎

**Corollary 5.21** *Let $p$ be a prime number and $a$ an integer. Then $a$ has a multiplicative inverse, mod $p$, if and only if $a \not\equiv 0 \bmod p$.*

PROOF: $a \not\equiv 0 \bmod p$ if and only if $GCD(a, p) = 1$ since $p$ is prime. ∎

Working modulo 40, there are 16 numbers which have multiplicative inverses. These are the numbers which are coprime to 40. Each is joined to its multiplicative inverse, mod 40, in the diagram below.

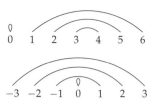

Figure 5.4: Each number modulo 7 is partnered with its additive inverse. Note that 0 is its own additive inverse. Additive inverses look more familiar with minimal representatives than with natural representatives.

[13] In the integers, multiplicative inverses are rare: the multiplicative inverse of 1 is 1, and the multiplicative inverse of $-1$ is $-1$. No other integers have multiplicative inverses among the integers.

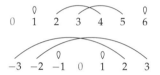

Figure 5.5: Every number modulo 7, except 0, has a multiplicative inverse. Note that 1 and 6 are their own multiplicative inverses; this is clearer with minimal representatives, where 1 and $-1$ are certainly their own multiplicative inverses since

$$1 \cdot 1 = 1 \text{ and } (-1) \cdot (-1) = 1.$$

Figure 5.6: Mod 40, most numbers do not have multiplicative inverses. And many are their own inverse, e.g. $19 \times 19 \equiv 1 \bmod 40$.

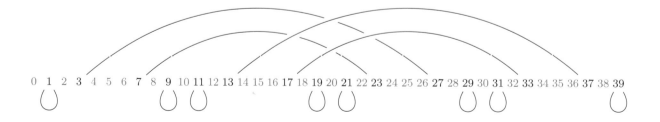

MULTIPLICATIVE INVERSES can be used to solve linear congruences.

**Problem 5.22** Solve the congruence $13x \equiv 20 \bmod 40$.

SOLUTION: At the bottom of the opposite page, we find that 37 is a multiplicative inverse[14] of 13. So multiply both sides of the congruence by 37 to obtain

$$x \equiv (37 \cdot 13)x \equiv 37 \cdot 20 \equiv (-3) \cdot 20 \equiv 20 \bmod 40.$$

The only solution to the congruence is $x \equiv 20 \bmod 40$. ✓

**Corollary 5.23** *Let m be a positive integer and a an integer satisfying* $GCD(a, m) = 1$. *Then for all integers x and y,*

$$ax \equiv ay \bmod m \text{ implies } x \equiv y \bmod m.$$

*In other words, one can "cancel a" if a is coprime to the modulus.*

PROOF: If $ax \equiv ay \bmod m$ and $GCD(a, m) = 1$, then let $b$ be the multiplicative inverse of $a$. Multiplying both sides by $b$, we find that

$ax \equiv ay \bmod m$ implies that $abx \equiv aby \bmod m$, so $x \equiv y \bmod m$. ∎

[14] One can find a multiplicative inverse of 13 by observing that

$$13 \times 3 = 39 \equiv -1 \bmod 40.$$

Hence

$$13 \times (-3) \equiv 1 \bmod 40.$$

So $-3$ is a multiplicative inverse of 13 mod 40.

MODULAR LIFE is easy when the modulus is prime. In this case our intuition for algebra carries into the modular world.

**Corollary 5.24 (Cancellation Property)** *Let p be a prime number. If* $a \not\equiv 0 \bmod p$ *then for all integers x and y,* $ax \equiv ay \bmod p$ *implies* $x \equiv y \bmod p$.

In the diagram below, we connect numbers to their multiplicative inverses, mod 41.

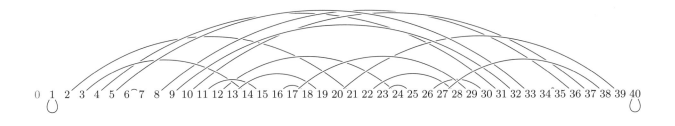

Figure 5.7: Mod 41, only two numbers ($\pm 1$) are connected to themselves, and only zero lacks a multiplicative inverse. A left-right symmetry is apparent.

POLYNOMIALS are the basic objects of algebra, and they arise in the modular worlds too. We work only with a **prime modulus** $p$ through the rest of the chapter. A **polynomial mod** $p$ is an expression of the form

$$a_0 + a_1 T + a_2 T^2 + \cdots + a_d T^d,$$

where the coefficients $a_0, \ldots, a_d$ are considered as integers mod $p$. Here $T$ is a "formal variable" – it will be treated for now just as a symbol; for addition it satisfies rules such as $T + T = 2T$. For multiplication it satisfies rules such as $T \cdot T = T^2$.

For example, if we work modulo 7, polynomials are considered the same if their coefficients are the same modulo 7.

$$1 + 7T + 15T^2 = 1 + T^2 \text{ mod } 7.$$

To add or subtract polynomials, add or subtract their coefficients.

$$(1 + 7T) + (2 + 3T + 5T^2) = 3 + 10T + 5T^2 \equiv 3 + 3T + 5T^2 \text{ mod } 7.$$

To multiply polynomials, use the distributive law.

$$(1 + 3T) \cdot (2 + 4T) = 2 + 6T + 4T + 12T^2 = 2 + 10T + 12T^2$$
$$\equiv 2 + 3T + 5T^2 \text{ mod } 7.$$

We use capital letters to stand for polynomials, e.g. $A = 1 + 3T^2 + 7T^3$. Working mod $p$, we define the **degree** of a polynomial $A$ to be the largest exponent that occurs with nonzero coefficient (mod $p$). So for the polynomial $A$, we have $\deg(A) = 3$ if we work mod 2 or 3 or 5. But we have $\deg(A) = 2$ if we work mod 7, since $A \equiv 1 + 3T^2 \text{ mod } 7$.

The **zero polynomial** is just 0; its degree is undefined since it has no nonzero coefficients (for every modulus). A **constant polynomial** mod $p$ is a polynomial which is zero or has degree 0. In other words, a constant polynomial is just a number in the world mod $p$. A **linear polynomial** mod $p$ is a polynomial of degree 1.

Polynomials can be evaluated by substitution and computation. If $x$ is a number, and $A$ is a polynomial, we write $A(x)$ for the number arising from substituting $x$ for $T$ everywhere. We work modulo $p$ throughout. For example, if $A = 1 + 3T^2 + 7T^3$ and $p = 7$ as before, we compute

$$A(2) = 1 + 3 \cdot 2^2 + 7 \cdot 2^3 = 1 + 12 + 7 \cdot 8 \equiv 1 + 12 = 13 \equiv 6 \text{ mod } 7.$$

If $A$ is a polynomial, then a **root** of $A$ mod $p$ is a number $x$ which satisfies $A(x) \equiv 0 \text{ mod } p$. Roots themselves live in the modular world; indeed, if $x \equiv y \text{ mod } p$ then $A(x) \equiv A(y) \text{ mod } p$. Hence $x$ is a root of $A$ if and only if $y$ is a root of $A$, mod $p$.

---

Given $A = 1 + 3T^2 + 7T^3$, we can simplify it modulo 2, 3, 5, and 7. The degree drops modulo 7 since $7 \equiv 0 \text{ mod } 7$.

| | |
|---|---|
| $A \equiv 1 + T^2 + T^3$ | mod 2 |
| $A \equiv 1 + T^3$ | mod 3 |
| $A \equiv 1 + 3T^2 + 2T^3$ | mod 5 |
| $A \equiv 1 + 3T^2$ | mod 7 |

ROOTS of polynomials, mod $p$, exhibit unusual phenomena which we study further in this and later chapters. In degree 1, roots of polynomials are solutions to linear congruences.

**Problem 5.25** Find the roots of the polynomial $3T - 5$, mod 7.

SOLUTION: We seek the solutions to the congruence $3x - 5 \equiv 0 \mod 7$. But this is equivalent to the congruence $3x \equiv 5 \mod 7$. We have already seen the solution in Problem 5.13: $x \equiv 4 \mod 7$. ✓

We can find roots of all linear polynomials, except when coincidences forbid it.

**Proposition 5.26** *Let $F = a + bT$ be a polynomial, and $p$ be a prime number. If $b \not\equiv 0 \mod p$, then there exists a unique root of $F$, mod $p$. If $b \equiv 0 \mod p$, and $a \not\equiv 0 \mod p$, then $F$ has no roots, mod $p$.*

PROOF: We seek solutions to the congruence $a + bx \equiv 0 \mod p$. This is equivalent to the congruence $bx \equiv -a \mod p$. If $b \not\equiv 0 \mod p$, then $b$ has a multiplicative inverse $c$ mod $p$, and the unique root of $F$ is $x \equiv -ac \mod p$.

If $b \equiv 0 \mod p$ and $a \not\equiv 0 \mod p$, then the congruence is equivalent to $a \equiv 0 \mod p$. This is always false, regardless of the value of $x$, and so $F$ has no roots mod $p$. ∎

IN DEGREE 2, roots of polynomials, mod $p$, begin to get interesting.

**Problem 5.27** Find the roots of the polynomial $3T^2 - 5$, mod 7.

SOLUTION: We seek the solutions to the congruence $3x^2 - 5 \equiv 0 \mod 7$, which is equivalent to $3x^2 \equiv 5 \mod 7$. Not having other methods at our disposal (yet), we search the 7 numbers in the modular world.

$$3(0)^2 \equiv 0, \quad 3(1)^2 \equiv 3, \quad 3(2)^2 \equiv 5, \quad 3(3)^2 \equiv 6,$$

$$3(4)^2 \equiv 6, \quad 3(5)^2 \equiv 5, \quad 3(6)^2 \equiv 3 \mod 7.$$

The solutions are $x \equiv 3 \mod 7$ and $x \equiv 5 \mod 7$. ✓

For a random prime modulus $p$, the polynomial $3T^2 - 5$ exhibits a 50% chance of having no roots, and a 50% chance of having two roots (and $p = 2, 3, 5$ are special cases); cases are tabulated in the margin.

More generally, given a polynomial $P$ with integer coefficients, one can ask the statistical question – for a random prime modulus $p$, how often does $P$ have no roots, one root, two roots, etc.? A general solution was given by Cebotarev's density theorem (1922); an implication, for example, is that the polynomial $x^5 - 5x + 12$ has five roots, modulo $p$, for about 10% of prime numbers $p$.

| Modulus | Number of roots |
|---|---|
| 2 | 0 |
| 3 | 0 |
| 5 | 1 |
| 7 | 0 |
| 11 | 1 |
| 13 | 1 |
| 17 | 1 |
| 19 | 1 |
| ⋮ | 1 |

Table 5.3: The number of roots of
$$17 + 42T$$
mod $p$, for various primes $p$. The exceptional primes are 2, 3, and 7; these are factors of 42, and so that polynomial drops in degree modulo 2, 3, and 7.

| Modulus | Number of roots |
|---|---|
| 2 | 1 |
| 3 | 0 |
| 5 | 1 |
| 7 | 2 |
| 11 | 2 |
| 13 | 0 |
| 17 | 2 |
| 19 | 0 |
| 23 | 0 |
| 29 | 0 |
| 31 | 0 |
| 37 | 0 |
| 41 | 0 |
| 43 | 2 |
| 47 | 0 |

Table 5.4: The number of roots of $3T^2 - 5$ mod $p$, for various primes $p$. For $p \geq 7$, the number of roots is either 0 or 2.

| No. of roots | No. primes | % of primes |
|---|---|---|
| No roots | 486 | 39.54 % |
| One root | 626 | 50.93 % |
| Two roots | 1 | 0 % |
| Three roots | 0 | 0 % |
| Four roots | 0 | 0 % |
| Five roots | 116 | 9.44 % |

Table 5.5: Proportion of primes $p < 10000$, for which the polynomial $x^5 - 5x + 12$ has no roots, one roots, etc., mod $p$. Cebotarev's density theorem gives the proportions 40%, 50%, 0%, 0%, 0%, and 10%, asymptotically.

POLYNOMIALS MOD $p$ form an alternate universe for number theory – one in which familiar properties of integers carry over, one which sheds light on the integers themselves, and one in which some of the greatest number theoretic conjectures become provable. For these reasons, we give a short tour.

We have seen that polynomials may be added, subtracted and multiplied; familiar commutative, associative, and distributive properties hold. The zero polynomial is the additive identity, and the number 1 is a polynomial too – a multiplicative identity. It makes sense to speak of divisibility of polynomials: if $A$ and $B$ are polynomials (mod $p$), then we say that $A \mid B$ if there exists a polynomial $C$ such that $B \equiv AC$ mod $p$. The two out of three principle for divisibility carries over. These are the *Beginning Principles* in Figure 4.16. The arithmetic of polynomials mod $p$ obeys the same basic principles as the arithmetic of integers obeys.

For integers, the absolute value provides a notion of size. An analogous notion can be defined for polynomials, by exponentiating the degree. **Fix a prime modulus** $p$ and define the **absolute value** of a nonzero polynomial $A$ by the formula

$$|A| = p^{\deg(A)}.$$

If $A = 0$, the degree is undefined, so define $|0| = 0$.[15]

**Theorem 5.28 (Degree-formula for products of polynomials)** *If $A$ and $B$ are nonzero polynomials mod $p$, then $deg(AB) = deg(A) + deg(B)$.*

PROOF: Let $d = \deg(A)$ and $e = \deg(B)$. Let $a_0, \ldots, a_d$ be the coefficients of $A$ and $b_0, \ldots, b_e$ the coefficients of $B$. Since $\deg(A) = d$ and $\deg(B) = e$, we have $a_d, b_e \not\equiv 0$ mod $p$. In other words, neither $a_d$ nor $b_e$ is a multiple of $p$. Since $p$ is prime, $a_d b_e$ cannot be a multiple of $p$ (see Corollary 2.13). In congruence notation, $a_d b_e \not\equiv 0$ mod $p$.

Computing the product of polynomials by the distributive law,

$$A \cdot B = (a_0 + \cdots + a_d T^d) \cdot (b_0 + \cdots + b_e T^e) = a_0 b_0 + \cdots + a_d b_e T^{d+e}.$$

The coefficients among the elided ($\cdots$) terms are often complicated; but at the ends the coefficients are the simple products $a_0 b_0$ and $a_d b_e$ as written above. In particular, since $a_d b_e \not\equiv 0$ mod $p$, the degree of $A \cdot B$ is $d + e$. ∎

**Corollary 5.29** *If $A$ and $B$ are polynomials mod $p$, then $|AB| = |A| \cdot |B|$.*

PROOF: If $A$ and $B$ are nonzero polynomials, then

$$|AB| = p^{\deg(AB)} = p^{\deg(A)+\deg(B)} = p^{\deg(A)} \cdot p^{\deg(B)} = |A| \cdot |B|.$$

If $A$ or $B$ are the zero polynomial, then $|A| = 0$ or $|B| = 0$ respectively. In this case $AB = 0$ and the result can be seen directly. ∎

Consider the polynomial $A = 3T^2 - 5$. If $p \neq 3$ then $\deg(A) = 2$. The absolute value depends more dramatically on the prime modulus. If $p = 2$ then $|A| = 2^2 = 4$. If $p = 7$ then $|A| = 7^2 = 49$. The usefulness of this construct will become apparent over time.

[15] Note that $|A|$ is always a natural number. This is important, since we know that every decreasing sequence of natural numbers must eventually terminate.

The proof requires $p$ to be prime, and indeed the theorem fails for composite $p$. For example, if we work modulo 6, then $\deg(1 + 2T) = 1$ and $\deg(1 + 3T) = 1$, but

$$(1 + 2T)(1 + 3T) \equiv 1 + 5T \text{ mod } 6.$$

Their product has degree 1 instead of the expected degree 2.

THE ABSOLUTE VALUE for polynomials (mod $p$) shares the most important basic properties with the absolute value for integers, Gaussian integers, and Eisenstein integers.

**Corollary 5.30** *If A and B are polynomials and $B \neq 0$, then $|A| \leq |AB|$. Equivalently, $\deg(A) \leq \deg(AB)$.*

Compare to the following property of integers: if $x$ and $y$ are integers and $y \neq 0$, then $|x| \leq |xy|$.

PROOF: Since $B$ is a nonzero polynomial, $|B| = p^{\deg(B)} \geq 1$. Hence $|A| = |A| \cdot 1 \leq |A||B| = |AB|$. ∎

**Proposition 5.31** $|A| = 1$ *if and only if there exists a polynomial B such that $A \cdot B \equiv 1 \bmod p$.*

PROOF: If $|A| = 1$, then $\deg(A) = 0$ and $A$ is a nonzero constant polynomial. Thus $A \equiv a$ for some integer $a$ with $a \not\equiv 0 \bmod p$. Thus $a$ has a multiplicative inverse $b$ mod $p$. Let $B = b$ and we find $AB = ab \equiv 1 \bmod p$.

Conversely, if $AB \equiv 1 \bmod p$, then $|AB| = 1$ and both $A$ and $B$ are positive integers. By the previous corollary, $|A| = |B| = 1$. ∎

The nonzero constant polynomials are the **units** in the world of polynomials mod $p$. They are analogous to $\pm 1$ among the integers, or $\pm 1, \pm i$ among the Gaussian integers, or $\pm 1, \pm \omega, \pm \omega^2$ among the Eisenstein integers.

For ordinary integers, and for Gaussian and Eisenstein integers, the absolute value satisfies the *triangle inequality*:

$$|x + y| \leq |x| + |y|.$$

For polynomials, an even stronger inequality holds.

**Proposition 5.32 (Ultrametric triangle inequality)** *If A and B are polynomials mod $p$, then*

$$|A + B| \leq \max(|A|, |B|).$$

PROOF: If $A$ or $B$ is zero, then the result is easy to check.
If $A$ and $B$ are nonzero, observe that

$$\deg(A + B) \leq \max(\deg(A), \deg(B)). \tag{5.1}$$

Indeed, one cannot produce higher powers of $T$ by addition. The highest power of $T$ in $A + B$ can be as high as the highest power that occurs in $A$ or the highest power that occurs in $B$.[16] The degree inequality (5.1) implies the inequality for absolute values:

$$|A + B| = p^{\deg(A+B)} \leq p^{\max(\deg(A),\deg(B))} = \max(|A|, |B|). \quad \blacksquare$$

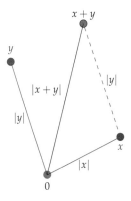

Figure 5.8: The triangle inequality for complex numbers holds, because one can construct a triangle with sides $|x|$, $|y|$, and $|x + y|$. This demonstrates the triangle inequality for Gaussian and Eisenstein integers.

[16] It is possible that cancellation *lowers* the degree in the sum.

DIVISION WITH REMAINDER is possible for polynomials mod $p$.

**Lemma 5.33** *Let B be a nonzero polynomial mod p. Suppose that $Q_1, Q_2$ and $R_1, R_2$ are polynomials, with $|R|_1 < |B|$ and $|R|_2 < |B|$. If $Q_1B + R_1 \equiv Q_2B + R_2$ mod $p$, then $Q_1 \equiv Q_2$ mod $p$ and $R_1 \equiv R_2$ mod $p$.*

PROOF: The congruence $Q_1B + R_1 \equiv Q_2B + R_2$ mod $p$ implies that

$$(Q_1 - Q_2)B \equiv (R_2 - R_1) \text{ mod } p.$$

Consider the degrees of each side. If $Q_1 \not\equiv Q_2$ mod $p$, then $\deg((Q_1 - Q_2)B) \geq \deg(B)$ by Corollary 5.30. But since $|R|_1 < |B|$ and $|R|_2 < |B|$, we have $\deg(R_2 - R_1) < \deg(B)$. This is a contradiction. Thus $Q_1 \equiv Q_2$ mod $p$, which implies $R_2 \equiv R_1$ mod $p$, proving the lemma. ∎

Compare to the following statement: Let $b$ be a nonzero natural number. If $q_1, q_2, r_1, r_2$ are natural numbers, and $r_1 < b$ and $r_2 < b$, then $q_1 b + r_1 = q_2 b + r_2$ if and only if $q_1 = q_2$ and $r_1 = r_2$.

**Lemma 5.34** *If A is a polynomial, then the number of polynomials P mod p satisfying $|P| < |A|$ is equal to $|A|$.*

PROOF: If $A = 0$, the result is clear. If $|A| = 1$ then $\deg(A) = 0$ and the only possible $P$ is 0; the result holds in this case too. If $\deg(A) > 0$, consider the choices needed to choose a polynomial $P$ satisfying $|P| < |A|$. If $\deg(A) = d$, then such a polynomial $P$ has the form

$$P = c_0 + c_1 T + c_2 T^2 + \cdots + c_{d-1} T^{d-1}.$$

Mod $p$, we have $p$ choices for $c_0$, $p$ choices for $c_1$, ..., etc. Hence there are $p^d = |A|$ choices for the polynomial $P$. ∎

Compare to the following statement: if $n$ is a natural number, then the number of natural numbers $x$ satisfying $x < n$ is equal to $|n|$.

| $a_0$ | $a_1$ | $a_2$ | $P$ | $\lvert P \rvert$ |
|---|---|---|---|---|
| 0 | 0 | 0 | 0 | 0 |
| 1 | 0 | 0 | 1 | 1 |
| 1 | 1 | 0 | $1+T$ | 2 |
| 0 | 1 | 0 | $T$ | 2 |
| 0 | 0 | 1 | $T^2$ | 4 |
| 0 | 1 | 1 | $T+T^2$ | 4 |
| 1 | 0 | 1 | $1+T^2$ | 4 |
| 1 | 1 | 1 | $1+T+T^2$ | 4 |

Table 5.6: The 8 polynomials $P$, mod 2, satisfying $|P| < 8$.

**Theorem 5.35 (Division with remainder for polynomials mod $p$)**
*If A and B are polynomials, and $B \neq 0$, then there exist polynomials Q, R such that*

$$A \equiv QB + R \text{ mod } p, \text{ and } |R| < |B|.$$

PROOF: If $A = 0$, let $Q = 0$ and $R = 0$, to verify the theorem. Otherwise, let $d = \deg(A)$ and $e = \deg(B)$. If $d < e$, define $Q = 0$ and $R = A$ to verify the theorem. If $d \geq e$ we consider two sets:

1. The set of all polynomials of the form $QB + R$, in which $|Q| \leq p^{d-e}$ and $|R| < |B|$. Equivalently,[17] $|Q| < p^{d-e+1}$ and $|R| < p^e$.

2. The set of all polynomials $P$ satisfying $|P| \leq |A|$. Equivalently, $|P| < p^{d+1}$.

[17] $|Q| < p^{d-e+1}$ is equivalent to $|Q| \leq p^{d-e}$, since absolute values must lie among the powers of $p$. For the same reason, $|P| \leq |A| = p^d$ is equivalent to $|P| < p^{d+1}$.

By the two lemmas, the first set contains $p^{d-e+1} \cdot p^e = p^{d+1}$ distinct (mod $p$) polynomials, one for each of the possible choices for $Q$ and $R$. The second set contains $p^{d+1}$ polynomials. By the triangle inequality, Proposition 5.32, the first set is contained in the second. Since both sets contain the same number of polynomials, every polynomial in the second set is also in the first set. Since $A$ is in the second set, it is in the first. ∎

The previous theorem guarantees that division with remainder is possible, though the given proof is nonconstructive. Fortunately, for those who wish to compute, the long division algorithm that is taught[18] in schools works as well for dividing polynomials mod $p$ as it does for dividing integers. In the margin we exhibit this process, dividing $T^4 + T^2 + 2$ by $T^2 + 2T + 1$, mod 3.

THE EUCLIDEAN ALGORITHM can now be carried out for polynomials mod $p$; the flow of Figure 4.16 applies to polynomials mod $p$, using the results of the previous few pages. The "primes" in the polynomial context are typically called **irreducible polynomials**. An irreducible polynomial mod $p$ is a polynomial $P$ such that

1. $P$ is not a unit; equivalently, $\deg(P) \neq 0$.

2. If $A$ and $B$ are polynomials, and $AB \equiv P$ mod $p$, then $A$ or $B$ is a constant polynomial, i.e., $\deg(A) = 0$ or $\deg(B) = 0$.

**Theorem 5.36 (Unique factorization of polynomials mod $p$)** *If $A$ is a nonzero polynomial, then $A$ can be decomposed uniquely (up to units) into irreducible polynomials.*

LINEAR POLYNOMIALS place a special role.

**Proposition 5.37** *If $\deg(P) = 1$, then $P$ is irreducible.*

PROOF: If $A$ and $B$ are polynomials and $P \equiv AB$ mod $p$, then $\deg(A) + \deg(B) = \deg(P) = 1$. The only two natural numbers which add to 1 are 0 and 1. Hence $A$ or $B$ has degree zero. ■

**Lemma 5.38** *Suppose that $x$ is a root of a polynomial $P$, mod $p$. Then the irreducible polynomial $(T - x)$ is a factor of $P$.*

PROOF: Since $\deg(T - x) = 1$, there exists a constant polynomial $R$ such that $P = Q \cdot (T - x) + R$; here $R = r$ for some integer $r$ mod $p$. Evaluation of the polynomial at $T = x$ yields
$$0 = P(x) = Q(x) \cdot (x - x) + r = r.$$
Hence $R = r = 0$. Therefore $P = Q \cdot (T - x)$ and evidently the irreducible linear polynomial $(T - x)$ is a factor of $P$. ■

We will apply the following theorem in chapters that follow.

**Theorem 5.39 (Roots are bounded by the degree)** *Let $P$ be a nonzero polynomial, mod $p$. The number of roots of $P$ (mod $p$) is less than or equal to the degree of $P$.*

PROOF: For each root $x$ of $P$, the irreducible polynomial $(T - x)$ is a factor of $P$. Therefore, if $P$ has $d$ distinct roots mod $p$, then $P$ has $d$ distinct[19] irreducible linear factors. Hence $\deg(P) \geq d$. ■

[18] Whether and how long division is taught varies greatly.

$$\begin{array}{r} T^2 + 1T + 1 \ \ \text{R} \ \ 1 \\ T^2 + 2T + 1 \overline{)T^4 + 0T^3 + 1T^2 + 0T + 2} \\ -\ \underline{T^4 + 2T^3 + 1T^2} \\ 1T^3 + 0T^2 + 0T \\ -\ \underline{1T^3 + 2T^2 + 1T} \\ 1T^2 + 2T + 2 \\ -\ \underline{1T^2 + 2T + 1} \\ 0T + 1 \end{array}$$

Figure 5.9: Long division of polynomials, mod 3. Dividing $T^4 + T^2 + 2$ by $T^2 + 2T + 1$ yields a quotient of $T^2 + T + 1$ and a remainder 1.

[19] For distinct numbers $r_1, r_2$ mod $p$, the polynomials $(T - r_1)$ and $(T - r_2)$ are coprime.

FOR EVERY THEOREM one has about prime numbers, one can seek a theorem about irreducible polynomials. For example, Euclid's proof[20] of the infinitude of primes (Theorem 2.3) adapts to polynomials mod $p$.

[20] The adaptation of Euclid's proof is straightforward and left to the exercises.

**Theorem 5.40 (Infinitude of irreducible polynomials mod $p$)** *The number of irreducible polynomials mod $p$ is infinite.*

The opposite page displays the irreducible polynomials mod 2, of degree up to 19, in "lexicographic order."[21]

While they are probably not familiar by name, their distribution should be reminiscent of the distribution of prime numbers. For prime numbers, recall that Gauss used the heuristic: the chance that a random $n \geq 2$ is prime is $1/\log(n)$. This heuristic suggests an estimate for the number of primes up to $x$ (denoted $\pi(x)$):

$$\pi(x) \approx \sum_{n=2}^{x} \frac{1}{\log(n)} \approx \int_{t=2}^{\infty} \frac{dt}{\log(t)}.$$

[21] We name and order them by transforming polynomials into binary expansions, then expressing them in base 10. For example, the irreducible polynomial $T^2 + T + 1$ is transformed into the binary expansion **b** 111 which represents the number 7. In the other direction, 61 has binary expansion **b** 111101, which stands for the irreducible polynomial

$$61 \to T^5 + T^4 + T^3 + T^2 + 1.$$

The Riemann hypothesis is equivalent to the statement that the absolute error is $O(\sqrt{x}\log(x))$, in the above approximation.

The analogous heuristic for polynomials is: the chance that a random mod $p$ polynomial of degree $d$ is irreducible is $1/d$. The degree is the appropriate analogue of $\log(n)$ in the polynomial context. This heuristic suggests an estimate for the number of irreducible monic[22] polynomials of degree $d$ (denoted $\pi(p;d)$):

$$\pi(p;d) \approx \frac{p^d}{d}.$$

[22] A polynomial is **monic** if its top-degree coefficient is 1. Every nonzero polynomial mod $p$ is a unit multiple of a monic polynomial. Counting only monic polynomials is like counting positive prime numbers (we don't count both 7 and $-7$ when we count primes).

Indeed, there are exactly $p^d$ monic polynomials of degree $d$, and according to the heuristic, each has a $1/d$ chance of being irreducible.

Remarkably, the analogue of the Riemann hypothesis for polynomials was proven by Gauss.[23] The absolute error in the above approximation $O(p^{d/2}/d)$, which is the analogue of $\sqrt{x}/\log(x)$.

**Theorem 5.41 (Prime number theorem for polynomials mod $p$)**
*Let $d$ be a positive integer. The number of irreducible monic polynomials of degree $d$, mod $p$, satisfies the estimate:*

$$\left|\pi(p;d) - \frac{p^d}{d}\right| \leq 2\frac{p^{d/2}}{d}.$$

[23] This analogue considers irreducible polynomials in aggregate – all polynomials of a given degree are counted at once. The original Riemann hypothesis is more refined, considering primes one by one. But a more refined Riemann hypothesis for polynomials – considering irreducible polynomials one at a time, in lexicographic order, was proven by Paul Pollack. See "Revisiting Gauss's analogue of the prime number theorem for polynomials over a finite field", in *Finite Fields and their Applications* vol. 16, 4 (2010).

The BIG question in number theory becomes an old settled theorem in the world of polynomials mod $p$. Some number theorists hold out hope that the methods of proof used in the world of polynomials mod $p$ can someday be adapted for the world of integers.

# THE MODULAR WORLDS 147

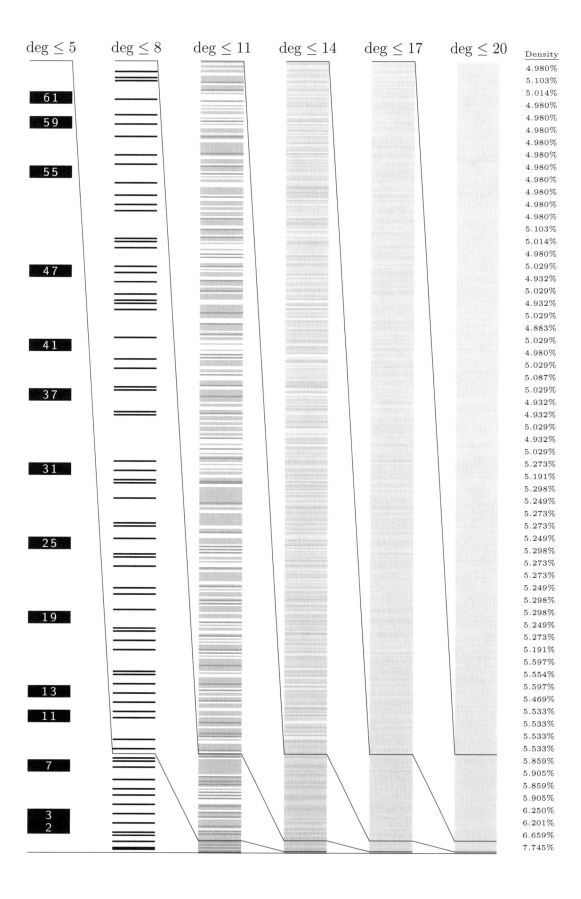

## Historical notes

Fermat and Euler and many others before[24] studied remainders after division, but Gauss changed our entire perspective on the matter. For division with remainder is an *operation* on numbers, but Gauss instead considers the *relation* given by *congruence* of numbers.[25]

A partial shift in this direction can be found in the work of Euler, who makes a detailed study of the *residues* of numbers. In his number theory text, published long after his death, Euler writes,

> Ita si sit $a = mb + r$, erit $r$ **residuum** ex divisione numeri $a$ per $b$ natum.

Euler goes on to discuss residue classes, writing

> 147. Simili modo divisor 5 has quinque numerorum **classes** suppeditat: I. $5m$, II. $5m + 1$, III. $5m + 2$, IV. $5m + 3$, V. $5m + 4$.[26]

The idea of residue classes may originate with Euler, but the language of congruence belongs to Gauss. Moreover, congruence is the very first topic of the *Disquisitiones*. Section 1.1 of the *Disquisitiones* begins,

> If a number $a$ divides the difference of the numbers $b$ and $c$, $b$ and $c$ are said to be *congruent relative to* $a$; if not, $b$ and $c$ are noncongruent. The number $a$ is called the *modulus*. If the numbers $b$ and $c$ are congruent, each of them is called a *residue* of the other. If they are noncongruent they are called *nonresidues*.[27]

The fundamental results on modular arithmetic and linear congruences can be found in the first two sections of the Disquisitiones, Art. 1-32. Moreover, within these articles, Gauss proves the uniqueness of prime decomposition we presented in Chapter 2. A passage exhibiting Gauss's instinct is below.

> In the same way that the root of the equation $ax = b$ can be expressed as $b/a$, we will designate by $b/a$ a root of the congruence $ax \equiv b$ and join to it the modulus of the congruence to distinguish it. Thus, e.g., $19/17 (\text{mod. } 12)$ denotes any number that is $\equiv 11 (\text{mod. } 12)$...
>
> One can work with these expressions very much as with common fractions.[28]

In this way Gauss anticipates the later abstract theory of rings and fields, where the meaning of fractions depends on context.

We mentioned in this chapter the four-square theorem of Lagrange – every natural number can be expressed as the sum of four squares. This result has a long history, going back at least to the Latin *Arithmetica* of Bachet de Méziriac (1621), a translation and adaptation of earlier manuscripts of Diophantus. In letters to Mersenne, and having read Bachet, Fermat mentions the result, first writing,

---

[24] See "Modular arithmetic before C.F. Gauss" by Maarten Bullynck, in *Historia Mathematica* **36** (2009).

[25] In 1797, Legendre uses notation like
$$\frac{a-b}{c} = entier = e$$
to mean that $(a-b)/c$ is an integer; this is equivalent to Gauss's $a \equiv b \mod c$, but not as streamlined.

[26] From Chapter V, "De residuis ex divisione natis" of Euler's "Tractatus de numerorum doctrina capita sedecim, quae supersunt," available from the Euler Archive as E792 at http://eulerarchive.maa.org/pages/E792.html. This was published in 1849, long after Euler's death in 1783.

[27] From *Disquisitiones Arithmeticae*, English edition translated by Arthur A. Clarke, published by Springer-Verlag in 1986, reprinted from the 1966 edition by Yale University Press.

[28] Art. 31 of the *Disquisitiones*, translated by Arthur A. Clarke.

Premierement, tout nombre est compose d'autant de quarres entiers qu'il a d'unites;[29]

As early as 1730, Euler became interested in this statement of Fermat; Euler proved that if $x$ and $y$ can be expressed as the sum of four squares, so too can their product. This reduces the proof to the case of prime numbers. After 40 years of attempts by Euler, Lagrange first proved the four-square theorem in 1770 (using ideas of Euler along the way). Euler rapidly simplified Lagrange's proof by 1773.[30]

Knowing the four square theorem immediately raises a deeper question: how many ways can one represent a natural number $n$ as the sum of four squares? The number of ways is denoted $r_4(n)$, where different orders and signs are counted separately.[31] A formula for $r_4(n)$ was given by Jacobi (1834),[32]

$$r_4(n) = \begin{cases} 8\sigma_1(n) & \text{if } n \text{ is odd;} \\ 24\sigma_1^o(n) & \text{if } n \text{ is even.} \end{cases}$$

Here $\sigma_1(n)$ denotes the sum of the divisors of $n$, and $\sigma_1^o(n)$ the sum of the odd divisors of $n$. About his formula for $r_4(n)$, Jacobi writes,

Ce théorème remarquable parait être assez difficile à démontrer par les méthods connues de la théorie des nombres. La démonstration fournie par la théorie des fonctions elliptiques est entièrement analytique.[33]

Jacobi's work is the first where we see the analytic theory of (elliptic) modular forms applied to number theoretic questions.

We have concluded this chapter with a brief introduction to the algebra of polynomials mod $p$ (for $p$ prime). This theory was meant to be Section 8 of Gauss's *Disquisitiones*; but Gauss did not complete the section in time for publication with the *Disquisitiones*, and in fact his notes for Section 8 (a manuscript from 1797) were only published posthumously in 1863. See the article by Frei for more on the history of this work of Gauss, titled the *Disquisitiones Generales de Congruentiis*.[34]

Gauss first understood the deep analogy between the polynomials mod $p$ and the ordinary integers – an analogy we have pursued in the factorization of polynomials. By Art. 336 of the *Disquisitiones Generales*, Gauss develops the Euclidean algorithm for polynomials mod $p$, and proves the unique decomposition of polynomials into irreducibles. By Art. 347, Gauss obtains a formula for the number of irreducible polynomials mod $p$, of a given degree $d$. This can be used to derive the analogue of the prime number theorem we present. A closer analogue and detailed treatment can be found in the recent work of Paul Pollack.[35]

---

[29] From a letter from Fermat to Mersenne, dated September 2, 1636. In "Oeuvres de Fermat", edited by Paul Tannery, published 1891.

[30] See Herbert Pieper's "On Euler's contributions to the Four-Squares theorem," in *Historia Mathematica* **20** (1993), upon which this paragraph is based.

[31] For example, $r_4(1) = 8$ since
$$1 = 1^2 + 0^2 + 0^2 + 0^2,$$
$$1 = (-1)^2 + 0^2 + 0^2 + 0^2,$$
$$1 = 0^2 + 1^2 + 0^2 + 0^2,$$
etc.

[32] A sketch was given in a one-page note, "Note sur la decomposition d'un nombre donne en quatre carrés", in *Journal für die reine und angewandte Mathematik* (1828).

[33] From Jacobi (1828) cited above.

[34] See Günther Frei's, "The unpublished section eight: on the way to function fields over a finite field," Chapter II.4 of "The shaping of arithmetic after C.F. Gauss's *Disquisitiones Arithmeticae*", Springer-Verlag (2007).

[35] "Revisiting Gauss's analogue of the prime number theorem for polynomials over a finite field," by Paul Pollack, in *Finite Fields and their Applications* vol. 16 **4** (2010).

*Exercises*

1. Translate the following into the language of congruences.

    (a) $3u - 5v$ is a multiple of 7.

    (b) 10 divides $x^2 - 1$.

    (c) If $x$ is an even number, then $x^2$ is a multiple of 4.

    (d) $x$ belongs to the sequence $\ldots, -7, -3, 1, 5, 9, 13, 17, \ldots$.

    (e) $x$ belongs to the sequence $-18, -8, 2, 12, 22, 32, \ldots$.

    (f) When $e$ is even, $10^e$ leaves a remainder 1 when divided by 11.

    (g) The linear Diophantine equation $11x + 13y = 1$ has a solution. (Many translations are possible.)

    (h) If $x$ is a multiple of 2 and of 3, then $x$ is a multiple of 6.

2. Simplify the following, using natural representatives.

    (a) $10 \times 9 \times 8 \times \cdots \times 1$, mod 7.

    (b) $37^5$ mod 5.

    (c) $1 + 2 + 3 + \cdots + 23 + 24 + 25$ mod 13.

    (d) $1^2 + 2^2 + 3^2 + \cdots + 101^2$ mod 3.

3. Let $N$ be a positive integer, with units digit $u$, tens digit $t$, hundreds digit $h$, thousands digit $s$, etc. Let $A$ be the alternating sum of these digits, i.e., $A = u - t + h - s + \cdots$. Demonstrate that $N \equiv A$ mod 11. Is 123456789 divisible by 11?

4. A palindromic number is one whose digits are the same forward as backwards, e.g., 13531 and 178871 are palindromic. Using the results of the previous problem, prove that a palindromic number with an even number of digits is divisible by 11.

5. Prove that $m$ is even if and only if there exists a number $x$ such that $x \not\equiv 0$ mod $m$ and $x + x \equiv 0$ mod $m$.

6. Solve the congruences $5x \equiv 11$ and $11y \equiv 5$, mod 37. If both are satisfied, simplify $xy$ mod 37.

7. Find a multiplicative inverse of 4, or prove that one does not exist, modulo 30, 31, 32, 33, 34, and 35.

8. Prove that the equation $x^2 + y^2 + z^2 = 8007$ has no solutions. (Hint: Work modulo 8). Demonstrate that there are infinitely many positive integers which *cannot* be expressed as the sum of three squares.[36]

[36] Legendre (1798) proved the **three-squares theorem**: a natural number $n$ can be expressed as the sum of three squares, unless it has the form $4^e \cdot (8k + 7)$ for some natural numbers $e, k$.

9. Prove that there are infinitely many primes in the sequence

$$5, 11, 17, 23, 29, 35, 41, \ldots.$$

   Hint: these are the numbers $x$ which satisfy $x \equiv 5 \bmod 6$. Try a proof in the style of Euclid, considering a quantity of the form $p_1 \cdots p_n - 1$.

10. Explain the left-right symmetry that appears in multiplicative inverse diagrams such as Figures 5.6 and 5.7.

11. Let $p$ be a prime number. The following exercises lead to a proof of Fermat's Little Theorem, which we prove by another method in the next chapter.

    (a) For any integer $k$ with $0 \leq k \leq p$, let $\binom{p}{k} = \frac{p!}{k!(p-k)!}$ denote the binomial coefficient. Prove that $\binom{p}{k} \equiv 0 \bmod p$ if $1 \leq k \leq p-1$.

    (b) Prove that for all integers $x, y$, $(x+y)^p \equiv x^p + y^p \bmod p$.

    (c) Prove that for all integers $x$, $x^p \equiv x \bmod p$. Hint: This is obvious for $x = 0$, and use induction.

    (d) Factor $T^p - T$ into irreducible polynomials mod $p$.

12. Let $F_n$ denote the Fibonacci sequence $0, 1, 1, 2, 3, 5, 8, 13, \ldots$. Each term is the sum of the previous two terms. Number the sequence so that $F_0 = 0$ and $F_1 = 1$ and $F_2 = 1$, etc. Prove that for all $n \geq 1$, $F_n \equiv 4^{n-1}(2^n - 1) \bmod 11$.

13. List all irreducible polynomials mod 3, of degree 2. Hint: Multiply and cross off, rather than testing each one.

14. Factor the polynomial $T^3 + 1$ into irreducibles, modulo 7.

15. What is the greatest common divisor of $T^6 + 1$ and $T^{15} + 1$, modulo 2? Use the Euclidean algorithm.

16. Let $P$ be an irreducible polynomial mod $p$. Let $P^*$ be the polynomial obtained by reversing the coefficients of $P$.[37] Demonstrate that $P^*$ is irreducible too. Hint: formally speaking, $P^*(T) = T^d \cdot P(T^{-1})$ where $d$ is the degree of $P$. Use this to construct a factorization of $P^*$ from a factorization of $P$.

17. Adapt Euclid's proof to demonstrate that there are infinitely many irreducible polynomials mod $p$ (for any fixed prime number $p$).

18. Rather than considering the integers mod $p$, one may consider the Gaussian or Eisenstein integers modulo a Gaussian or Eisenstein prime. Find representatives for the Eisenstein integers mod 2, and create addition and multiplication tables (mod 2) for these representatives. Try the same with the Gaussian integers mod 3.

[37] E.g., if $P(T) = T^3 + T + 1$ (with coefficients 1,0,1,1), then $P^*(T) = T^3 + T^2 + 1$ (with coefficients 1,1,0,1).

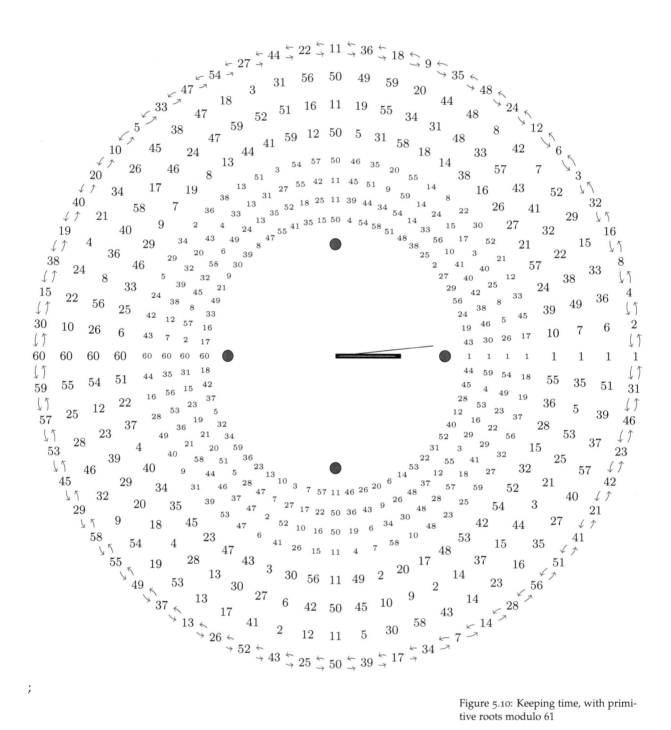

Figure 5.10: Keeping time, with primitive roots modulo 61

# 6
# Modular Dynamics

REPETITION is at the heart of arithmetic. Multiplication is repeated addition. Exponentiation is repeated multiplication. So to understand these operations, one should understand *dynamics* of numbers – how numbers change when they are repeatedly operated upon.

Below we draw the dynamics of "addition of 6, modulo 21". We use natural representatives throughout this chapter.

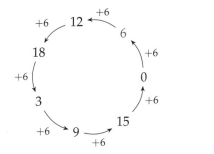
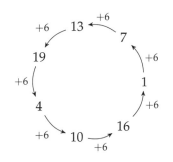
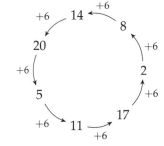

Figure 6.1: The graph of addition of 6, modulo 21. Every arrow denotes the result of adding 6. The graph contains three 7-cycles – loops with seven numbers.

Repeated multiplication can give more complicated dynamics. Below we draw the dynamics of "multiplication by 6, modulo 21".

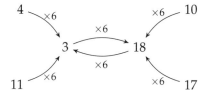
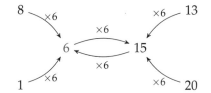
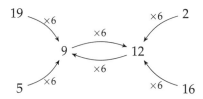

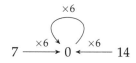

Figure 6.2: The graph of multiplication by 6, modulo 21. Every arrow denotes a multiplication by 6. The graph contains three 2-cycles, (3, 18) and (6, 15) and (9, 12), and one fixed point: 0.

The dynamics of multiplication allow us to compute exponents.

**Problem 6.1** Compute $12^{1000}$ mod 20.

SOLUTION: The dynamics of multiplication by 12, modulo 20 are graphed below, and tabulated in the margin.

$$1 \xrightarrow{\times 12} 12 \xrightarrow{\times 12} 4$$
$$\times 12 \uparrow \qquad \downarrow \times 12$$
$$16 \xleftarrow{\times 12} 8$$

| $e$ | $12^e$ mod 20 |
|---|---|
| 0 | 1 |
| 1 | 12 |
| 2 | 4 |
| 3 | 8 |
| 4 | 16 |
| 5 | 12 |
| 6 | 4 |
| 7 | 8 |
| 8 | 16 |

The powers of 12 repeat in cycles of length 4. For positive exponents $e$, we find that

$$12^e = \begin{cases} 12 & \text{if } e = 1 \bmod 4 \\ 4 & \text{if } e = 2 \bmod 4 \\ 8 & \text{if } e = 3 \bmod 4 \\ 16 & \text{if } e = 0 \bmod 4. \end{cases}$$

Since $1000 = 0 \bmod 4$, we find that $12^{1000} = 16 \bmod 20$. ✓

In this chapter we closely examine the dynamics of addition and multiplication in modular arithmetic. We already have most of the tools we need to study the simpler dynamics of addition. The next theorem verifies that the dynamics of addition involve cycles of equal length.

**Proposition 6.2 (Cycle length for addition mod $m$)** *Let $m$ be a positive integer. Let $a$ be an integer. Then the dynamics of addition of $a$, mod $m$, consist of $m/\ell$ cycles of length $\ell$, where the cycle length is given by*

$$\ell = \frac{m}{GCD(a, m)}.$$

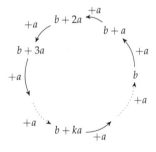

PROOF: Suppose we begin with a number $b$, and repeatedly add $a$, modulo $m$. The dynamics are displayed in the margin.

After $k$ steps, we arrive at the number $b + ka$ mod $m$. Eventually, we must return to $b$, since

$$b + ma \equiv b + 0 = b \bmod m.$$

The length of the cycle is the smallest positive integer $\ell$ for which $b + \ell a \equiv b \bmod m$. Such an equation holds if and only if $\ell a \equiv 0 \bmod m$. In other words, $\ell$ is the smallest positive integer for which $\ell a$ is a multiple of $m$. Hence $\ell a = LCM(a, m)$ and

$$\ell = \frac{LCM(a, m)}{a} = \frac{am}{a\,GCD(a, m)} = \frac{m}{GCD(a, m)}. \blacksquare$$

THE DYNAMICS of addition of 6, modulo 21, illustrate Proposition 6.2.

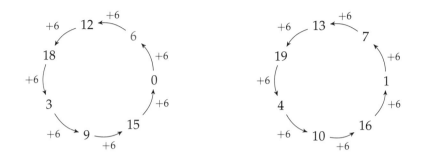

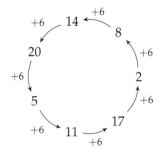

Figure 6.3: Since $21/GCD(6,21) = 21/3 = 7$, the cycle lengths are equal to 7. The 21 numbers are partitioned into three 7-cycles.

When $GCD(a, m) = 1$, the theorem simplifies.

**Corollary 6.3** *Let $m$ be a positive integer. Let $a$ be an integer and suppose $GCD(a, m) = 1$. Then the dynamics of addition of $a$, mod $m$, consist of a single cycle of length $m$.*

PROOF: The cycle length is $\ell = m/GCD(a, m) = m/1 = m$. ■

Since $GCD(5, 21) = 1$, addition of 5 modulo 21 exhibits a single cycle in its dynamics.

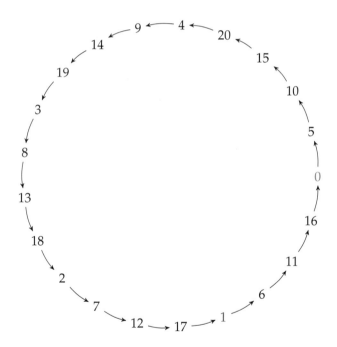

Figure 6.4: Every number, modulo 21, appears in a single cycle. The appearance of 1 corresponds to the fact that 5 has a multiplicative inverse, modulo 21. Notice that

$$5 \times 17 \equiv 1 \bmod 21.$$

Correspondingly, if one begins at zero, and travels counterclockwise 17 steps, one finds the number 1.

In fact, one may reason that since the cycle has "full length", every number including 1 must appear in the cycle. Hence $5 \times b \equiv 1 \bmod 21$ for some $b$. Hence, the full length of the cycle reflects the fact that 5 has a multiplicative inverse, mod 21.

ADDITIVE DYNAMICS are straightforward, modulo $m$. They are described by one or more cycles, all of the same length. The length is predictable from the formula $\ell = m/GCD(a,m)$.

MULTIPLICATIVE DYNAMICS can be subtle, modulo $m$. The root of this subtlety is that – in general – the process of "multiplying by $a$, mod $m$" is not reversible. But if one avoids nonreversible processes then the dynamics become similar to the additive case.

**Definition 6.4** Let $m$ be a positive integer. Let $\Phi(m)$ be the set of numbers between 0 and $m-1$ which are coprime with $m$. The **totient** of $m$ is the number $\phi(m)$ of elements in the set $\Phi(m)$.

**Problem 6.5** Describe the set $\Phi(10)$ and the number $\phi(10)$.

SOLUTION: Among the numbers $0,1,2,3,4,5,6,7,8,9$, the numbers coprime to 10 are
$$\Phi(10) = \{1,3,7,9\}.$$
Hence $\phi(10) = 4$; there are four numbers in the above set. ✓

We will study the totient in more detail in the next chapter. For now, we mention just the easiest result.

**Proposition 6.6** *A positive integer $p$ is prime if and only if $\phi(p) = p - 1$.*

PROOF: If $p$ is prime, then among the numbers between 1 and $p-1$, all are coprime with $p$. Thus, among the numbers from 0 to $p-1$, exactly $p-1$ are coprime with $p$. Conversely, if all of the numbers between 1 and $p-1$ are coprime with $p$, then in particular, none can divide $p$ except for 1. Hence $p$ is prime. ∎

The numbers in $\Phi(m)$ have a dynamic interpretation.

**Proposition 6.7** *Given an integer $a$ between 0 and $m-1$, we have $a \in \Phi(m)$ if and only if multiplication by $a$, mod $m$, is reversible.*

PROOF: This is a restatement of Theorem 5.20. ∎

**Proposition 6.8** *If $a \in \Phi(m)$ and $b \in \Phi(m)$, then there exists $c \in \Phi(m)$ such that $a \cdot b \equiv c \mod m$.*

PROOF: Since multiplication by $a$ and multiplication by $b$ are reversible, mod $m$, it must be the case that multiplication by $ab$ is also reversible mod $m$.

Let $c$ be the natural representative of $ab$, mod $m$. Modulo $m$, multiplication by $ab$ and multiplication by $c$ give the same result. Hence multiplication by $c$ is also reversible mod $m$. Thus $ab \equiv c \mod m$ and $c \in \Phi(m)$. ∎

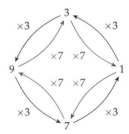

Figure 6.5: Multiplication by 3, in $\Phi(10)$, is reversed by multiplication by 7.

If we restrict the dynamics of multiplication to the set $\Phi(m)$, we find a reversible process.

CONSIDER THE DYNAMICS of multiplication, modulo 21. But let us restrict our attention to those numbers coprime to 21.

$$\Phi(21) = \{1, 2, 4, 5, 8, 10, 11, 13, 16, 17, 19, 20\}, \quad \phi(21) = 12.$$

The dynamics of multiplication by 5 are displayed below.

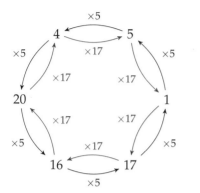
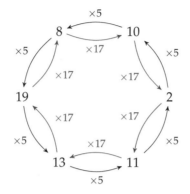

Figure 6.6: Dynamics of multiplication by 5, on the 12 elements of $\Phi(21)$. The process is reversible, since each number has exactly one arrow pointing towards it. The reverse process is given by multiplication by 17.

The cycle lengths are more difficult to describe in the multiplicative case than in the additive case. But equality of cycle lengths still holds after restriction to $\Phi(m)$.

**Lemma 6.9 (Dynamics of multiplication mod $m$)** *Let $m$ be a positive integer, and suppose that $GCD(a, m) = 1$. Then the dynamics of multiplication by $a$, mod $m$, within $\Phi(m)$, consist only of cycles, all of the same length.*

PROOF: Consider the sequence of powers of $a$,

$$1 = a^0, a, a^2, a^3, a^4, a^5, a^6, \ldots.$$

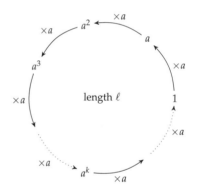

Modulo $m$, this seemingly infinite sequence of numbers lies among the finite set $\Phi(m)$. Hence there must be a repetition. Let $e$ and $f$ (with $e < f$) be two exponents at which a repetition occurs: $a^e \equiv a^f \bmod m$. We may rewrite this as $a^e \equiv a^e \cdot a^{f-e} \bmod m$.

By Corollary 5.23, we find that $1 \equiv a^{f-e} \bmod m$: a positive power of $a$ is congruent to 1 modulo $m$; let $\ell$ be the smallest such power. Thus the powers of $a$ belong to a cycle of length $\ell$.

Now let $b$ be any element of $\Phi(m)$. Starting from $b$, the dynamics of multiplication by $a$, modulo $m$ give a sequence

$$b, b \cdot a, b \cdot a^2, b \cdot a^3, \ldots, b \cdot a^k, \ldots$$

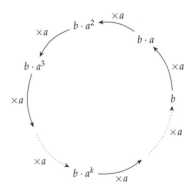

Since $a^\ell \equiv 1 \bmod m$, we find that $a^\ell \cdot b \equiv b \bmod m$, so $b$ belongs to a cycle of length *at most* $\ell$ steps. If this cycle returned to $b$ in fewer steps, i.e. $b \cdot a^k \equiv b \bmod m$ for some $0 < k < \ell$, then Corollary 5.23 would give $a^k \equiv 1 \bmod m$, contradicting the minimality of $\ell$.

Hence $b$ belongs to a cycle of length $\ell$ too. The dynamics of multiplication by $a$ are composed of cycles, all of the same length $\ell$. ∎

A WONDERFUL THEOREM of Euler and Fermat[1] now follows from the dynamics of multiplication.

**Theorem 6.10 (The Fermat-Euler Theorem)** *Let $m$ be a positive integer. Let $a$ be an integer coprime to $m$. Then*

$$a^{\phi(m)} \equiv 1 \bmod m.$$

*In particular, if $p$ is a prime number, and $a \not\equiv 0 \bmod p$, then*

$$a^{p-1} \equiv 1 \bmod p.$$

PROOF: Consider the dynamics of multiplication by $a$, on the set $\Phi(m)$, mod $m$. Let $\ell$ be the length of the resulting cycles (all cycles have the same length). One cycle begins at 1, and after multiplying by $a^\ell$, returns to 1. In other words, $a^\ell \equiv 1 \bmod m$. On the other hand, let $c$ be the *number* of cycles in the dynamics. Since each number in $\Phi(m)$ belongs to some cycle, we find that

$$\ell \cdot c = \phi(m).$$

It follows that

$$a^{\phi(m)} = (a^\ell)^c \equiv 1^c = 1 \bmod m.$$

This proves the general result.

When $p$ is a prime number, $\phi(p) = p - 1$, and $GCD(a, p) = 1$ if and only if $a \not\equiv 0 \bmod p$. Thus the latter statement is a special case of the former. ∎

This theorem allows us to rapidly compute some exponents in modular arithmetic.

**Problem 6.11** Simplify $2^{1001}$ mod 15.

SOLUTION: Since $GCD(2, 15) = 1$, the Fermat-Euler theorem applies. Note that $\Phi(15) = \{1, 2, 4, 7, 8, 11, 13, 14\}$ so $\phi(15) = 8$. Hence $2^8 \equiv 1 \bmod 15$. Since 1000 is a multiple of 8, we find that

$$2^{1001} = 2^{1000} \cdot 2 = (2^8)^{125} \cdot 2 \equiv 1 \cdot 2 = 2 \bmod 15. \checkmark$$

Do not attempt to apply the theorem, unless $GCD(a, m) = 1$.

**Problem 6.12** Simplify $3^8$ mod 15.

SOLUTION: Consider the dynamics of multiplication by 3, modulo 15.

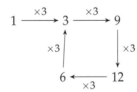

We find that $3^4 \equiv 3^8 \equiv 6 \bmod 15$. $\checkmark$

[1] The general case, with $m$ a positive integer, is a result due to Euler. The specific case, with $p$ a prime number, is called **Fermat's Little Theorem**.

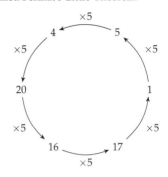

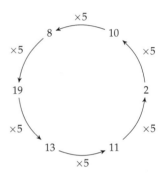

Figure 6.7: Dynamics of multiplication by 5, on the 12 elements of $\Phi(21)$. Here $\phi(21) = 12$. There are two cycles, each of length six. Here, $\ell = 6$ and $c = 2$. Note that $5^6 \equiv 1 \bmod 21$.

If instead, we attempted to use the Fermat-Euler theorem, we would incorrectly deduce that $3^8 \equiv 1 \bmod 15$. The theorem does not apply, since $GCD(3, 15) \neq 1$.

**Problem 6.13** Demonstrate that the equation $x^6 + y^{12} = 700003$ has no integer solutions.

SOLUTION: If the equation had integer solutions, then it would have solutions modulo 7. Reducing modulo 7 yields
$$x^6 + y^{12} \equiv 3 \bmod 7.$$

By Fermat's Little Theorem, we find that
$$x^6 \equiv \begin{cases} 0 \bmod 7 & \text{if } x \equiv 0 \bmod 7; \\ 1 \bmod 7 & \text{if } x \not\equiv 0 \bmod 7. \end{cases}$$

Similarly,
$$y^{12} = (y^6)^2 \equiv \begin{cases} 0 \bmod 7 & \text{if } y \equiv 0 \bmod 7; \\ 1 \bmod 7 & \text{if } y \not\equiv 0 \bmod 7. \end{cases}$$

Hence
$$x^6 + y^{12} \equiv (0 \text{ or } 1) + (0 \text{ or } 1) = (0 \text{ or } 1 \text{ or } 2) \bmod 7.$$

In particular, $x^6 + y^{12} \not\equiv 3 \bmod 7$. Since the congruence has no solutions, the original equation has no integer solutions. ✓

**Corollary 6.14** *Let m be a positive integer, and a an integer coprime to m. Let x and y be natural numbers. Then*
$$x \equiv y \bmod \phi(m) \text{ implies } a^x \equiv a^y \bmod m.$$

PROOF: Assume without loss of generality that $x \geq y$, and let $t = x - y$. If $x \equiv y \bmod \phi(m)$, then $t = n \cdot \phi(m)$ for some natural number $n$. Therefore, by the Fermat-Euler theorem,
$$a^t = a^{n \cdot \phi(m)} = \left(a^{\phi(m)}\right)^n \equiv 1^n = 1 \bmod m.$$

It follows that
$$a^x = a^t \cdot a^y \equiv 1 \cdot a^y = a^y \bmod m. \quad \blacksquare$$

When the base is coprime to the modulus $m$, the exponent effectively belongs to a different modular world. When the base lives in the world of numbers modulo $m$ (and is coprime to $m$), the exponent lives in a world of numbers modulo $\phi(m)$.

**Problem 6.15 (Revisited)** Simplify $2^{1001} \bmod 15$.

SOLUTION: Since $\phi(15) = 8$, and $1001 \equiv 1 \bmod 8$, we have $2^{1001} \equiv 2^1 = 2 \bmod 15$. ✓

SINCE ZEBRAS HAVE STRIPES, an animal without stripes must not be a zebra. Since primes $p$ have the property labelled (FLT)[2] below, a number not satisfying (FLT) must not be a prime.

[2] FLT stands for Fermat's Little Theorem.

$$\textbf{(FLT)} \quad \text{If } a \not\equiv 0 \bmod p, \text{ then } a^{p-1} \equiv 1 \bmod p.$$

**Definition 6.16** Let $n$ be a positive integer. If $a \not\equiv 0 \bmod n$, and $a^{n-1} \not\equiv 1 \bmod n$, then we say $a$ **witnesses** the nonprimality of $n$.

Witnesses can demonstrate that a number $n$ is *not* prime, without finding any particular factor of $n$.

**Problem 6.17** Does 2 witness the nonprimality of 91?

SOLUTION: We show that $2^{90} \not\equiv 1 \bmod 91$. We compute $2^{90}$ by a successive squaring method. We fill in a table of powers of 2, in which the exponents are powers of 2, by squaring each row to find the next.

Here we note that $16 \times 16 = 256 \equiv -17 \bmod 91$ and $(-17) \times (-17) \equiv 16 \bmod 91$. This aids our computation of $2^{90}$, using the binary expansion of 90:

$$90 = 64 + 16 + 8 + 2.$$

| $2^{2^e}$ | mod 91 |
|---|---|
| $2^1$ | 2 |
| $2^2$ | 4 |
| $2^4$ | 16 |
| $2^8$ | $-17$ |
| $2^{16}$ | 16 |
| $2^{32}$ | $-17$ |
| $2^{64}$ | 16 |

We compute

$$\begin{aligned} 2^{90} &= 2^{64} \times 2^{16} \times 2^8 \times 2^2 \\ &\equiv 16 \times 16 \times (-17) \times 4 \\ &\equiv (-17) \times (-17) \times 4 \\ &\equiv 16 \times 4 = 64 \bmod 91. \end{aligned}$$

We find that $2^{90} \equiv 64 \bmod 91$. Since $64 \not\equiv 1 \bmod 91$, we find that 91 cannot be a prime number. ✓

WITNESSES are not the most reliable test for primality. If a witness claims that a number is not prime, one can be certain that the number is not prime. But sometimes witnesses suggest a number might be prime and we cannot be certain.

**Problem 6.18** Does 3 witness the nonprimality of 91?

SOLUTION: We may compute $3^{90} \bmod 91$, in the same manner as we computed $2^{90} \bmod 91$. We find that

$$3^{90} = 3^{64} \times 3^{16} \times 3^8 \times 3^2 \equiv 1 \bmod 91.$$

| $3^{2^e}$ | mod 91 |
|---|---|
| $3^1$ | 3 |
| $3^2$ | 9 |
| $3^4$ | 81 |
| $3^8$ | 9 |
| $3^{16}$ | 81 |
| $3^{32}$ | 9 |
| $3^{64}$ | 81 |

This is consistent with the primality of 91. The witness 3 cannot be sure that 91 is not prime. ✓

RAISING A NUMBER to a large exponent seems like an inefficient[3] way to test whether a number is prime. **Pingala's algorithm**,[4] demonstrated below, gives a rapid way to compute large exponents.

**Problem 6.19** Simplify $3^{666}$ mod 667.

SOLUTION: Assemble the table below; the left columns are obtained by repeatedly dividing by two or subtracting one, according to whether the number is even or odd.

[3] We can see that 91 is not prime quickly by finding the factor 7; this seems better than raising a number to the 90th power!

[4] A similar algorithm was used by Piṅgala (India, c. 200 BCE) to enumerate possible meters for poetic verse.

|  | Exponent $e$ | Binary | $3^e$ mod 667 |  |
|---|---|---|---|---|
|  | 666 | b 1010011010 | 660 | Square |
| Divide by 2 | 333 | b 101001101 | 188 |  |
| Subtract 1 then divide by 2 | 332 |  | 285 | Square then multiply by 3 |
|  | 166 | b 10100110 | 187 | Square |
| Divide by 2 | 83 | b 1010011 | 39 |  |
| Subtract 1 then divide by 2 | 82 |  | 13 | Square then multiply by 3 |
|  | 41 | b 101001 | 512 |  |
| Subtract 1 then divide by 2 | 40 |  | 393 | Square then multiply by 3 |
|  | 20 | b 10100 | 547 | Square |
| Divide by 2 | 10 | b 1010 | 353 | Square |
| Divide by 2 | 5 | b 101 | 243 |  |
| Subtract 1 then divide by 2 | 4 | b | 81 | Square then multiply by 3 |
|  | 2 | b 10 | 9 | Square |
| Divide by 2 | 1 | b 1 | 3 |  |

The rightmost column is built from bottom up. Every time the exponent is doubled, the value of $3^e$ is squared, mod 667. Every time the exponent is increased by 1, the value of $3^e$ is multiplied by 3, mod 667. We find the value $3^{666} \equiv 660$ mod 667. So 667 is not prime. ✓

Tracking binary expansions, we find one bit lost for each arrow. The bits can be used to construct the rightmost column; from left to right, each 0-bit tells us to square, and each 1-bit tells us to multiply by 3 then square. With Proposition 0.17, we can count bits[5] to mark time.

**Proposition 6.20** *Pingala's algorithm for computing $a^e$ mod $m$ requires at most $2\lfloor \log_2(e) \rfloor$ multiplications, each modulo $m$.*

Figure 6.8: Pingala's algorithm, applied to computing $3^{666}$ mod 667.

[5] The number of bits in $e$ is $\lfloor \log_2(e) \rfloor + 1$. But the leftmost bit is always 1, so only $\lfloor \log_2(e) \rfloor$ arrows can occur. A squaring requires one multiplication, and a multiplication-by-$a$-then-squaring requires two multiplications.

WITNESSES can very quickly inform us if a number is not prime, thanks to Pingala's algorithm. But for some composite numbers, it is hard to find a witness. A **Carmichael number** is a composite number $N$ which satisfies

$$\mathrm{GCD}(a, N) = 1 \text{ implies } a^{N-1} \equiv 1 \bmod N.$$

For Carmichael numbers, finding a witness for nonprimality is as hard as finding a factor. Remarkably such numbers exist; more remarkably, there are infinitely many such numbers.[6] For example, 41041 is a Carmichael number. When $\mathrm{GCD}(a, 41041) = 1$, we have $a^{41040} \equiv 1 \bmod 41041$.

The first five witnesses cannot be certain that 41041 is *not* prime. But alas, $7 \mid 41041$.

Fortunately, witnesses can be sharpened a bit if we look beyond the "stripes" of Fermat's Little Theorem.

**Proposition 6.21 (Roots Of One property of primes)** *Let $p$ be a prime number. Then*

**(ROO)**  $x^2 \equiv 1 \bmod p$ *implies* $x \equiv \pm 1 \bmod p$.

PROOF: If $x^2 \equiv 1 \bmod p$, then $x^2 - 1 \equiv 0 \bmod p$, so $(x+1)(x-1) \equiv 0 \bmod p$. By Corollary 2.13, $x + 1 \equiv 0$ or $x - 1 \equiv 0 \bmod p$. Hence $x \equiv -1$ or $x \equiv 1 \bmod p$. ■

As primes have property (ROO) above, numbers which do not have property (ROO) are not primes. For example, $19^2 = 361 \equiv 1 \bmod 40$, but $19 \not\equiv \pm 1 \bmod 40$. This implies 40 is not prime.

Witnesses use Fermat's Little Theorem (property FLT of primes) to examine the primality of numbers. We can improve witnesses, by searching for violations of (ROO) within Pingala's algorithm. On the opposite page, we have carried out Pingala's algorithm to compute

$$2^{41040} \equiv 1 \bmod 41041.$$

From this, the witness 2 states that 41041 might be prime. But examining the rightmost column in Pingala's algorithm we find that

$$27182^2 \equiv 1 \bmod 41041.$$

This is a dramatic violation of (ROO), and hence 41041 cannot be prime after all.

We have made the witness 2 more **perceptive** without adding significantly to our computations; it now witnesses the nonprimality of 41041 after all. This approach is called the **Miller-Rabin primality test**. To test whether a number $N$ is prime, we search for violations of (FLT) **and** violations of (ROO) as we perform squaring in Pingala's algorithm.

[6] See "There are infinitely many Carmichael numbers" by W.R. Alford, Andrew Granville, and Carl Pomerance, in *Annals of Mathematics*, **140** (1994). The Carmichael numbers are

$$561, 1105, 1729, 2465, 2821, \ldots$$

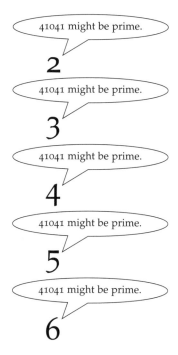

| Exponent $e$ | Binary | $2^e$ mod 41041 | |
|---|---|---|---|
| 41040 | **b** 1010000001010000 | 1 | ⎫ Square |
| 20520 | **b** 101000000101000 | 1 | ⎬ Square |
| 10260 | **b** 10100000010100 | 1 | ⎬ Square |
| 5130 | **b** 1010000001010 | 27182 | ⎬ Square |
| 2565 | **b** 101000000101 | 27994 | ⎭ |
| 2564 | | 13997 | ⎫ Square then multiply by 2 |
| 1282 | **b** 10100000010 | 8122 | ⎬ Square |
| 641 | **b** 1010000001 | 17140 | ⎭ |
| 640 | | 8570 | ⎫ Square then multiply by 2 |
| 320 | **b** 101000000 | 22551 | ⎬ Square |
| 160 | **b** 10100000 | 8570 | ⎬ Square |
| 80 | **b** 1010000 | 22551 | ⎬ Square |
| 40 | **b** 101000 | 8570 | ⎬ Square |
| 20 | **b** 10100 | 22551 | ⎬ Square |
| 10 | **b** 1010 | 1024 | ⎬ Square |
| 5 | **b** 101 | 32 | ⎭ |
| 4 | | 16 | ⎫ Square then multiply by 2 |
| 2 | **b** 10 | 4 | ⎬ Square |
| 1 | **b** 1 | 2 | ⎭ |

Figure 6.9: Pingala's algorithm, applied to computing $2^{41040}$ mod 667.

The Miller-Rabin test greatly improves the reliability of witnesses.

**Proposition 6.22** *If $N < 25,326,001$, and $N$ is not prime, then either 2, 3, or 5 will witness the nonprimality of $N$ via the Miller-Rabin test.*

Going further, if we use the perceptive witnesses $2, 3, 5, 7, 11$, we can quickly and accurately check primality of numbers up to 2 trillion. If none of these five witnesses observe the non-primality of $N$, and $N$ is smaller than 2 trillion, then $N$ is prime.

In practice, the Miller-Rabin test can be used to have *confidence* if not certainty in the primality of numbers. If $N$ is a large number (e.g., 1000 digits), then choose $w$ witnesses at random. If none of the witnesses observes the nonprimality of $N$, then the probability that $N$ is not prime drops to 1 in $4^w$. With just twenty witnesses, our certainty of primality can rise to 99.999999999909 percent.

WE RETURN now to the dynamics of multiplication mod $p$, assuming here that $p$ is a prime number. If $GCD(a, p) = 1$, its **cycle-length** $\ell(a)$ is the length of the cycle of powers of $a$; equivalently, $\ell(a)$ is the smallest positive integer for which $a^{\ell(a)} \equiv 1 \bmod p$. Equivalently, the solutions $x$ to $a^x \equiv 1 \bmod p$ consist precisely of the multiples of $\ell(a)$: $x = 0, \ell(a), 2\ell(a), 3\ell(a), \ldots$.

Define the **cycle-number** $c(a)$ to be the number of cycles obtained in the dynamics of multiplication by $a$ mod $p$, amongst the numbers $\Phi(p) = \{1, \ldots, p-1\}$. We have seen that

$$\ell(a) \cdot c(a) = (p-1).$$

**Lemma 6.23** *Let $\lambda$ be a positive integer. Then the number of elements of $\Phi(p)$ of cycle-length $\lambda$ is either 0 or $\phi(\lambda)$.*

PROOF: Suppose that this number is not zero, so there exists $a \in \Phi(p)$ of cycle-length $\lambda$. Note that $(a^e)^\lambda = (a^\lambda)^e \equiv 1 \bmod p$, for every exponent $e$. As $e$ varies between 0 and $\lambda - 1$, this provides $\lambda$ distinct roots to the polynomial $T^\lambda - 1$. By Theorem 5.39, **every** root of $T^\lambda - 1$ must lie among the powers of $a$, mod $p$. In particular, every element of $\Phi(p)$ of cycle-length $\lambda$ must lie among the powers of $a$, mod $p$.

So we are left to examine the cycle-length of powers of $a$. If $e$ is a positive integer, then $(a^e)^f \equiv 1 \bmod p$ if and only if $ef$ is a multiple of $\lambda$. This occurs if and only if $f$ is a multiple of $\lambda / GCD(e, \lambda)$. From this we find that

$$\ell(a^e) = \frac{\ell(a)}{GCD(e, \lambda)}.$$

Consequently, $\ell(a^e) = \lambda$ if and only if $GCD(e, \lambda) = 1$. Thus we find $\phi(\lambda)$ possible exponents $e$ for which $a^e$ has cycle-length $\lambda$. ∎

**Lemma 6.24 (Totient sum formula)** *Let $N$ be a positive integer. Then the sum of the totients of the divisors of $N$ equals $N$ itself.*

For example, if $N = 12$, then we find

$$\phi(1) + \phi(2) + \phi(3) + \phi(4) + \phi(6) + \phi(12) = 1 + 1 + 2 + 2 + 2 + 4 = 12.$$

PROOF: On one hand, consider the set of fractions with denominator $N$ and numerator between 1 and $N$. On the other hand, consider the set of **reduced** fractions $a/d$ in which $1 \leq a \leq d$ and $d \mid N$. The first set has $N$ elements. The second set has $\phi(d)$ elements for each possible denominator $d \mid N$. Every fraction in the first set can be expressed uniquely as an element of the second set, and vice versa, proving the lemma. ∎

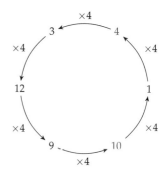

Figure 6.10: The powers of 4 mod 13. Here $\ell(4) = 6$ and $c(4) = 2$. The number of elements of $\Phi(13)$ of cycle-length 6 is $\phi(6) = 2$. The two elements are highlighted in red above, and both belong to the cycle generated by 4.

Here the exposition and slick proof are based on Chapter 4 of the text of Ronald Graham and Donald Knuth, "Concrete Mathematics" (1989).

Figure 6.11: Illustrating the sum of the totients of the divisors of 12.

**Definition 6.25** We say that $a$ is a **primitive root** if $\ell(a) = p - 1$, i.e., if multiplication by $a$ mod $p$ yields a single cycle of length $p - 1$.

The following theorem is due to Gauss.

**Theorem 6.26 (Existence of primitive roots mod $p$)** *If $p$ is a prime number, then there exists a primitive root mod $p$. Their number is $\phi(p-1)$.*

PROOF: The possible cycle-lengths are the divisors of $p - 1$. For each divisor $\lambda \mid (p - 1)$, there are either 0 or $\phi(\lambda)$ elements of cycle-length $\lambda$. By Lemma 6.23, we have

$$(0 \text{ or } \phi(\lambda_1)) + \cdots + (0 \text{ or } \phi(\lambda_s)) = p - 1.$$

Here, there is one term for each divisor $\lambda_i$ of $p - 1$, and each term is 0 or $\phi(\lambda_i)$, according to the number of elements of cycle-length $\lambda$. The totient sum formula implies that every term must **equal** $\phi(\lambda_i)$, since otherwise the left side would be less than the right. Hence, for every divisor $\lambda$ of $p - 1$, there exist $\phi(\lambda)$ elements of $\Phi(p)$ of cycle-length $\lambda$. In particular, for $\lambda = p - 1$, we find $\phi(p - 1)$ elements of cycle-length $p - 1$. Since $\phi(p - 1) \geq 1$, the result is proven. ∎

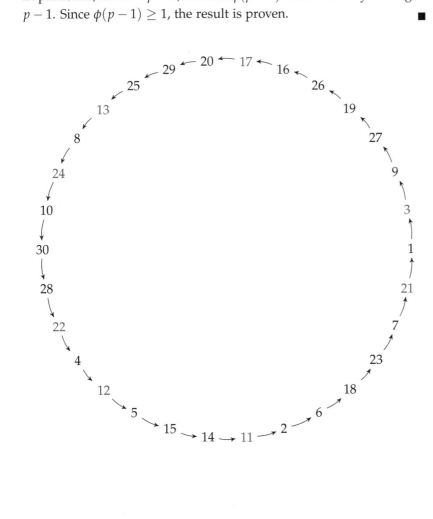

Figure 6.12: Modulo 31, 3 is a primitive root. Multiplication by 3 yields a cycle of maximal length, $\ell(3) = 30$. The other primitive roots are highlighted in red. There are $\phi(30) = 8$ primitive roots modulo 31.

CRYPTOGRAPHY is the study of secure communications over insecure channels. We consider the model scenario, in which Alice wishes to communicate privately with Bob, while Eve attempts to eavesdrop. We imagine Alice and Bob at their computers, their communications transmitted over a long wire, with Eve tapped in.

Security requires **authentication** – that Alice and Bob can be sure they are communicating to each other and not an imposter. It requires Alice and Bob to agree on a **cipher** – a way of encoding and decoding their messages back and forth. And it requires the cipher to be unbreakable by Eve.

Here we focus on the question of how Alice and Bob can agree on a cipher while separated at a great distance, without allowing Eve to decode their communication later. An approach was given by Diffie and Hellman in 1976.[7] To agree on a cipher while separated, Alice and Bob must somehow share a secret without ever having communicated the secret directly to each other. The **Diffie-Hellman protocol** provides a method based on modular arithmetic.

The protocol begins when Alice or Bob initiates communication. The initiator chooses a prime number[8] $p$ and a primitive root $g$ mod $p$, and sends $p$ and $g$ across the line. The Miller-Rabin test can be used to find such a large prime, and Theorem 6.26 guarantees the existence of many primitive roots $g$.

[7] See "New Directions in Cryptography," by Whitfield Diffie and Martin E. Hellman, in *IEEE Transactions on Information Theory*, vol. IT-22 **6** (1976). Their paper begins with the precient "We stand today on the brink of a revolution in cryptography" before laying the challenges and some solutions to problems of cryptography in the digital age.

[8] For security, a large (e.g., 2048 bits) $p$ is taken, and one in which $p - 1$ has a large prime factor too.

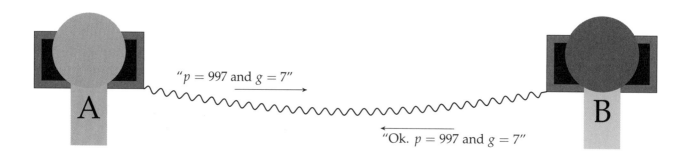

At this point, Alice and Bob generate[9] secret numbers $a$ and $b$. They will not tell these secrets to anyone, not even each other. Alice computes $g^a$ modulo $p$ and calls it $A$. Bob computes $g^b$ modulo $p$ and calls it $B$. They send $A$ and $B$ to each other.

[9] $a$ and $b$ can be randomly chosen between 1 and $p-1$. A good random number generator is advisable!

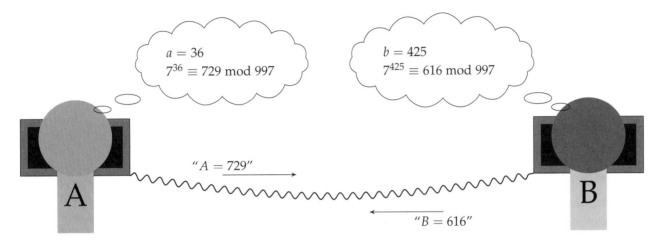

Now Alice computes $B^a$, modulo $p$. Bob computes $A^b$ modulo $p$.[10] From how $A$ and $B$ were computed, we find

$$B^a \equiv (g^b)^a = g^{ba} = g^{ab} = (g^a)^b = A^b.$$

Therefore, both Alice and Bob have computed the same number. It is their special secret, a number called $S$.

Alice and Bob may use their secret $S$ as the **key** for a cipher, to encode and decode their future communications. Eve, having listened in, knows the numbers $p$, $g$, $A$, and $B$. If she knew $a$ and $b$, she could figure out the secret $S$. To figure out $a$, for example, Eve would need to solve the congruence $g^a \equiv A \bmod p$ for $a$. This is the **discrete logarithm problem**. The extreme difficulty[11] of solving the discrete logarithm problem keeps Alice and Bob's secret safe from Eve.

[10] Note that $A$ and $B$ are openly communicated, but only Alice knows $a$ and only Bob knows $b$!

[11] The best computers (that the public knows about) are unable to solve the discrete logarithm problem, when the primes are sufficiently large and well chosen. Poor choices and infiltrated computers threaten secrecy regardless.

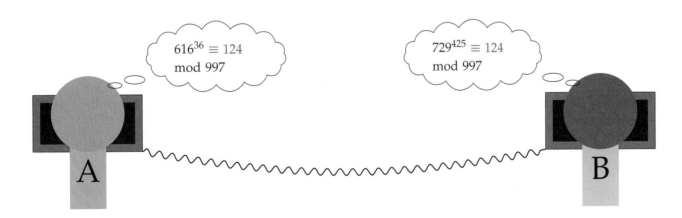

## Historical notes

Our dynamic approach to modular arithmetic is inspired by Gauss, Art. 45 of the *Disquisitiones*, and his approach to Fermat's Little Theorem more broadly. In Art. 50, Gauss writes the theorem (calling it Fermat's theorem), "$a^{p-1} - 1$ will always be divisible by $p$ when $p$ is a prime and does not divide $a$."

As Gauss notes, Fermat did not supply a proof, though he stated the result in 1640.

> Tout nombre premier mesure infailliblement une des puissances $-1$ de quelque progression que ce soit, et l'exposant de la dite puissance est sous-multiple du nombre premier donné $-1$; et, après qu'on a trouvé la première puissance qui satisfait à la question, toutes celles dont les exposants sont multiples de l'exposant de la première satisfont tout de même à la question.[12]

In this correspondence, Fermat begins with interest in Mersenne primes; he demonstrates that the factors of $2^q - 1$ have the form $2kq + 1$ for some positive integer $k$. This is an obvious aid in factoring Mersenne prime candidates – looking for factors of $2^q - 1$, one must proceed through an arithmetic progression of step size $2q$.[13]

A proof of Fermat's Little Theorem came in the work of Euler (1736), who first employed binomial coefficients. Euler's statement is much closer to Gauss's above.

> Significante $p$ numerum primum, formula $a^{p-1} - 1$ semper per $p$ diuidi poterit, nisi $a$ per $p$ diuidi queat.[14]

The dynamic approach, implicit in work of Fermat, leads to a later (1761) proof by Euler.[15] Euler and Gauss study the nature of *geometric progressions* $1, a, a^2, a^3$, etc., when divided by a prime $p$. Gauss discusses the *period*[16] of the residues which occur – this is the cycle length we consider.

In a third paper (1763)[17], Euler demonstrates the general "Fermat-Euler" theorem; in our language, $\text{GCD}(a, m) = 1$ implies $a^{\phi(m)} \equiv 1 \bmod m$.

The primitive root theorem is far deeper than Fermat's Little Theorem. Gauss proves the existence of a primitive root in Art. 55 of the *Disquisitiones*:

> There always exist numbers with the property that no power less than the $p - 1$st is congruent to unity, and there are as many of them between 1 and $p - 1$ as there are numbers less than $p - 1$ and relatively prime to $p - 1$.

After giving two proofs of this fact, Gauss mentions in Art. 56,

> This theorem furnishes an outstanding example of the need for circumspection in number theory so that we do not accept fallacies as certainties.

[12] Letter from Fermat to Frénicle, 18 October 1640. From Fermat's *Oeuvres*, XLIV.

[13] For example,
$$2^{37} - 1 = 223 \times 616\,318\,177.$$
Finding this factorization (by hand) is much easier if one looks for factors within the arithmetic progression $1, 75, 149, 223, \ldots$.

[14] From §3 of Euler's "Theorematum quorundam ad numeros primos spectantium demonstratio" in *Comm. acad. Petrop.* **8** (1736). Available as E54 at the Euler Archive http://eulerarchive.maa.org/. A similar proof using binomial coefficients can be found in the exercises of the previous chapter.

[15] See "Theoremata circa residua ex divisione potestatum relicta," in *Novi Commentarii academiae scientiarum Petropolitanae* **7** (1761). Available as E262 at the Euler Archive http://eulerarchive.maa.org/

[16] See Art. 46 of the *Disquisitiones*.

[17] See "Theoremata arithmetica nova methodo demonstrata," in *Novi Commentarii academiae scientiarum Petropolitanae* **8** (1763). Available as E271 at the Euler Archive http://eulerarchive.maa.org/. Also see Art. 83 of the *Disquisitiones*.

For the existence of a primitive root was stated without proof before Gauss, and an incorrect proof was given earlier by Euler. The proof we give, based on the algebra of polynomials mod $p$, can also be found in Gauss's unpublished Section 8 of the *Disquisitiones*.

The original motivation for Fermat's Little Theorem was the factorization of large numbers (Mersenne prime candidates). From this standpoint, it may not be surprising that it yields some of the best primality tests today. Pingala's algorithm belongs to an entirely separate mathematical tradition.[18] Pingala worked on the mathematics of poetic verse – his *Chandaḥ-sūtra* (300–100BCE) enumerates the ways of arranging long and short syllables in a progression of finite length. His concise rule reads

> When halved, [record] two. When unity [is subtracted, record] *śūnya*. When *śūnya*, [multiply by] two; when halved, [it is] multiplied [by] so much [i.e., squared].[19]

The number of ways of choosing $n$ syllables to be long or short equals $2^n$. Pingala's algorithm gives a method for computing $2^n$, by repeatedly subtracting one and halving from $n$, and carrying out doubling and squaring from 1 up to $2^n$. His algorithm adapts well to computers which store the exponent $n$ in binary.

The Miller-Rabin test was introduced by Miller (1976)[20] and adapted by Rabin (1980).[21] Assuming the generalized Riemann hypothesis (GRH), Miller gave a test for the primality of a number $N$ with runtime[22]

$$O(\log(N))^4 \cdot \log\log\log(N)).$$

Rabin considered the same algorithm as Miller, but chose witnesses at random. Rabin estimated the power of the resulting *probabilistic* primality test. Using a "medium-sized computer" (in 1980!), Rabin verified all Mersenne primes $2^p - 1$ with $p \leq 500$, in about 10 minutes. The Miller-Rabin test is frequently used to certify large primes, e.g., in the OpenSSL cryptography library.

Since the generalized Riemann hypothesis is far from proven, Miller's results do not give a proof that primality testing can be carried out (deterministically) with runtime $O(\log(N)^e)$ for any exponent $e$. This was an open problem, essentially since the beginning of computational complexity theory in the 1960s, and it was solved by Agrawal, Kayal, and Saxena in 2004.[23] The AKS primality test they developed is not so practical, but is proven to have runtime $\tilde{O}(\log(N)^6)$ (using an improvement of C. Pomerance and H.W. Lenstra, Jr.).

---

[18] The six *vedāṅga* – limbs of Veda – can be loosely translated as phonetics, prosody (poetic verse), grammar, etymology, ritual practice, and astronomy. It is unsurprising to a Western audience that astronomical studies led to deep mathematical investigations. But mathematics occurred elsewhere – geometry in ritual practice and combinatorics in the studies of prosody and grammar.

[19] Translation from p.56 of Kim Plofker, "Mathematics in India" (2009). The term *śūnya* refers to a symbol – in later texts, it become the standard term for zero.

[20] Gary L. Miller, "Riemann's Hypothesis and Tests for Primality", Journal of Computer and System Sciences, 13 (3) (1976).

[21] Michael O. Rabin, "Probabilistic algorithm for testing primality", Journal of Number Theory, 12 (1) (1980).

[22] See Theorem 3 of Miller's article; Miller applied the Riemann hypothesis for Dirichlet L-functions to obtain his best runtime estimate. Without this assumption, Miller proves the runtime is $O(N^{1/7})$. Both are improvements over the previous record of $O(N^{1/4})$ using the Pollard rho algorithm.

[23] See "PRIMES is in P" by Agrawal, Kayal, and Saxena, in *Annals of Mathematics* 160 (2004). Kayal and Saxena were undergraduates at the time of their discovery.

## Exercises

1. Describe $\Phi(24)$. What is $\phi(24)$?

2. Demonstrate that the equation $x^{10} + y^{10} - z^{10} = 5$ has no integer solutions. (Hint: work modulo 11)

3. In the following, simplify using natural representatives.
$$3^{23} \bmod 10, \quad 7^{100} \bmod 16, \quad 2^{20} \bmod 21,$$
$$5^{2015} \bmod 24, \quad 3^{2015} \bmod 24, \quad 101^{100^{99}} \bmod 7.$$

4. Does 3 witness the nonprimality of 65? Of 100? Of 121?

5. Each circle in the opening figure displays dynamics of multiplication, modulo 61. Some rays from the origin contain one number,[24] and some two numbers. Why is this?

    [24] For example, the westward ray contains only the number 60.

6. Suppose that $p$ is prime and $p \equiv 2 \bmod 3$. Prove that every integer is a cube modulo $p$. In other words, prove that for every integer $x$, there exists an integer $a$ such that $a^3 \equiv x \bmod p$. Hint/example: $(x^{11})^3 \equiv x \bmod 17$. Generalize.

7. Let $a$ and $b$ be coprime integers, with $a > 1$ and $b > 1$. Prove that there exists an integer $x$ such that $1 \leq x \leq ab - 1$, $x \equiv 1 \bmod a$ and $x \equiv -1 \bmod b$. Show that $x^2 \equiv 1 \bmod ab$ and $x \not\equiv \pm 1 \bmod ab$. Thus $ab$ fails to have property (ROO) from Proposition 6.21.

8. Consider the Fibonacci sequence $0, 1, 1, 2, 3, 5, 8, \ldots$. If 0 is the zeroth Fibonacci number, what is the 1000th Fibonacci number, modulo 3?

9. Consider the sequence $1, 2, 5, 26, \ldots$ defined recursively by the rule
$$a_0 = 1, a_n = a_{n-1}^2 + 1 \text{ for all positive integers } n.$$
What is $a_{1000}$ modulo 11?

10. Consider the sequence given by
$$a_0 = 3, a_1 = 3^3, a_2 = 3^{3^3}, \ldots, a_n = 3^{a_{n-1}} \text{ for all positive integers } n.$$
What is the last digit of $a_{1000}$?

11. The Fermat numbers are defined by $F_n = 2^{2^n} + 1$, for natural numbers $n$. When $n = 0, 1, 2, 3, 4, 5, 6$, the corresponding Fermat numbers are
$$3, 5, 17, 257, 65537, 4294967297, 18446744073709551617.$$
Fermat suspected[25] these were prime. The following exercises relate to the primality of Fermat numbers.

    [25] From his letter to Frenicle, 1640, "I do not have the exact proof, but I have excluded such a large number of divisors by infallible proofs, and I have such a strong insight, which is the foundation for my thought, that it would be difficult for me to retract it." (Translation from French by Amanda Bergeron and David Jao, from the Euler Archive at http://eulerarchive.maa.org/docs/other/fermat1.pdf.

(a) Let $p = 2^m + 1$ for some positive integer $m$. Prove that if $p$ is prime, then $m$ is a power of 2. Hint: if $m = (2k+1)r$ for some natural numbers $k, r$, then prove that $2^r + 1$ is a factor of $2^m + 1$.

(b) Demonstrate that $3, 5, 17, 257, 65537$ are prime. Brute force suffices for the first four; use the Miller-Rabin test to verify that $65537$ is prime. (The perceptive witnesses 2 and 3 suffice to certify primality of numbers up to a million.)

(c) Use witnesses[26] to prove that $F_5$ and $F_6$ are not prime.[27]

(d) Prove that if $p$ is prime and $p | F_n$, then $p - 1$ is a multiple of $2^{n+1}$. Hint: use Fermat's Little Theorem, and consider powers of 2 modulo $p$. (This is also a theorem of Euler. Lucas demonstrated that $p - 1$ is in fact a multiple of $2^{n+2}$ in 1878.)

(e) Why might 2 be a bad witness for determining whether $F_n$ is prime?

[26] A computer will help! The Python command pow(a,b,m) computes $a^b$ mod $m$ in natural representatives.

[27] Among the Fermat numbers, $3, 5, 17, 257, 65537$ are prime, and $F_5, \ldots, F_{32}$ are known to be composite. In 1732, Euler discovered that $4294967297 = 641 \cdot 6700417$. See http://eulerarchive.maa.org/pages/E026.html for more.

12. Suppose that $p$ is an odd prime, and $q$ is a prime factor of $2^p - 1$. Prove that $q \equiv 1 \bmod 2p$.

13. A **Sophie Germain prime**[28] is a prime number $p$ for which $2p + 1$ is also prime. Prove that if $p$ is a Sophie Germain prime then $2p + 1$ is a divisor of $2^p - 1$ or $2p + 1$ is a divisor of $2^p + 1$.

[28] Germain proved that if $p$ is a prime number for which $2p + 1$ is prime, then the Diophantine equation $x^p + y^p = z^p$ has no solutions $(x, y, z)$ for which none of $x, y, z$ are multiples of $p$. This is the "first case" of Fermat's Last Theorem for $p$.

14. Find Alice and Bob's secret number $S$, if $g = 3$, $p = 17$, $A = 8$, and $B = 7$.

15. Suppose that $g^a \equiv A \bmod p$, $g$ is a primitive root modulo $p$, and $p = 2^n + 1$ is a Fermat prime.[29] The following exercises provide a method for finding $a$, given $A$ and $g$, i.e., solving the discrete logarithm problem modulo $p$. In choosing a prime $p$ for Diffie-Hellman key exchange, it is important that $p - 1$ has a large prime factor; Fermat primes are as bad as possible!

[29] In particular, $n$ is also a power of 2, but this will not be needed in what follows.

(a) The desired exponent $a$ has a binary expansion,
$$a = a_0 + 2a_1 + 4a_2 + \cdots + 2^{n-1}a_{n-1},$$
where each bit $a_0, \ldots, a_n$ is either zero or one. Demonstrate that if $g^a \equiv A \bmod p$, then
$$A^{2^{n-1}} \equiv \begin{cases} 1 \bmod p & \text{if } a_0 = 0; \\ -1 \bmod p & \text{if } a_0 = 1. \end{cases}$$
This yields $a_0$.

(b) Let $h$ be a multiplicative inverse of $g$, mod $p$. Let $B \equiv Ah^{a_0} \bmod p$. Demonstrate that $B^{2^{n-2}} \bmod p$ can be used to compute $a_1$. Continue this to describe an algorithm to find $a_2, \ldots, a_{n-1}$.

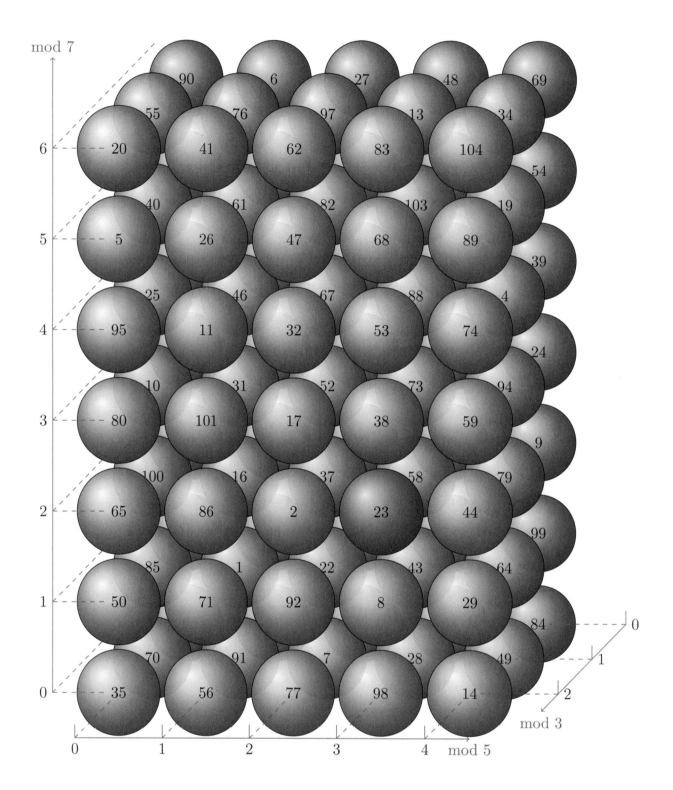

# 7
# *Assembling the Modular Worlds*

THE MODULAR WORLDS are shadows of the world of integers. To know an integer $n$, it suffices to know *all* of its remainders, after all possible divisions. For example, the number 2 is characterized by the fact that it leaves a remainder 0 after division by 2, and it leaves a remainder 2 after division by every number bigger than 2. Each modular world provides a piece of information about the world of integers, and the pieces altogether give a complete picture.

But the modular worlds are not wholly independent pieces. If $n \equiv 4 \bmod 6$, then it cannot be the case that $n \equiv 11 \bmod 15$. If $n \equiv 1 \bmod 2$, then $n \equiv 1$ or $3 \bmod 4$. Congruence for one modulus constrains congruences for other moduli. In this chapter, we study the connections across modular worlds. These connections span two directions.

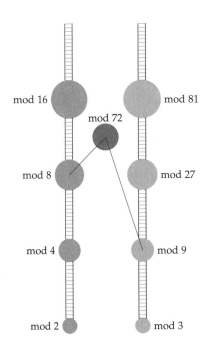

VERTICAL CONNECTIONS link the worlds of numbers mod $p$, $p^2$, $p^3$, $p^4$, etc., when $p$ is a prime number. The congruence of a number modulo a high power of $p$ determines its congruence modulo lower powers of $p$. Conversely, variants of *Hensel's Lemma* allow us to "lift" congruences modulo powers of $p$ to congruences modulo higher powers of $p$.

HORIZONTAL CONNECTIONS allow us to bring together congruences from coprime moduli. A congruence modulo 3 and a congruence modulo 5 can be assembled into a congruence modulo 15. Or, in the other direction, a congruence modulo 15 can be disassembled into a congruence modulo 3 and another modulo 5. The *Chinese Remainder Theorem* enables this assembly and disassembly, and that is where we will begin.

From Chapter 3, Problem 26, of the *Sunzi suan jing* (Mathematical classic of Master Sun, c. 300CE), we adapt the following.

**Problem 7.1** There are an unknown number of things, but fewer than 100. If we count by threes, there is a remainder of 2; if we count by fives, there is a remainder 3; if we count by sevens, there is a remainder 2. Find the number of things.

SOLUTION: First, we find a number $x$ for which $x \equiv 2 \bmod 3$ and $x \equiv 0 \bmod 5$ and $x \equiv 0 \bmod 7$. Such a number must be a multiple of 35 (since it is a multiple of 5 and 7). Among the numbers $35, 70, 105, \ldots$, we find that $35 \equiv 2 \bmod 3$. So let $x = 35$.

Second, we find a number $y$ for which $y \equiv 0 \bmod 3$ and $y \equiv 3 \bmod 5$ and $y \equiv 0 \bmod 7$. Such a number must be a multiple of 21. Among the numbers $21, 42, 63, 84, 105, \ldots$, we find that $63 \equiv 3 \bmod 5$. So let $y = 63$.

Third, we find a number $z$ for which $z \equiv 0 \bmod 3$ and $z \equiv 0 \bmod 5$ and $z \equiv 2 \bmod 7$. Such a number must be a multiple of 15; Among the numbers $15, 30, 45, 60, 75, 90, 105, \ldots$, we find that $30 \equiv 2 \bmod 7$. So let $z = 30$.

The sum $x + y + z$ has the desired remainders.

$$128 = x + y + z \equiv 2 + 0 + 0 \bmod 3,$$
$$128 = x + y + z \equiv 0 + 3 + 0 \bmod 5,$$
$$128 = x + y + z \equiv 0 + 0 + 2 \bmod 7.$$

| Number | mod 3 | mod 5 | mod 7 |
|---|---|---|---|
| $x = 35$ | 2 | 0 | 0 |
| $y = 63$ | 0 | 3 | 0 |
| $z = 30$ | 0 | 0 | 2 |
| 128 | 2 | 3 | 2 |
| 105 | 0 | 0 | 0 |
| 23 | 2 | 3 | 2 |

Hence the number of things could be $35 + 63 + 30 = 128$, if it were fewer than 100.

But observe that we may change the number of things by any multiple of $3 \times 5 \times 7$, without changing the remainders. The resulting solutions are $128 + 105n$ for integers $n$. Since there are fewer than 100 things, there must be $128 - 105 = 23$ things. ✓

TO ACCOMODATE MANY MODULAR WORLDS, we introduce a bracket notation. Instead of writing a system of congruences, $N \equiv 2 \bmod 3$ and $N \equiv 3 \bmod 5$, we write simply $N \equiv [2,3] \bmod [3,5]$. The previous problem could be restated: find a number $N$ between 0 and 99 such that

$$N \equiv [2,3,2] \bmod [3,5,7].$$

To solve this problem, we used Master Sun's method to transform the system of congruences modulo $3, 5, 7$ into a single congruence $N \equiv 128 \bmod 105$, from which we found that $N = 23$.

**Theorem 7.2 (Chinese Remainder Theorem)** *Suppose that d and e are coprime positive integers. There is a one-to-one correspondence between*

- *the set of pairs $[a, b]$ with $0 \leq a < d$ and $0 \leq b < e$;*
- *the set of numbers $N$ with $0 \leq N < de$;*

*such that the solutions to $x \equiv [a, b] \mod [d, e]$ are the same as the solutions to $x \equiv N \mod de$.*

PROOF: Gather $de$ pigeons, and label them by numbers $N$ between 0 and $de - 1$. Arrange $de$ pigeonholes, in $d$ columns and $e$ rows, and label them by brackets $[a, b]$ according to their column and row.

Given a pigeon labeled $N$, there exist unique integers $a, b$ such that $0 \leq a < d$ and $0 \leq b < e$ and $N \equiv [a, b] \mod [d, e]$. Send the pigeon $N$ to the pigeonhole $[a, b]$ accordingly.

If two pigeons labeled $M$ and $N$ landed in the same pigeonhole $[a, b]$, then we would find

$$M \equiv [a, b] \mod [d, e] \text{ and } N \equiv [a, b] \mod [d, e].$$

Then $M - N$ would be a multiple of $d$ and a multiple of $e$. Since $d$ and $e$ are coprime, $M - N$ is a multiple of $de$. But since $M$ and $N$ are between 0 and $de - 1$, this implies $M = N$.

Hence no two pigeons land in the same pigeonhole. Since there are the same number ($de$) of pigeons as pigeonholes, this gives a one-to-one correspondence and the result follows. ∎

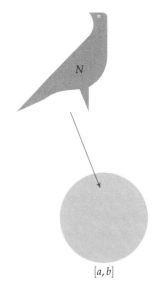

Figure 7.1: Pigeons fly into their pigeonholes. A version of the **pigeonhole principle** is used in the proof: if there are the same number of pigeons as pigeonholes, and no two pigeons occupy the same pigeonhole, then each pigeonhole is occupied by exactly one pigeon. Pigeon images are based on a design by Dorota, of Emu Gallery.

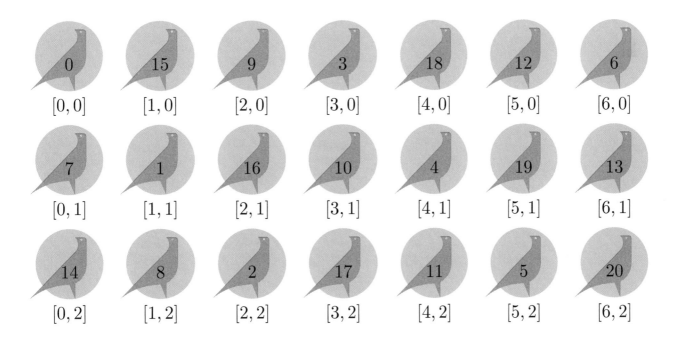

GIVEN A NUMBER $N$ between 0 and 20, it is straightforward to find the corresponding bracket $[a,b]$ mod $[7,3]$. Simply compute the remainder after dividing $N$ by 7 and by 3. For example, the number 17 mod 21 corresponds the bracket $[3,2]$ mod $[7,3]$.

Reversing this process is precisely as hard as solving a linear Diophantine equation.

**Problem 7.3** Find a number $N$ between 0 and 34, such that $N \equiv [6,2]$ mod $[7,5]$.

SOLUTION: If $N = [6,2]$ mod $[7,5]$, then $N = 7x + 6$ and $N = 5y + 2$, for some integers $x, y$. Thus a solution $N$ gives a solution to the equation $7x + 6 = 5y + 2$, or equivalently, $7x - 5y = -4$. We have discussed the solutions to linear Diophantine equations in the first part of this text, and so we trust the reader can find the solution $x = 3$ and $y = 5$. If $x = 3$ then $N = 7x + 6 = 27$.

By the Chinese Remainder Theorem, $N \equiv 27$ mod 35 if and only if $N \equiv [6,2]$ mod $[7,5]$. ✓

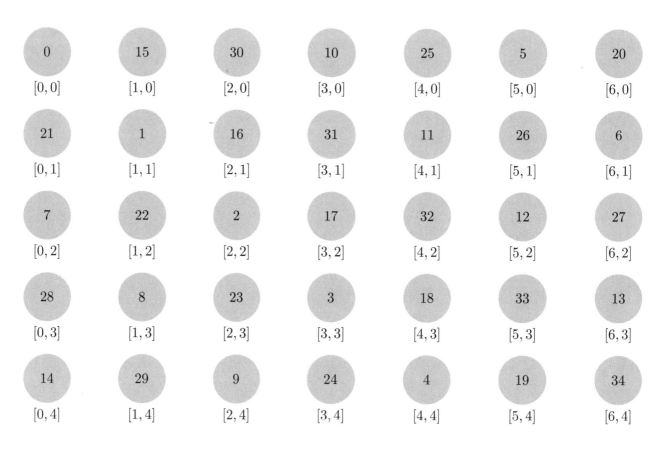

Figure 7.2: The numbers from 0 to 34, modulo $[7,5]$. Highlighted is the solution $27 \equiv [6,2]$ mod $[7,5]$.

The Chinese Remainder Theorem can be used to disassemble a problem, solve the pieces, and reassemble the result.

**Problem 7.4** Solve the congruence $x^2 \equiv 1 \mod 21$.

SOLUTION: By the Chinese Remainder Theorem, $x^2 \equiv 1 \mod 21$ if and only if $x^2 \equiv [1,1] \mod [3,7]$. So we may disassemble the original problem into two pieces:

$$x^2 \equiv 1 \mod 7 \text{ and } x^2 \equiv 1 \mod 3.$$

Since 3 and 7 are prime, there are exactly two square roots of 1 mod 3, and two square roots of 1 mod 7. We find that

$$x \equiv 1 \text{ or } 6 \mod 7 \text{ and } x \equiv 1 \text{ or } 2 \mod 3.$$

Two options modulo 7, and two options modulo 3, give four options modulo $[7,3]$.

$$x \equiv [1,1] \text{ or } [6,1] \text{ or } [1,2] \text{ or } [6,2] \mod [7,3].$$

Each solution, mod $[7,3]$ corresponds[1] to a solution mod 21 by the Chinese Remainder Theorem. We find four solutions

$$x \equiv 1 \text{ or } 13 \text{ or } 8 \text{ or } 20 \mod 21.$$

[1] Turn the page back and find the pigeons.

The same mode of reasoning yields a general result.

**Proposition 7.5** *Suppose that d and e are coprime positive integers. Then the solutions to the congruence $x^2 \equiv a \mod de$ are in one-to-one correspondence[2] with pairs of solutions $(u,v)$ to the congruences $u^2 \equiv a \mod d$ and $v^2 \equiv a \mod e$.*

[2] Here we count the solutions $x$ with $0 \leq x < de$, and the solutions $(u,v)$ with $0 \leq u < a$ and $0 \leq v < b$.

**Corollary 7.6** *Suppose that d and e are coprime positive integers. A number a is a square, modulo de, if and only if a is a square modulo d and a is a square modulo e.*

PROOF: There is at least one solution to the congruence $x^2 \equiv a \mod de$, if and only if there is at least one pair of solutions to the congruences $x^2 \equiv a \mod d$, $x^2 \equiv b \mod e$. ∎

**Corollary 7.7** *Let m be a positive integer with at least 2 distinct odd prime factors. Then there are at least 4 solutions to the congruence $x^2 \equiv 1 \mod m$.*

PROOF: Since $m$ has at least 2 distinct odd prime factors, we can write $m = de$ for coprime integers $d,e$ greater than 2. There are at least two solutions to the congruence $x^2 \equiv 1 \mod d$, and at least two for the congruence $x^2 \equiv 1 \mod e$. Indeed, 1 and $d-1$ are solutions to the first, and 1 and $e-1$ are solutions to the second. Hence there are at least four solutions to the congruence $x^2 \equiv 1 \mod m$. ∎

This corollary demonstrates that the property we called (ROO) in Proposition 6.21 is very frequently violated for non-prime numbers. The corollary is behind the effectiveness of the Miller-Rabin test in identifying non-prime numbers.

TOTIENTS OF COMPOSITE NUMBERS can be computed by the Chinese remainder theorem. The reason for this is that the set $\Phi(N)$ can be described in terms of solving an equation modulo $N$. Given an integer $x$, the following two conditions are equivalent.

- $x \in \Phi(N)$, i.e., $0 \leq x < N$ and $\mathrm{GCD}(x, N) = 1$;

- $0 \leq x < N$ and there exists $y$ such that $xy \equiv 1 \bmod N$.

In other words, $\Phi(N)$ consists of numbers between 0 and $N - 1$ which have multiplicative inverses, modulo $N$. To find multiplicative inverses, one may disassemble and reassemble.

**Problem 7.8** Find a multiplicative inverse of 13, modulo 100.

SOLUTION: We wish to find $y$ such that $13y \equiv 1 \bmod 100$. Equivalently, by the Chinese Remainder Theorem, $13y \equiv [1, 1] \bmod [4, 25]$. Multiplicative inverses, modulo 4 and 25 can be found relatively quickly[3] by guessing:

$$13 \cdot 1 \equiv 1 \bmod 4 \text{ and } 13 \cdot 2 = 26 \equiv 1 \bmod 25.$$

[3] Or more slowly and surely by the Euclidean algorithm.

So the multiplicative inverse of 13, mod 100, satisfies

$$y \equiv [1, 2] \bmod [4, 25].$$

To reassemble, we list numbers which are 2 mod 25, between 0 and 100: these are $2, 27, 52, 77$, among which only $77 \equiv 1 \bmod 4$. We find $77 \equiv [1, 2] \bmod [4, 25]$ and so 77 is the sought-after multiplicative inverse of 13, modulo 100:

$$77 \times 13 = 1001 \equiv 1 \bmod 100. \quad \checkmark$$

**Theorem 7.9 (The totient is a multiplicative function)** *Let $d$ and $e$ be positive coprime integers. Then $\phi(de) = \phi(d)\phi(e)$.*

PROOF: Let $x$ be a number between 0 and $de - 1$. Then solving the equation $xy \equiv 1 \bmod de$ is equivalent to solving the pair of equations $xy \equiv [1, 1] \bmod [d, e]$ by the Chinese remainder theorem. Hence $x \in \Phi(de)$ if and only if $x$ has a multiplicative inverse modulo $d$ and modulo $e$. In other words, $x \in \Phi(de)$ if and only if $x \equiv [a, b] \bmod [d, e]$ for integers $a \in \Phi(d)$ and $b \in \Phi(e)$.

The Chinese remainder therefore gives a one-to-one correspondence between $\Phi(de)$ and pairs $[a, b]$ in which the first term is in $\Phi(d)$ and the second term is in $\Phi(e)$. Therefore $\phi(de) = \phi(d)\phi(e)$. ∎

To illustrate the theorem, we highlight the set $\Phi(36)$ below.

| 0 | 28 | 20 | 12 | 4 | 32 | 24 | 16 | 8 |
|---|---|---|---|---|---|---|---|---|
| [0,0] | [1,0] | [2,0] | [3,0] | [4,0] | [5,0] | [6,0] | [7,0] | [8,0] |
| 9 | 1 | 29 | 21 | 13 | 5 | 33 | 25 | 17 |
| [0,1] | [1,1] | [2,1] | [3,1] | [4,1] | [5,1] | [6,1] | [7,1] | [8,1] |
| 18 | 10 | 2 | 30 | 22 | 14 | 6 | 34 | 26 |
| [0,2] | [1,2] | [2,2] | [3,2] | [4,2] | [5,2] | [6,2] | [7,2] | [8,2] |
| 27 | 19 | 11 | 3 | 31 | 23 | 15 | 7 | 35 |
| [0,3] | [1,3] | [2,3] | [3,3] | [4,3] | [5,3] | [6,3] | [7,3] | [8,3] |

The numbers in $\Phi(36)$ are those $N$ which are congruent to a member of $\Phi(9)$ mod 9 and to a member of $\Phi(4)$ mod 4. We have

$$\Phi(9) = \{1, 2, 4, 5, 7, 8\} \text{ and } \Phi(4) = \{1, 3\}.$$

We find the elements of $\Phi(36)$ in rows 1 and 3, and in columns 1, 2, 4, 5, 7, 8, accordingly. There are $2 \times 6 = 12$ elements of $\Phi(36)$:

$$\Phi(36) = \{1, 29, 13, 5, 25, 17, 19, 11, 31, 23, 7, 35\}.$$

Figure 7.3: The numbers from 0 to 35, modulo [9,4]. Those in $\Phi(36)$ are highlighted; they occur in rows numbered by $\Phi(4)$ and columns numbered by $\Phi(9)$.

COMPUTING TOTIENTS now boils down to prime powers. When $p$ is prime, $\phi(p) = p - 1$. The following takes care of all prime powers.

**Proposition 7.10** *Let $p$ be a prime number, and let $e$ be a positive integer. Then $\phi(p^e) = p^e - p^{e-1}$.*

PROOF: A number $n$ is coprime to $p^e$ precisely when $n$ is **not** a multiple of $p$. Hence $\Phi(p^e)$ consists of those numbers, between 0 and $p^e - 1$, which are not multiples of $p$.

We count the multiples of $p$ between 0 and $p^e - 1$.

$$\left.\begin{array}{ccccc} 0 & p & 2p & \cdots & p^e - p \\ 0 & 1 & 2 & \cdots & p^{e-1} - 1 \\ 1^{\text{st}} & 2^{\text{nd}} & 3^{\text{rd}} & \cdots & (p^{e-1})^{\text{th}} \end{array}\right\} \begin{array}{l} \text{Divide by } p. \\ \text{Add 1.} \end{array}$$

There are $p^{e-1}$ multiples of $p$ between 0 and $p^e - 1$. Hence there are $(p^e - p^{e-1})$ elements of $\Phi(p^e)$. ∎

We can compute the totient of any number that we can factor, using the two proven facts.

1. If $GCD(m,n) = 1$ then $\phi(mn) = \phi(m)\phi(n)$.

2. If $p$ is prime and $e \geq 1$, then $\phi(p^e) = p^e - p^{e-1}$.

We carry out two examples.

**Problem 7.11** Compute the totient $\phi(100)$.

SOLUTION: Since $100 = 2^2 \cdot 5^2$, and $GCD(2^2, 5^2) = 1$, we find that

$$\begin{aligned}\phi(100) &= \phi(2^2)\phi(5^2) \\ &= (2^2 - 2^1) \cdot (5^2 - 5^1) \\ &= (4-2) \cdot (25-5) = 2 \cdot 20 = 40.\end{aligned}$$ ✓

**Problem 7.12** Compute the totient $\phi(10!)$.

SOLUTION: We factor $10!$ into primes.

$$\begin{aligned}10! &= 1 \cdot 2 \cdot 3 \cdot 4 \cdot 5 \cdot 6 \cdot 7 \cdot 8 \cdot 9 \cdot 10 \\ &= 1 \cdot 2 \cdot 3 \cdot (2^2) \cdot 5 \cdot (2 \cdot 3) \cdot 7 \cdot (2^3) \cdot (3^2) \cdot (2 \cdot 5) \\ &= 2^{1+2+1+3+1} \cdot 3^{1+1+2} \cdot 5^{1+1} \cdot 7 \\ &= 2^8 \cdot 3^4 \cdot 5^2 \cdot 7.\end{aligned}$$

Now we compute the totient, using our two facts.

$$\begin{aligned}\phi(2^8 \cdot 3^4 \cdot 5^2 \cdot 7) &= \phi(2^8 \cdot 3^4) \cdot \phi(5^2 \cdot 7) \\ &= \phi(2^8) \cdot \phi(3^4) \cdot \phi(5^2) \cdot \phi(7) \\ &= (2^8 - 2^7) \cdot (3^4 - 3^3) \cdot (5^2 - 5) \cdot (6) \\ &= 128 \cdot 54 \cdot 20 \cdot 6 \\ &= 829440.\end{aligned}$$ ✓

It is crucial not to misapply the multiplicativity of the totient. When $m$ and $n$ are *not* coprime, it is exceedingly rare for $\phi(mn)$ to equal $\phi(m)\phi(n)$. For example, if you compute

$$\phi(4 \cdot 6) \stackrel{?}{=} \phi(4) \cdot \phi(6) = (4-2) \cdot 2 = 4,$$

you would be incorrect! The better method separates the primes.

$$\phi(4 \cdot 6) = \phi(2^3 \cdot 3) = \phi(2^3) \cdot \phi(3) = (2^3 - 2^2) \cdot (3-1) = 4 \cdot 2 = 8.$$

The corresponding eight numbers are

$$\Phi(4 \cdot 6) = \Phi(24) = \{1, 5, 7, 11, 13, 17, 19, 23\}.$$

Once we can compute totients, we can compute many exponents in modular arithmetic using Euler's Theorem.

**Problem 7.13** What are the last two digits of $7^{7^7}$?

SOLUTION: To know the last two digits of a number is to simplify the number modulo 100. On the opposite page, we found that $\phi(100) = 40$. Therefore, if $x \equiv 7^7 \bmod 40$, then $7^{7^7} \equiv 7^x \bmod 100$.

To simplify $7^7 \bmod 40$, we note that $7^2 = 49 \equiv 9 \bmod 40$, and $7^4 = (7^2)^2 \equiv 9^2 \equiv 1 \bmod 40$. Therefore

$$7^7 = 7^4 \cdot 7^2 \cdot 7^1 \equiv 1 \cdot 9 \cdot 7 = 63 \equiv 23 \bmod 40.$$

Therefore,
$$7^{7^7} \equiv 7^{23} \bmod 100.$$

| $2^e$ | $7^{2^e} \bmod 100$ |
|---|---|
| 1 | 7 |
| 2 | 49 |
| 4 | 1 |
| 8 | 1 |
| 16 | 1 |

At this point, Fermat and Euler cannot help us. We are left with the technique of successive squaring.

From the table in the margin, we find

$$7^{23} = 7^{16} \cdot 7^4 \cdot 7^2 \cdot 7^1 \equiv 7^3 \equiv 43 \bmod 100.$$

Hence the last two digits of $7^{7^7}$ are 43. ✓

We can compute $\phi(n)$ as soon as we know the prime decomposition of $n$. For every prime power $p^e$, note that $\phi(p^e) = p^e - p^{e-1} < p^e$; it follows from Theorem 7.9 that $\phi(n) < n$ for every integer $n > 1$. In other words, $\phi(n)/n < 1$ for every integer $n > 1$. As $n$ grows very large, the average value of $\phi(n)/n$ approaches $6/\pi^2$. But the totient-ratio $\phi(n)/n$ itself does not tend towards its average – its structure is far more complex.[4]

[4] To be precise, it is known that
$$\lim_{N \to \infty} \frac{1}{N} \sum_{n=1}^{N} \frac{\phi(n)}{n} = \frac{6}{\pi^2}.$$

On the other hand,
$$\liminf \frac{\phi(n)}{n} = 0,$$
$$\limsup \frac{\phi(n)}{n} = 1.$$

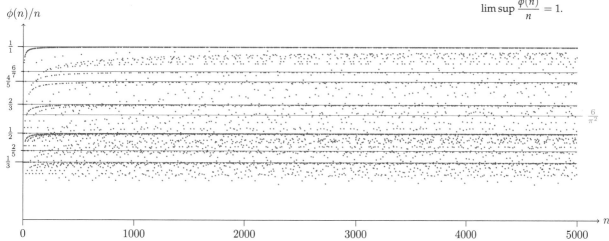

The Chinese Remainder Theorem allows us to understand the world modulo $mn$ in terms of the worlds modulo $m$ and modulo $n$ separately, **if** $\text{GCD}(m,n) = 1$. But some numbers cannot be broken as a product of two coprime numbers – these are the prime powers. So now we study the vertical connections between modular worlds: the connections between the worlds modulo $p^e$ as $e$ varies and $p$ is a fixed prime.

Before specializing to prime powers, we examine the connection between the worlds modulo $m$ and modulo $n$, assuming $m$ divides $n$. The first observation is that a congruence modulo $n$ uniquely determines a congruence modulo $m$.

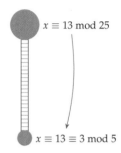

If $m \mid n$, and $x \equiv a \bmod n$, then $x \equiv a \bmod m$.

Indeed, if $x \equiv a \bmod n$ then $n \mid x - a$. Since $m \mid n$ and $n \mid (x - a)$, we find that $m \mid (x - a)$. Hence $x \equiv a \bmod m$ too. We say that the congruence $x \equiv a \bmod n$ **descends** to the congruence $x \equiv a \bmod m$.

In the other direction, a congruence modulo $m$ constrains but does not uniquely determine a congruence modulo $n$. An example will illustrate this.

**Problem 7.14** What numbers modulo 12 satisfy $x \equiv 2 \bmod 4$?

SOLUTION: If $x \equiv 2 \bmod 4$, then $x$ belongs to the sequence

$$\ldots, -18, -14, -10, -6, -2, 2, 6, 10, 14, 18, 22, \ldots.$$

Modulo 12, every number in this sequence is congruent to 2, 6, or 10. Thus $x \equiv 2 \bmod 4$ implies that $x \equiv 2$ or $6$ or $10 \bmod 12$. ✓

The more general observation is the following.

If $m \mid n$ and $x \equiv b \bmod m$ then
$$x \equiv b \text{ or } b + m \text{ or } \cdots \text{ or } b + \left(\tfrac{n}{m} - 1\right)m \bmod n.$$

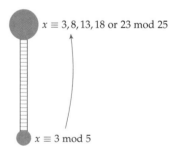

We find that the congruence $x \equiv b \bmod m$ **lifts** to $m/n$ possible congruences modulo $n$.

**Problem 7.15** Solve the congruence $x^2 \equiv 1 \bmod 49$.

SOLUTION: If $x^2 \equiv 1 \bmod 49$, then by descending we find that $x^2 \equiv 1 \bmod 7$. Thus $x \equiv 1$ or $6 \bmod 7$. Modulo 49, we find the **possibilities**:[5]

$$x \equiv 1, 8, 15, 22, 29, 36, 43 \text{ or } 6, 13, 20, 27, 34, 41, 48 \bmod 49.$$

Going through these 14 possibilities, one by one, the only $x$ satisfying $x^2 \equiv 1 \bmod 49$ are
$$x \equiv 1 \text{ or } 48 \bmod 49. \checkmark$$

[5] Using this descending/lifting method, we have reduced our search from 49 possibilities to 14 possibilities. We will have a better method soon.

MULTIPLICATIVE INVERSES can be lifted up a tower of prime powers. If $p$ is prime, then $GCD(x, p) = 1$ if and only if $GCD(x, p^e) = 1$ for all positive exponents $e$. Thus, if $x$ has a multiplicative inverse modulo $p$, then $x$ has a multiplicative inverse modulo all powers of $p$.

Here we give a recipe for *lifting* a multiplicative inverse mod $p$ to an inverse mod $p^2$, mod $p^4$, mod $p^8$ etc., accelerating up the ladder.

**Lemma 7.16 (Lifting multiplicative inverses)** *Suppose that* $xy \equiv 1$ *mod* $p^e$. *Let* $r$ *be the integer satisfying* $xy = 1 + rp^e$. *Define*[6]

$$z = y - yrp^e.$$

*Then* $xz \equiv 1$ *mod* $p^{2e}$.

PROOF: We compute directly,

$$\begin{aligned} xz &= xy - xyrp^e, \\ &= 1 + rp^e - (1 + rp^e)rp^e, \\ &= 1 + rp^e - rp^e - r^2p^{2e} = 1 - r^2p^{2e}. \end{aligned}$$

Hence $xz \equiv 1$ mod $p^{2e}$. ∎

**Problem 7.17** Find the multiplicative inverse of 3, modulo $2^{12}$.

SOLUTION: Modulo 2, 4, and 8, we find that $3 \times 3 \equiv 1$ mod 8. We lift further up the powers of 2 using the lemma. With $x = 3$, $y = 3$, $p = 2$ and $e = 3$, the equation $xy = 1 + rp^e$ is satisfied if $r = 1$:

$$3 \times 3 = 9 = 1 + 1(8),$$

Define

$$z = y - yrp^e = 3 - 3(1)(2^3) = 3 - 24 = -21.$$

We find that $3 \cdot (-21) \equiv 1$ mod $2^6$. Simplifying modulo $2^6 = 64$,

$$3 \cdot 43 \equiv 1 \text{ mod } 64.$$

We have *lifted* the multiplicative inverse of 3 from the world mod 8 to the world mod 64. We can go further, setting $x = 3$, $y = 43$, $p = 2$, and $e = 6$. The equation $xy = 1 + rp^e$ is satisfied if $r = 2$:

$$3 \cdot 43 = 129 = 1 + 2(64).$$

Now define

$$z = y - yrp^e = 43 - 43(2)2^6 = -5461.$$

We find that

$$3 \cdot (-5461) \equiv 1 \text{ mod } 2^{12}.$$

Simplifying modulo $2^{12} = 4096$ yields

$$3 \cdot 2731 \equiv 1 \text{ mod } 4096.$$

[6] Where does this $z$ come from? We seek a $z$ such that $z \equiv y$ mod $p^e$ and $xz \equiv 1$ mod $p^{2e}$. The first condition implies $z = y + tp^e$ for some $t$. The second condition yields

$$x(y + tp^e) \equiv 1 \text{ mod } p^{2e}.$$

As $xy = 1 + rp^e$, we find

$$1 + rp^e + xtp^e \equiv 1 \text{ mod } p^{2e}.$$

This will be satisfied as long as $r + xt \equiv 0$ mod $p^e$. Setting $t = -yr$ solves this congruence, since $y$ is a multiplicative inverse of $x$ modulo $p^e$.

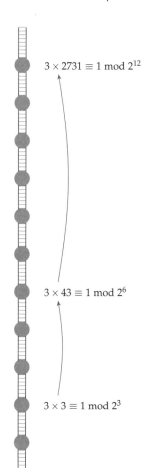

SQUARE ROOTS can also be lifted up a tower of prime powers, though a few more subtleties occur. Let $a$ be an integer; a **square root** of $a$, modulo $m$, is a solution to the congruence $x^2 \equiv a \bmod m$.

**Lemma 7.18** *Let $p$ be a prime number and $e$ a positive integer. The square roots of 1, modulo $p^e$, are*

$$\begin{cases} 1 & \text{if } p = 2 \text{ and } e = 1; \\ 1 \text{ or } p^e - 1 & \text{if } p \text{ is an odd prime}; \\ 1 \text{ or } 3 & \text{if } p = 2 \text{ and } e = 2; \\ 1 \text{ or } p^{e-1} - 1 \text{ or } p^{e-1} + 1 \text{ or } p^e - 1 & \text{if } p = 2 \text{ and } e \geq 3. \end{cases}$$

| $2^e$ | Square roots of 1 mod $2^e$ |
|---|---|
| 2 | 1 |
| 4 | 1 and 3 |
| 8 | 1, 3, 5, 7 |
| 16 | 1, 7, 9, 15 |
| 32 | 1, 15, 17, 31 |
| 64 | 1, 31, 33, 63 |

PROOF: First suppose $p$ is an odd prime. If $x^2 \equiv 1 \bmod p^e$ then

$$(x+1)(x-1) \equiv 0 \bmod p^e.$$

Hence $(x+1)(x-1)$ is a multiple of $p^e$. Since $p > 2$, it cannot be the case[7] that both $x+1$ and $x-1$ are multiples of $p$. Thus, either $x+1$ or $x-1$ must be a multiple of $p^e$. Therefore $x \equiv \pm 1 \bmod p^e$.

[7] Numbers that differ by two cannot both be multiples of $p$, if $p > 2$.

Now suppose $p = 2$. If $e = 1$ or $e = 2$, the lemma can be checked directly, so take $e \geq 3$ in what follows. If $x^2 \equiv 1 \bmod 2^e$, then

$$(x+1)(x-1) \equiv 0 \bmod 2^e.$$

Let $f$ and $g$ be the exponents of 2 in the prime decompositions of $x+1$ and $x-1$, respectively. Then $x+1$ is a multiple of $2^f$ and $x-1$ is a multiple of $2^g$, and $f + g \geq e$. If both $f$ and $g$ were greater than 1, we would find that $x+1$ and $x-1$ would be multiples of 4. But numbers that differ by 2 cannot both be multiples of 4!

Hence $f \leq 1$ or $g \leq 1$, which implies that $g \geq e-1$ or $f \geq e-1$. We find that $x+1$ or $x-1$ must be a multiple of $2^{e-1}$. Thus $x+1$ or $x-1$ must be congruent to 0 or $2^{e-1} \bmod 2^e$. We find the solutions:

$$x \equiv 1 \text{ or } x \equiv -1 \text{ or } x \equiv 2^{e-1} + 1 \text{ or } x \equiv 2^{e-1} - 1 \bmod 2^e. \blacksquare$$

**Proposition 7.19** *Suppose that $p$ is a prime and $\mathrm{GCD}(a, p) = 1$. If $a$ has a square root modulo $p^e$, then the number of square roots of $a$ equals the number of square roots of 1: either 1, 2, or 4.*

PROOF: Let $\epsilon$ be a square root of 1, and suppose that $x^2 \equiv a \bmod p^e$. Then $x\epsilon$ is a square root of $a$ too.[8] To prove the proposition, we check that all square roots of $a$ can be obtained from $x$ in this way.

[8] Note that
$$(x\epsilon)^2 = x^2\epsilon^2 \equiv x^2 \equiv a \bmod p^e.$$

Suppose that $y^2 \equiv a \bmod p^e$ too. Since $\mathrm{GCD}(a, p) = 1$, there exists $b$ such that $ab \equiv 1 \bmod p^e$. It follows[9] that $yxb$ is a square root of 1. Moreover, $y \cdot yxb = x \cdot y^2 b \equiv x \cdot ab \equiv x \bmod p^e$. Therefore $y$ arises from $x$ by multiplying by a square root of 1. $\blacksquare$

[9] Note that
$$(yxb)^2 = y^2 x^2 b^2 \equiv aab^2 \equiv 1 \bmod p^e.$$

LIFTING SQUARE ROOTS can be accomplished by the following recipe. We consider the case of an odd prime $p$ first.

**Theorem 7.20 (Lifting square roots)** *Suppose that $p$ is an odd prime, $e \geq 1$, $\mathrm{GCD}(a, p) = 1$, and $x^2 \equiv a \bmod p^e$. Let $r$ be the integer for which $x^2 = a + rp^e$. Let $b$ be a multiplicative inverse[10] of $2x$, modulo $p^e$. Define*

$$z = x - brp^e.$$

*Then $z^2 \equiv a \bmod p^{2e}$.*

[10] $b$ can be found by lifting multiplicative inverses mod $p$, $p^2$, up to $p^e$.

PROOF: Since $b$ is a multiplicative inverse of $2x$, modulo $p^e$, we have $2xb = 1 + sp^e$ for some integer $s$. Now we compute,

$$\begin{aligned} z^2 &= (x - brp^e)^2, \\ &= x^2 - 2xbrp^e + b^2r^2p^{2e}, \\ &= x^2 - (1 + sp^e)rp^e + b^2r^2p^{2e}, \\ &= (a + rp^e) - rp^e - rsp^{2e} + b^2r^2p^{2e} \equiv a \bmod p^{2e}. \end{aligned}$$ ∎

**Corollary 7.21** *Suppose that $p$ is an odd prime and $\mathrm{GCD}(a, p) = 1$. Then $a$ is a square modulo $p$ if and only if $a$ is a square modulo every power of $p$.*

When $p = 2$, lifting can require shifting.

**Proposition 7.22** *Suppose that $e \geq 3$, $a$ is odd, and $x^2 \equiv a \bmod 2^e$. Then either*

$$x^2 \equiv a \bmod 2^{e+1} \quad or \quad \left(x + 2^{e-1}\right)^2 \equiv a \bmod 2^{e+1}.$$

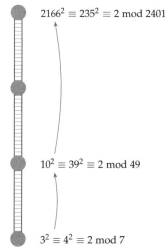

Figure 7.4: Lifting the square roots of 2, modulo powers of 7. The lifting accelerates, modulo 7, $7^2$, then $7^4$, and could be quickly continued to $7^8$, $7^{16}$, etc.

PROOF: We may write $x^2 = a + r2^e$ for some integer $r$. Note that $x$ is odd since $a$ is odd. If $r$ is even, then $x^2 \equiv a \bmod 2^{e+1}$ as needed. If $r$ is odd, then we must shift $x$ to lift it. We compute

$$\begin{aligned} \left(x + 2^{e-1}\right)^2 &= x^2 + 2x2^{e-1} + 2^{2e-2}, \\ &= a + r2^e + x2^e + 2^{2e-2}, \\ &= a + (x + r)2^e + 2^{2e-2} \equiv a \bmod 2^{e+1}. \end{aligned}$$

The last step requires a bit of justification: when $e \geq 3$, $2e - 2 = e + (e - 2) \geq e + 1$. Since $x$ is odd and $r$ is odd, $(x + r)2^e$ is a multiple of $2^{e+1}$ as well. The result follows. ∎

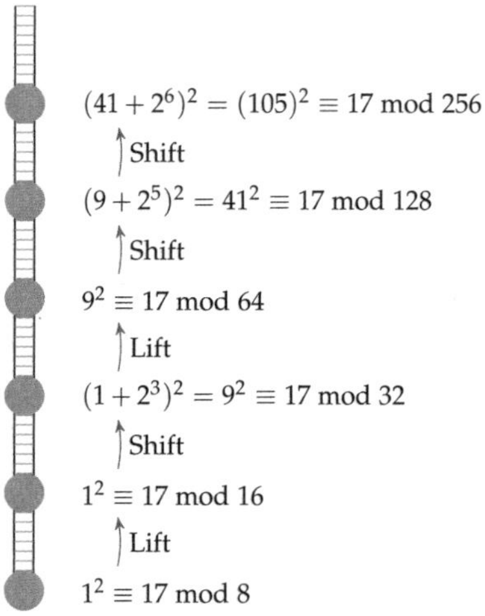

**Corollary 7.23** *Let $a$ be an odd number. Then $a \equiv 1 \bmod 8$ if and only if $a$ is a square modulo $8, 16, 32, 64, \ldots$.*

PROOF: Observe that $1^2 \equiv 3^2 \equiv 5^2 \equiv 7^2 \equiv 1 \bmod 8$. Hence 1 is the only square modulo 8. If $a \equiv 1 \bmod 8$, then we may shift and lift to obtain square roots of $a$ modulo all higher powers of 2. ∎

Figure 7.5: Lifting and shifting, to find square roots of 17, modulo powers of 2.

We finish this chapter with a cryptographic application of the Chinese remainder theorem – the famous public-key cryptosystem known as **RSA**.[11]

While RSA can be used for various purposes, we focus here on the common task of sending a short encrypted message. How might Bob send a short private message to Alice, if they are separated by distance and Eve has tapped the wire.

[11] RSA is named for its discoverers, Rivest, Shamir, and Adleman. They first described their cryptosystem in "A method for obtaining digital signatures and public-key cryptosystems," in *Communications of the ACM* **21** (2) (1978).

Unlike Diffie-Hellman, RSA is asymmetric. RSA gives a method whereby *anyone* can encrypt a message but *only Alice* can decrypt it later. It is like a mathematical padlock, which anyone can push closed but only Alice knows the opening combination.

To receive encrypted messages, Alice begins (often in advance) by cooking up two large[12] prime numbers called $p$ and $q$. The numbers $p$ and $q$ are called Alice's **private keys**. These she never reveals to anyone, not even her friend Bob. But she multiplies them together to form $N = pq$ and makes the number $N$ public to everyone. She also publicizes an auxiliary number called $e$. The pair $(N, e)$ is called Alice's **public key**.

[12] 1024 bit primes should do, and are easy enough to find with the Miller-Rabin test.

When Bob wishes to send an encrypted message $m$ (a small natural number – smaller than $N$ at least) to Alice, he computes the **ciphertext** $c = m^e \bmod N$ using Alice's public key $(N,e)$. This number $c$, the simplification of $m^e \bmod N$, is in effect a scrambled form of $m$. Bob sends $c$ to Alice over the wire.

To recover the message $m$ from the ciphertext, Alice uses her private keys $p$ and $q$. First she computes $\phi(N)$ by the formula

$$\phi(N) = \phi(p)\phi(q) = (p-1)\cdot(q-1).$$

From this, she computes a multiplicative inverse[13] $d$ of $e$ modulo $\phi(N)$, via the Euclidean algorithm.

$$d \cdot e \equiv 1 \bmod \phi(N).$$

Alice keeps $\phi(N)$ and $d$ as secret as $p$ and $q$!

When Alice receives the ciphertext $c$, she computes $c^d \bmod N$. This remarkably recovers Bob's message $m$ by Euler's Theorem:

$$c^d \equiv (m^e)^d \equiv m^{ed} \equiv m^1 = m \bmod N.$$

In this way, Alice decrypts the message and finds $m$.

Decryption seems[14] to require knowledge of $\phi(N)$; as knowledge of $\phi(N)$ requires knowledge[15] of the factors $p$ and $q$, breaking the code seems as hard for Eve as factoring the large number $N$.

This asymmetry between public knowledge and private knowledge, provides a robust[16] mechanism for short encrypted messages over insecure channels. To secure a channel of communication, one can use RSA to encrypt a *symmetric key* which is then used to encrypt later ongoing communication. This is how RSA is used in practice,[17] e.g., when your email is transmitted from a server to your computer.

[13] It is important, for this purpose, that $\mathrm{GCD}(e,\phi(N)) = 1$. But Alice is in charge of choosing $e$, so this is not difficult to guarantee in advance. The value $e = 65537$ is commonly used, for pragmatic reasons of security and speed.

[14] Factorization of $N$ reveals the private key from the public key; but breaking RSA – finding $m$ from knowledge of $c, N, e$ in the equation $m^e \equiv c \bmod N$ – might be significantly easier than factorization.
[15] See the exercises on this.
[16] As with all security protocols, robustness depends on all aspects of implementation.
[17] E.g., in the TLS handshake protocol.

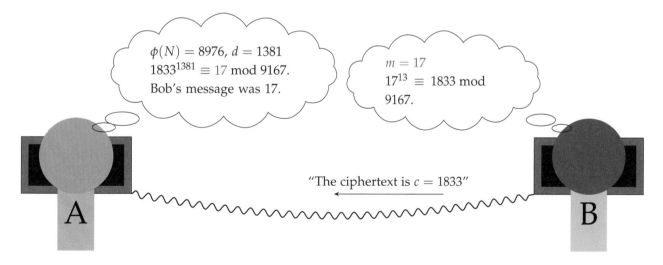

## Historical notes

The "Chinese remainder theorem" is a name of Western origin, for a theorem which relates to remainder problems found across cultures and times. There are two histories – one of remainder problems, and another of the naming of the theorem.

We first discuss the naming. Alexander Wylie, a Protestant missionary and British sinologist, participated in an exchange of scientific ideas between China and Western Europe from 1847 until 1877. A history of Wylie can be found in the work of Siu and Chan.[18]

With collaborators, Wylie translated Western mathematics into Chinese, and in a series of newspaper articles he introduced Chinese mathematics to a Western audience. In these *Jottings*, Wylie introduces a problem-solving method as follows.

> One of the most remarkable of these is the 大衍 Ta-yen, "Great Extension," a rule for the resolution of indeterminate problems. This rule is met with in embryo in Sun Tsze's[19] Arithmetical Classic under the name of 物不知數 Wuh-puh-chi-soo, "Unknown Numerical Quantities," where after a general statement in four lines of rhyme the following question is proposed:
>
> Given an unknown number, which when divided by 3, leaves a remainder of 2; when divided by 5, it leaves 3; and when divided by 7, leaves 2; what is the number?
>
> This is followed by a special rule for working out the problem, in terms sufficiently concise and elliptical, to elude the comprehension of the casual reader.[20]

L. Dickson, in his "History of the Theory of Numbers,"[21] states that the Chinese method of solution became known in the West due to Wylie, and uses the term "Chinese remainder theorem" in his 1929 "Introduction to the theory of numbers."[22] This appears to be the first instance of the name in its modern usage.

Before a theorem was stated as such, there was a long tradition of remainder problems – in China (e.g., in Problem 3.26 of the 孙子算经 *Sunzi suan jing* "Mathematical classic of Master Sun," mentioned by Wylie above), in India where a general solution was described in the *Aryabhatiya* of Āryabhata and commentary by Bhāskara I, in medieval Islam, and in Europe.[23] In particular, one finds the same moduli (3,5,7) of Sunzi's remainder problem also in a problem given by Abu Mansur (c. 1000) in the *al-Takmila fi'l-Hisab*, and in a problem from Fibonacci's *Liber Abaci*.

Presented as a theorem, one finds the Chinese Remainder Theorem in the work of Euler and Gauss. Gauss devotes Art. 30–36 of the *Disquisitiones* to the solution of remainder problems. The solution to

[18] See "On Alexander Wylie's *Jottings on the science of the Chinese arithmetic*," by Man-Keung Siu and Yip-Cheung Chan, presented at *History and Pedagogy of Mathematics*, DCC, Daejeon, Korea (2012).

[19] Sun Tsze ("Master Sun", roughly 220–420CE) is often spelled Sun Tzu or Sunzi in other sources. He should not be confused with the Sun Tzu whom Sima Qian credits with writing "The Art of War" around 500BCE.

[20] From "Jottings on the science of the Chinese arithmetic," by Alexander Wylie (1852). See p.175 of the collection *Chinese researches* of Wylie's writings, published in 1897 (after Wylie's death).

[21] See Chapter II of the "History of the Theory of Numbers," Volume II, by Leonard Eugene Dickson (1919), reprinted by the American Mathematical Society (1999). Dickson references the "Chinese problem of remainders."

[22] See p.11 of Dickson's "Introduction to the theory of numbers," published by the University of Chicago Press, 1929.

[23] See "Historical development of the Chinese remainder theorem," by Shen Kangsheng, in *Archive for History of Exact Sciences* vol. 38 4 (1988) for historical examples translated into modern notation. For later examples in Europe, see Maarten Bullynck's "Modular arithmetic before C.F. Gauss: Systematizations and discussions on remainder problems in 18th-century Germany," in *Historia Mathematica* vol. 36 1 (2009).

systems of linear congruences is given in Art. 32 of the *Disquisitiones*, before Gauss provides a calendric example for illustration,

> This happens in the problem of chronology when we seek to determine what Julian year it is whose indiction, golden number, and solar cycle are given.[24]

Remainder problems are inevitable when considering calendrics, in any system where one tracks time by measuring one's place on multiple cycles.

The *totient* $\phi(n)$, was introduced by Euler (1763),[25] though the name originates with Sylvester (1879)[26] and the symbol $\phi$ was chosen by Gauss in Art. 38 of the *Disquisitiones*. In §20 of Euler's work, Euler describes the number of positive integers, less than a number and coprime with it. For a prime $p$, he observes the answer is $p - 1$, for prime powers $p^n$, the answer is $p^{n-1}(p - 1)$, and for products of distinct primes $pq$, he finds the answer $(p - 1)(q - 1)$. Euler's Theorem 5 demonstrates that $\phi(ab) = \phi(a)\phi(b)$ (in our notation), when $\gcd(a, b) = 1$, and his Corollary 3 relates $\phi(N)$ to the prime decomposition of $N$.

The idea of *lifting* congruences, up a tower of prime powers, is now associated with *Hensel's Lemma* after an article by Kurt Hensel (1904).[27] A specific form of Hensel's Lemma is found on page 81,

> Ist die Diskriminante von $F(x)$ durch $p$ nicht teilbar, so folgt schon aus jeder Zerlegung des *ersten* Näherungswertes
> 
> $$F_0(x) \equiv f_0(x)g_0(x) \pmod{p}$$
> 
> für den Modul p eine eindeutig bestimmte Zerlegung[28]
> 
> $$F(x) = f(g)g(x) \quad (p)$$

In this paper, Hensel introduced the $p$-adic numbers. The $p$-adic numbers simultaneously encode congruences modulo $p$, $p^2$, $p^3$, etc., for *all* powers of a prime $p$. By introducing basic $p$-adic analysis, Hensel proves very general lifting results.

The specific form of Hensel's Lemma written above can also be found in Art. 373–375 of the unpublished Section 8 of Gauss's *Disquisitiones*. There Gauss writes,

> From this one sees that if the function $X$ does not have equal factors with respect to the modulus $p$, it can be decomposed into factors mod $p^k$ in a similar way as mod $p$.[29]

Gauss's result implies most of the lifting results of this chapter. For example, lifting square roots corresponds to lifting a factorization of $x^2 - a$ mod $p$ to a factorization modulo $p^k$. Note that when $p = 2$, Gauss's result does not apply; for example $x^2 - 1$ factors as $(x - 1)(x - 1)$, mod 2. In such cases, the stronger results of Hensel apply to lift from square roots modulo 8 to higher powers of 2.

[24] See the end of Art. 36 of the *Disquisitiones*. Translation from Arthur A. Clarke.

[25] See "Theoremata arithmetica nova methodo demonstrata", in *Novi Commentarii academiae scientiarum Petropolitanae* 8 (1763). Available as E271 at the Euler Archive http://eulerarchive.maa.org. This is the same paper in which Euler generalizes Fermat's Little Theorem to nonprime moduli.

[26] J.J. Sylvester, "On certain ternary cubic-form equations," in *American Journal of Mathematics* 2 (1879).

[27] See "Neue Grundlagen der Arithmetik" by K. Hensel, in *Journal für die reine und angewandte Mathematik* 127 (1904).

[28] The notation $(p)$ refers to equality as $p$-adic numbers, or equivalently, congruence mod all powers of $p$.

[29] Translation from Frei, "The unpublished Section 8", in "The shaping of arithmetic after C.F. Gauss's *Disquisitiones Arithmeticae*," (2007). The case where there are equal factors is more difficult, and Hensel's more general results cover this case as well.

## Exercises

1. Find a number $x$ between 0 and 90, such that $x \equiv [4, 8] \bmod [7, 13]$.

2. List all numbers $x$ such that $0 \leq x < 100$ and $x \equiv [3, 1] \bmod [4, 5]$.

3. Prove that if $n > 2$ then $\phi(n)$ is even.

4. Find all numbers $n$ such that $\phi(n) = 6$.

5. Prove that if $n$ and $e$ are positive integers, then $\phi(n^e) = n^{e-1} \cdot \phi(n)$. Use this to find a general expression for $\phi(10^e)$.

6. Use the Chinese remainder theorem to find all of the solutions to $x^2 + 1 = 0$, modulo 1313.

7. What are the last two digits of $3^{1000}$?

8. Find a positive integer $x$ such that the last three digits of $7^{7^x}$ are 007.

9. Find the multiplicative inverse of 3 modulo $5^8$.

10. What are the square roots of 3, modulo $11^4$?

11. Is 17 a square modulo 104? (Use the Chinese remainder theorem.)

12. Prove that if $p$ is prime, and $a \equiv b \bmod p^2 - p$, then $a^a \equiv b^b \bmod p$.

13. Fix a prime number $p$ in what follows. The *p-adic norm* of an integer $x$, denoted $|x|_p$ is defined to be $p^{-e}$, if $p^e$ is the power of $p$ appearing in the prime decomposition of $x$. For example,

$$|50|_2 = \frac{1}{2} \text{ and } |50|_3 = 1 \text{ and } |50|_5 = \frac{1}{25}.$$

The $p$-adic norm of zero is defined to be zero: $|0|_p = 0$. The *p-adic distance* between two integers $x$ and $y$ is defined to be $|x - y|_p$.

(a) Prove that $|x|_p \leq 1$ for all integers $x$, and $|x|_p = 1$ if and only if $x = \pm 1$.

(b) Prove the ultrametric triangle inequality:[30]

$$|x + y|_p \leq \max\{|x|_p, |y|_p\}.$$

[30] Compare to Proposition 5.32.

(c) Let $a$ be an integer. Describe the set of integers $x$ such that $|x - a|_p < p^{-e}$, using the language of congruences.

(d) Suppose that $x$ and $a$ are integers, and $x^2 \equiv a \bmod p$. Prove that for every positive real number $\epsilon$, there exists an integer $y$ such that $|y^2 - a|_p < \epsilon$.

14. The algorithm for lifting square roots mod $p^e$ to those mod $p^{2e}$ (when $p$ is odd) is related to a much older[31] algorithm for approximating square roots of positive real numbers. Let $a$ be a positive real number in what follows.

    [31] The first written example seems to be in the work of Heron of Alexandria. For more detail, see (the last page of) "Square root approximations in Old Babylonian mathematics: YBC 7289 in Context" by David Fowler and Eleanor Robson, in *Historia Mathematica* **25** (1998).

    (a) Suppose that $x$ is a positive real number, and define
    $$x' = \frac{1}{2}\left(x + \frac{a}{x}\right).$$
    Let $\epsilon = |x - \sqrt{a}|$ be the error in the approximation of $x$ to $\sqrt{a}$. Let $\epsilon' = |x' - \sqrt{a}|$ be the corresponding error for $x'$. Assume moreover that $x > 1/2$. Prove that $\epsilon' < \epsilon^2$.

    (b) Compare this approximation algorithm to the lifting of square roots modulo powers of $p$ (when $p$ is an odd prime). The comparison is more straightforward using the $p$-adic norm.

15. Suppose that $N$ is the product of two distinct odd primes, $N = pq$. If $N = 8633$ and $\phi(N) = 8448$ then what are $p$ and $q$? How do $N$ and $\phi(N)$ determine the private keys $p$ and $q$ in RSA? (Alice must keep $\phi(N)$ private!)

16. In 2012, researchers[32] discovered a significant problem with some implementations of RSA in the wild. A "key overlap" occurs when two independent parties use private keys $p_1, q_1$ and $p_2, q_2$, and there is a coincidence between them, e.g., $p_2 = p_1$ or $q_2 = q_1$ or $p_1 = q_2$ or $q_1 = p_2$. Their discovery was based on a large database of public keys (products $pq$ for many different people).

    [32] See Lenstra et al., "Ron was wrong, Whit is right" at http://eprint.iacr.org/2012/064.pdf and the independent work of Nadia Heninger et al., "Mining your Ps and Qs: Detection of widespread weak keys in network devices" at https://factorable.net/weakkeys12.extended.pdf.

    (a) How might one quickly discover such key overlaps, and why do they represent a security problem?

    (b) If $p$ and $q$ are randomly chosen prime numbers between $2^{1023}$ and $2^{1024}$, estimate the probability that $p = q$. Among a million such prime numbers, independently and randomly chosen, estimate the chance of a key overlap.[33]

    [33] You may use the heuristic that a number $n$ has a $1/\log(n)$ chance of being prime. To estimate the probability of a key overlap, you may wish to read about the "birthday paradox".

17. Let $p = 47$ and $q = 59$, $N = pq = 2773$, and $e = 157$.[34]

    [34] These keys are taken from the Section VIII of the original paper by Rivest, Shamir, and Adleman.

    (a) Compute a multiplicative inverse of $d$, modulo $\phi(N)$.

    (b) Every two-letter string (including A-Z and spaces) can be converted to a number-message between 0 and 2626, by replacing a space by 00, 'A' by 01, 'B' by 02, etc.. For example, 'ME' becomes 1305. Encrypt the two-letter string 'HI' by computing its number-message $m$, and the ciphertext $m^e$ mod $N$.[35]

    [35] The Python command pow(m,e,N) computes $m^e$ modulo $N$.

    (c) Decrypt the sequence of ciphertexts $0802, 2179, 2657, 1024$ to find a message.

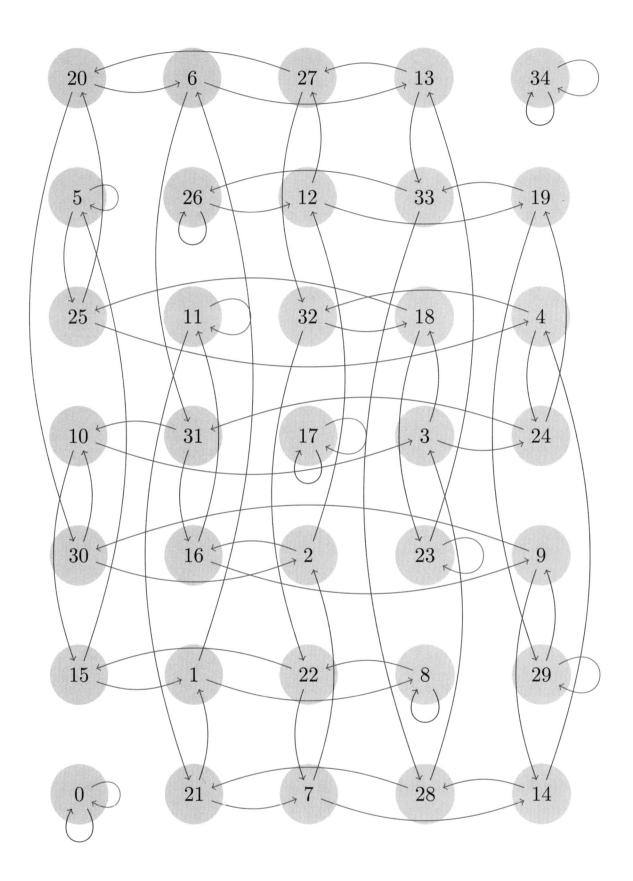

# 8
# Quadratic Residues

LINEAR CONGRUENCES can be solved easily, since linear congruences are really linear Diophantine equations in disguise. For example, the congruence $5x \equiv 7 \bmod 13$ states nothing more or less than $5x = 7 + 13y$ for some integer $y$. The latter equation can be solved using techniques from the first chapter of the text.

QUADRATIC CONGRUENCES are more difficult to solve. For example, the congruence $5x^2 \equiv 7 \bmod 13$ states that $5x^2 = 7 + 13y$ for some integer $y$. The Euclidean algorithm will not help us solve such an equation; it is fundamentally different. Finding integers $(x, y)$ which satisfy $5x^2 = 7 + 13y$ is the same as finding grid-points on the parabola $y = \frac{5}{13}x^2 - \frac{7}{13}$. It is a nonlinear problem, through and through.

Gauss's solution of quadratic congruences marks the birth of modern number theory. His *Quadratic Reciprocity*, conjectured in a different form by Euler and almost proven by Legendre, determines whether the congruence $x^2 \equiv a \bmod p$ has a solution, when $p$ is an odd prime. More general quadratic congruences follow quickly, and thanks to the Chinese remainder theorem and lifting techniques, one may solve congruences with composite moduli.

To understand the congruence $x^2 \equiv a \bmod p$, we explore the **squares** mod $p$, often called **quadratic residues** mod $p$. For example, the squares modulo 7 are 0, 1, 2, 4. These are the numbers, mod 7, which *result* from squaring.

$$0^2 = 0, \ 1^2 = 1, \ 2^2 = 4, \ 3^2 = 2, \ 4^2 = 2, \ 5^2 = 4, \ 6^2 = 1 \bmod 7.$$

The quadratic residues are those numbers $a$ mod $p$, for which the congruence $x^2 \equiv a \bmod p$ has a solution.

In this chapter, we study quadratic residues and prove *Quadratic Reciprocity*. We approach this great result through a deep study of dynamics, using an idea of Zolotarev.

Figure 8.1: The parabola $5x^2 = 7 + 13y$ intersects the grid points $(2,1)$ and $(-2,1)$. This reflects the fact that $5x^2 \equiv 7 \bmod 13$ has solutions with $x = \pm 2$.

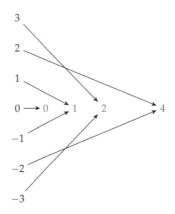

Figure 8.2: Arrows denote squaring modulo 7. Squares modulo 7 are highlighted in red.

PARTNERING can simplify a repeated operation. From the first chapter, we recall the technique for addition.

**Problem 8.1 (Repeated from Chapter 0)** What is $1 + 2 + 3 + \cdots + 47 + 48 + 49$?

SOLUTION: Partner the numbers from 1 to 49 by pairing $x$ with $y$ when $x + y = 50$.

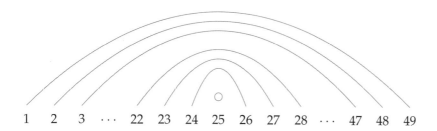

We find 24 pairs – each pair summing to 50 – and 25 is a **lonely number**.[1] Hence

[1] In partnering operations, we call a number $x$ lonely if $x$ is its own partner.

$$1 + 2 + 3 + \cdots + 47 + 48 + 49 = (24 \times 50) + 25 = 1200 + 25 = 1225.$$

✓

The utility of partnering is that each partnered pair sums to 50; so instead of adding 49 different numbers, we add the same number to itself: an act of multiplication. In the end, one must treat lonely numbers individually.

We use the method of partnering, for multplying a list of numbers in the modular context.

**Theorem 8.2 (Wilson's Theorem)** *If $p$ is a prime number, then*

$$(p-1)! \equiv -1 \bmod p.$$

PROOF: Recall that $\Phi(p) = \{1, 2, 3, \ldots, p-1\}$. Partner these numbers by pairing $x$ with $y$ when $xy \equiv 1 \bmod p$. Notice that every number $x \in \Phi(p)$ has a unique partner, since every element of $\Phi(p)$ has a unique multiplicative inverse, mod $p$.

The lonely numbers are those $x \in \Phi(p)$ for which $x^2 = xx \equiv 1 \bmod p$. By Proposition 6.21, the lonely numbers are 1 and $p - 1$.

Now to multiply the numbers in $\Phi(p)$, multiply in partners; each partnership $x, y$ yields 1 as its product. Only the lonely numbers contribute to the product.

$$1 \times 2 \times 3 \times \cdots \times (p-1) \equiv 1 \times (p-1) \equiv -1 \bmod p. \blacksquare$$

The partnering of Wilson's Theorem is displayed below. Only the lonely numbers 1 and $p-1$ contribute to the product, leaving the product equal to $p-1$, or $-1 \mod p$.

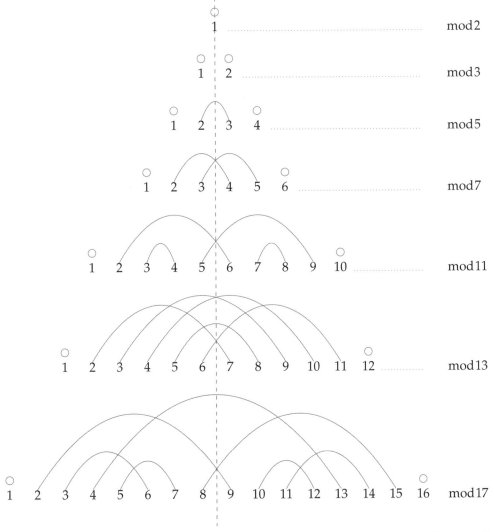

"Wilson's" Theorem can be found in the *Opuscula* of Ibn al-Haytham, c. 1040CE; see R. Rashed's "The development of Arabic Mathematics" for translation and discussion of the relevant sections of Ibn al-Haytham. A proof is found in Lagrange, *Démonstration d'un Théoreme nouveau concernant les nombres premiers*, published in 1771. Lagrange found the result in work of Waring, who in turn attributed the result to his student Wilson. Lagrange's proof was more difficult than ours – his difficulty is not surprising since modular arithmetic was not yet developed to the point where such a partnering might seem natural.

Figure 8.3: Each number is paired with its multiplicative inverse, mod $p$, in the proof of Wilson's Theorem. The left-right symmetry is due to the fact that the partnership

$$xy \equiv 1 \mod p$$

implies the partnership

$$(p-x)(p-y) \equiv (-x)(-y) \equiv 1.$$

WILSON'S THEOREM might be a mere curiosity, but we use it to analyze *squareness* in modular arithmetic, at least when the modulus is *prime*. To start our analysis, we begin by enumerating the squares.

**Proposition 8.3** *Let p be an odd prime number. Then among the set $\Phi(p)$, half the numbers are squares mod p and half are non-squares.*

PROOF: Consider the squaring mod $p$ function on the set $\Phi(p)$. This is displayed in the margin when $p = 11$.

Observe that this function defines a two-to-one correspondence. In other words, for each output there are precisely two inputs. Indeed, if $x$ and $p - x$ are input into the squaring mod $p$ function, then their outputs are the simplifications of $x^2$ mod $p$ and $(p - x)^2$ mod $p$. But $(p - x)^2 \equiv (-x)^2 = x^2$, so both inputs yield the same output.

No more than two inputs yield the same output. For if $x^2 \equiv y^2$ mod $p$, then $(x + y)(x - y) \equiv 0$ mod $p$, which implies $x \equiv \pm y$ mod $p$. Within the set $\Phi(p)$, this implies that $y = p - x$.

Since there are two inputs for each output, there are half as many outputs as inputs. Half of the elements of $\Phi(p)$ are squares. ∎

But **which** half? It is difficult to see a pattern.

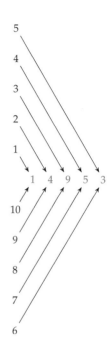

Figure 8.4: Arrows denote squaring modulo 11. Squares modulo 11, within $\Phi(11)$, are highlighted in red.

| Modulus $p$ | The numbers in $\Phi(p)$, with squares in red |
|---|---|
| 2 | 1 |
| 3 | 1,2 |
| 5 | 1,2,3,4 |
| 7 | 1,2,3,4,5,6 |
| 11 | 1,2,3,4,5,6,7,8,9,10 |
| 13 | 1,2,3,4,5,6,7,8,9,10,11,12 |
| 17 | 1,2,3,4,5,6,7,8,9,10,11,12,13,14,15,16 |
| 19 | 1,2,3,4,5,6,7,8,9,10,11,12,13,14,15,16,17,18 |
| 23 | 1,2,3, 4,5,6,7,8, 9,10,11,12,13,14,15,16,17,18,19,20,21,22 |
| 29 | 1,2,3,4,5,6,7,8,9,10,11,12,13,14,15,16,17,18,19,20,21,22,23,24,25,26,27,28 |

A NEW PARTNERSHIP distinguishes squares from nonsquares.

**Definition 8.4** *Let p be a prime number. Let a, x, and y be elements of $\Phi(p)$. We declare x and y to be a-**partners** if $xy \equiv a$ mod p.*

The partners of Wilson's Theorem were 1-partners. Below, we display the $a$-partners, modulo 11, for all $a$ between 1 and 10.

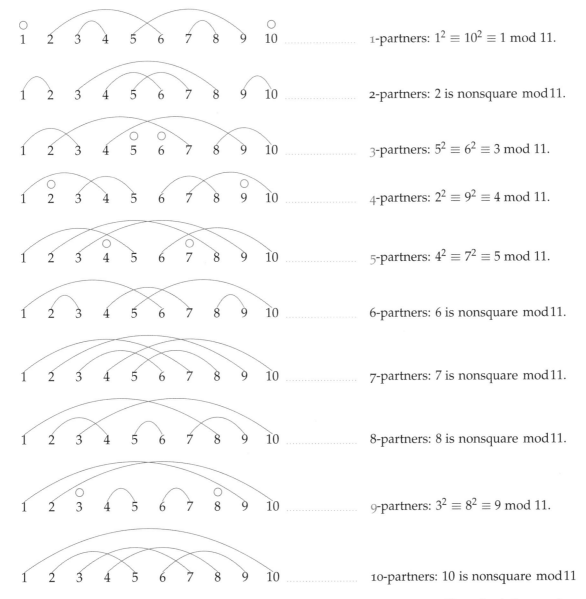

Figure 8.5: A diagram of $a$-partnership, mod 11. Notice that when $a$ is a square, mod 11, there are two lonely numbers, corresponding to the two square roots of $a$. Compare to the diagram of arrows on the previous page.

Observe first that every number in $\Phi(p)$ has an $a$-partner. Indeed, if $x \in \Phi(p)$, then consider the multiplicative inverse $y \in \Phi(p)$ of $x$. We find that $xy \equiv 1 \bmod p$ and so $xya \equiv a \bmod p$. Hence $(ya)$ is the $a$-partner of $x$, mod $p$.

When $a$-partnering, the lonely numbers are those $x$ which satisfy

$$x^2 = xx \equiv a \bmod p.$$

If $a$ is nonsquare, mod $p$ then there can be no lonely numbers after $a$-partnership. If $a$ is square, mod $p$, then there are two lonely numbers: the two *square roots* of $a$, mod $p$.

**Theorem 8.5 (Euler's Criterion for squareness mod $p$)** *Let $p$ be an odd prime number, and let $a$ be an integer coprime to $p$. Then,*

- *$a$ is square mod $p$, if and only if $a^{(p-1)/2} \equiv 1 \bmod p$.*

- *$a$ is nonsquare mod $p$, if and only if $a^{(p-1)/2} \equiv -1 \bmod p$.*

PROOF: Since we care about congruence mod $p$, we may take $a$ between 1 and $p - 1$. If $a$ is nonsquare mod $p$, then there are no lonely numbers when matching elements of $\Phi(p)$ by $a$-partnership.

Each number in $\Phi(p)$, together with its partner, has product equal to $a$, mod $p$. Since there are $p - 1$ numbers in the set $\Phi(p)$, there are $(p-1)/2$ pairs. Hence multiplying the numbers of $\Phi(p)$ yields

$$(p-1)! \equiv a^{\frac{p-1}{2}} \bmod p.$$

By Wilson's Theorem, $(p - 1)! \equiv -1 \bmod p$, and so when $a$ is nonsquare mod $p$,

$$a^{\frac{p-1}{2}} \equiv -1 \bmod p.$$

If $a$ is a square, mod $p$, then there are two lonely numbers, say $x$ and $y$, such that $x^2 \equiv y^2 \equiv a$. The two numbers $x$ and $y$ must be negatives of one another: $y \equiv -x \bmod p$. Hence

$$xy \equiv x \cdot (-x) = -x^2 \equiv -a \bmod p.$$

Among the $p - 1$ numbers in $\Phi(p)$, only $x$ and $y$ are lonely, and the remaining $p - 3$ numbers fit into $(p-3)/2$ partnerships. Each partnership contributes $a$ to the product, and $x$ and $y$ contribute $-a$ to the product. We multiply the numbers of $\Phi(p)$ using these partners and the lonely numbers:

$$(p-1)! \equiv a^{(p-3)/2} \cdot xy \equiv a^{(p-3)/2} \cdot (-a) = -a^{\frac{p-1}{2}}.$$

By Wilson's Theorem, $(p - 1)! \equiv -1 \bmod p$ and so when $a$ is a square mod $p$,

$$a^{\frac{p-1}{2}} \equiv 1 \bmod p. \qquad \blacksquare$$

Since 7 is nonsquare mod 11, there are no lonely numbers after 7-partnership. There are $(11-1)/2 = 5$ partnerships, each with product 7. Hence

$$7 \cdot 7 \cdot 7 \cdot 7 \cdot 7 = 7^5 \equiv 10! \bmod 11.$$

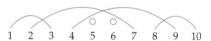

Since 3 is square mod 11, there are two lonely numbers after 3-partnership. The lonely numbers are 5 and 6, since

$$5^2 \equiv 6^2 \equiv 3 \bmod 11.$$

The remaining numbers fit into 4 partnerships, each with product 3. Hence

$$3 \cdot 3 \cdot 3 \cdot 3 \cdot (5 \cdot 6) = -1 \cdot 3^5 \equiv 10! \bmod 11.$$

Euler's Criterion refines Fermat's Little Theorem. When $p$ is a prime number and $GCD(a, p) = 1$, Fermat's Little Theorem guarantees that

$$a^{p-1} \equiv 1 \bmod p.$$

Consider the number $h \equiv a^{(p-1)/2}$ arising in Euler's criterion. Then,

$$h^2 = a^{2 \cdot (p-1)/2} = a^{p-1} \equiv 1 \bmod p.$$

Since $p$ is prime, the condition $h^2 \equiv 1 \bmod p$ implies that $h \equiv \pm 1$. Euler's criterion states that the number $h$ is 1 when $a$ is a square, and $-1$ when $a$ is a nonsquare, mod $p$.

Reciprocity laws relate questions of squareness, mod $p$, to properties of $p$ modulo another number. The first example is a corollary of Euler's criterion.

**Corollary 8.6 (Reciprocity for $-1$)** *Let $p$ be an odd prime. Then $-1$ is a square mod $p$, if and only if $p \equiv 1 \bmod 4$.*

Proof: By Euler's criterion, $-1$ is a square modulo $p$ if and only if $(-1)^{(p-1)/2} \equiv 1 \bmod p$. For any exponent $n$,

$$(-1)^n = \begin{cases} 1 & \text{if } n \text{ is even;} \\ -1 & \text{if } n \text{ is odd.} \end{cases}$$

Since $p \neq 2$, note that $1 \not\equiv -1 \bmod p$. Hence $-1$ is a square modulo $p$ if and only if $(p-1)/2$ is even. This occurs if and only if $(p-1)$ is a multiple of 4. Equivalently, $p \equiv 1 \bmod 4$. ∎

If one traces through the proof of reciprocity for $-1$, via Wilson's theorem, then one finds an explicit *square root* of $-1$ mod $p$, when $p \equiv 1 \bmod 4$. Indeed, the $(-1)$-partnerships join even representatives with odd representatives, mod $p$.

Figure 8.6: A diagram of all $(-1)$-partnerships, modulo 11. The even subproduct is the product of all green numbers, and the odd subproduct is the product of all blue numbers.

Consider two subproducts within the product defining $(p-1)!$,

the even subproduct: $E = 2 \times 4 \times 6 \times \cdots \times (p-3) \times (p-1)$;
the odd subproduct: $O = 1 \times 3 \times 5 \times \cdots \times (p-4) \times (p-2)$.

On one hand, we find that

$$E \times O = (p-1)! \equiv -1 \bmod p.$$

On the other hand, multiplying each even representative by $-1$ yields

$$O = E \times (-1)^{\frac{p-1}{2}}.$$

Putting these together, we find that

$$E^2 \equiv (-1)^{\frac{p-1}{2}} \cdot (-1) \bmod p.$$

Thus if $p \equiv 1 \bmod 4$, then $E^2 \equiv -1 \bmod p$. So the even subproduct is a square root of $-1$ modulo $p$.

Here we use reciprocity for $-1$ to prove a famous theorem.

**Theorem 8.7 (Fermat's Christmas Theorem)** *Let $p$ be a prime number with $p \equiv 1 \bmod 4$. Then the Diophantine equation $x^2 + y^2 = p$ has a solution. In other words, $p$ can be expressed as the sum of two squares.*

This theorem is sometimes called Fermat's Christmas Theorem, since Fermat announced it in a letter to Mersenne on December 25, 1640.

Later (Theorem 10.15) we give a proof using quadratic forms. Here we use a "geometry of numbers" method due to Minkowski.

**Proposition 8.8 (Minkowski's theorem in the plane)** *Consider a grid of parallelograms in the plane, with the origin at a grid-point, and a circle centered at the origin. If the area of the circle is greater than 4 times the area of a parallelogram, then the circle contains a grid-point besides the origin.*

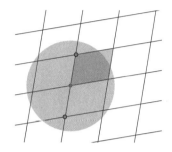

PROOF: Use the grid-lines through the origin to dissect the circle into four pieces. Move these four pieces to the four interior corners of a parallelogram twice as large as the original, as shown below.

Figure 8.7: The area of the circle is larger than four times the area of the parallelogram. Hence the circle contains a grid-point besides the origin (highlighted in green).

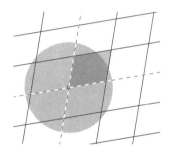 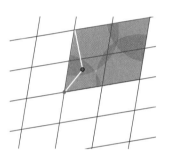

Since the area of the circle is larger than the area of the doubled parallelogram, there must be an overlap among the four pieces of circle. Place a dot at the overlap.

In coordinates,
$$P = Q + 2\lambda,$$
for some point $\lambda$ in the grid. Thus
$$Q' = -Q = 2\lambda - P.$$
The midpoint $M$ can then be located,
$$M = \frac{P + Q'}{2} = \lambda.$$

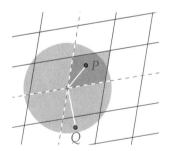 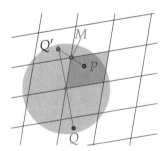

The dot corresponds to (at least) two points, $P$ and $Q$, in the original circle. Let $Q'$ be the point in the circle opposite to $Q$. Then the midpoint $M$ of $\overline{PQ'}$ is the point we seek. On one hand, as the midpoint of a line segment within the circle, $M$ too lies within the circle. On the other hand, $M$ must be a grid-point, by the algebraic argument in the margin. ∎

Note that we used only two geometric properties of the circle – that it was centrally symmetric (opposites of points are again in the circle) and convex (line segments with endpoints in the circle are fully contained in the circle). Minkowski's theorem extends to general regions with those two properties. Formally speaking, we also used measurability, but this follows from convexity.

Fermat's Christmas Theorem follows from clever choices.

Proof: Let $p$ be a prime number with $p \equiv 1 \bmod 4$. Let $u$ be an integer satisfying $u^2 \equiv -1 \bmod p$ (by reciprocity for $-1$). Consider the set of points $(x,y)$ such that $x \equiv uy \bmod p$. This set of points is the set of corners of a grid of parallelograms,[2] among which one has corners at
$$(0,0), \quad (u,1), \quad (p,0), \quad (p+u,1).$$
Each parallelogram has area $p$. Place a circle of area $4p$ at the origin.

[2] Indeed, every solution $(x,y)$ to $x \equiv uy \bmod p$ can be expressed uniquely as
$$(x,y) = s(u,1) + t(p,0)$$
for integers $s$ and $t$. Explicitly, if $x \equiv uy \bmod p$, then set $s = y$ and
$$t = \frac{x - yu}{p}.$$

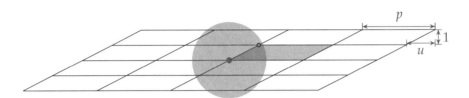

Minkowski's Theorem (Proposition 8.8) implies that the circle contains a grid-point $(x,y)$ besides the origin.[3] By construction, this point $(x,y)$ satisfies two conditions,[4]
$$x^2 + y^2 \leq \frac{4}{\pi} p \text{ and } x \equiv uy \bmod p.$$
Since $u^2 \equiv -1 \bmod p$, we find $x^2 \equiv -y^2 \bmod p$, and so
$$x^2 + y^2 \leq \frac{4}{\pi} p \text{ and } x^2 + y^2 \text{ is a multiple of } p.$$
As $(x,y)$ is not the origin, $x^2 + y^2 > 0$. Since $1 < 4/\pi < 2$,
$$0 < x^2 + y^2 < 2p \text{ and } x^2 + y^2 \text{ is a multiple of } p.$$
But the only multiple of $p$ strictly between $0$ and $2p$ is $p$. Hence
$$x^2 + y^2 = p. \qquad\blacksquare$$

[3] Strictly speaking, we should make the circle have area greater than $4p$ to apply the theorem. But at area equal to $4p$, we are safe if we allow the possibility that a grid-point lies on the edge of the circle.

[4] If the area of a circle is $4p$, then its squared radius satisfies $\pi r^2 = 4p$ and so $r^2 = 4p/\pi$.

Fermat's Christmas Theorem was first proven by Euler. Euler sketched an argument in a letter[5] to Goldbach, in 1747; this argument, with a slight (and acknowledged) gap, was published by Euler in 1758,[6] and a full proof appeared a bit later in 1760.[7]

A proof by Lagrange, using binary quadratic forms, was simplified by Gauss in Art. 182 of the *Disquisitiones Arithmeticae*. We will encounter Lagrange's proof in a later chapter.

Minkowski's Theorem is fundamental in the "Geometry of Numbers."[8] The proof of Fermat's Christmas Theorem, given here and using Minkowski's theorem, is influenced by notes of Matthew Baker and Peter L. Clark. It can also be found in §III.7.2 of the text of J.W.S. Cassells, "An Introduction to the Geometry of Numbers" (first published, 1959).

[5] The text of the letter is available as OO829 at the Euler Archive, http://eulerarchive.maa.org/correspondence/letters/000829.pdf.

[6] See "De numeris, qui sunt aggregata duorum quadratorum", in *Novi Commentarii academiae scientiarum Petropolitanae* 4 (1758), available as E228 at the Euler Archive http://eulerarchive.maa.org/.

[7] See "Demonstratio theorematis Fermatiani omnem numerum primum formae $4n+1$ esse summam duorum quadratorum", in *Novi Commentarii academiae scientiarum Petropolitanae* 5 (1760), available as E241 at the Euler Archive http://eulerarchive.maa.org/.

[8] Cf. H. Minkowski, "Geometrie der Zahlen" published by B.G. Teubner, 1896.

THE LEGENDRE SYMBOL encodes the squareness of numbers, modulo prime numbers. As a definition, we take the following.

**Definition 8.9** Let $p$ be an odd prime number. Let $a$ be an integer. Define the **Legendre symbol** by [9]

$$\left(\frac{a}{p}\right) = \begin{cases} 1 & \text{if } a \text{ is a nonzero square } \mod p, \\ 0 & \text{if } a \equiv 0 \mod p, \\ -1 & \text{if } a \text{ is nonsquare } \mod p. \end{cases}$$

[9] The pronunciation "$a$ on $p$" was suggested was suggested by Pete L. Clark on MathOverflow, Feb. 16, 2010. Others say "$a$ over $p$".

Euler's criterion gives a formula for the Legendre symbol.[10]

$$\left(\frac{a}{p}\right) \equiv a^{\frac{p-1}{2}} \mod p.$$

[10] When $p$ is odd, $0, 1, -1$ are distinct mod $p$, so knowing the Legendre symbol mod $p$ suffices.

From this, the Legendre symbol is multiplicative in its "numerator".

$$\left(\frac{ab}{p}\right) \equiv (ab)^{\frac{p-1}{2}} = a^{\frac{p-1}{2}} b^{\frac{p-1}{2}} \equiv \left(\frac{a}{p}\right) \cdot \left(\frac{b}{p}\right) \mod p.$$

Consequently the product of two nonsquares is a square, mod $p$. For example, 3 and 5 are nonsquare, mod 7, and we find that $3 \cdot 5 \equiv 1$, and 1 is certainly square, mod 7.

We have little knowledge of the Legendre symbol; we have only one reciprocity law so far:

$$\left(\frac{-1}{p}\right) = \begin{cases} 1 & \text{if } p \equiv 1 \mod 4, \\ -1 & \text{if } p \equiv 3 \mod 4. \end{cases}$$

Repartitioning $(p-1)!$ yields a reciprocity law for $\left(\frac{2}{p}\right)$.

**Theorem 8.10 (Reciprocity for 2)** *When $p$ is an odd prime number,*

$$\left(\frac{2}{p}\right) = \begin{cases} 1 & \text{if } p \equiv 1 \text{ or } 7 \mod 8, \\ -1 & \text{if } p \equiv 3 \text{ or } 5 \mod 8. \end{cases}$$

PROOF: We relate three subproducts[11] of $(p-1)!$.

[11] Two of these were used in the analysis of reciprocity for $-1$.

the half subproduct: $H = 1 \times 2 \times 3 \times \cdots \times \frac{p-3}{2} \times \frac{p-1}{2}$;

the even subproduct: $E = 2 \times 4 \times 6 \times \cdots \times (p-3) \times (p-1)$;

the odd subproduct: $O = 1 \times 3 \times 5 \times \cdots \times (p-4) \times (p-2)$.

Doubling each term of the half subproduct yields each term of the even subproduct. By Euler's criterion, we find

$$\left(\frac{2}{p}\right) \cdot H = 2^{\frac{p-1}{2}} \cdot H = E. \tag{8.1}$$

Next we consider the effect of multiplication by $-1$, mod $p$. In the figures below, even terms are green, odd are blue, and the half subproduct is shaded above.

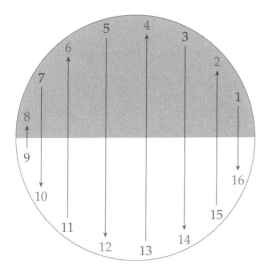
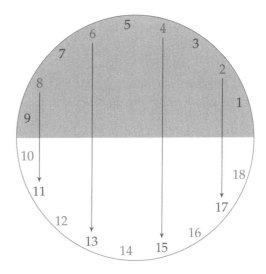

If we multiply every odd term by $-1$, the odd subproduct becomes the even subproduct. Therefore,

$$(-1)^{\frac{p-1}{2}} \cdot O = E. \tag{8.2}$$

Figure 8.8: On the left, each odd element of $\Phi(17)$ is multiplied by $-1$, mod 17. On the right, each even element of $\Phi(19)$ in the half subproduct is multiplied by $-1$, mod 19.

Finally, if we multiply every *even* term within the half subproduct by $-1$, the half subproduct becomes the odd subproduct. Therefore,

$$(-1)^{\lfloor \frac{p-1}{4} \rfloor} \cdot H = O. \tag{8.3}$$

The exponent arises from enumerating the even numbers between 1 and $(p-1)/2$. The equations (8.1), (8.2), (8.3) together give us

$$\left(\frac{2}{p}\right) = (-1)^{\frac{p-1}{2}} \cdot (-1)^{\lfloor \frac{p-1}{4} \rfloor}.$$

Every odd prime can be expressed as $8k+1$, $8k+3$, $8k+5$, or $8k+7$, for some integer $k$. The results are tabulated below.

| $p$ | $\frac{p-1}{2}$ | $\lfloor \frac{p-1}{4} \rfloor$ | $\left(\frac{2}{p}\right)$ |
|---|---|---|---|
| $8k+1$ | $4k$ | $2k$ | $1$ |
| $8k+3$ | $4k+1$ | $2k$ | $-1$ |
| $8k+5$ | $4k+2$ | $2k+1$ | $-1$ |
| $8k+7$ | $4k+3$ | $2k+1$ | $1$ |

This verifies the claimed formula for $\left(\frac{2}{p}\right)$. ∎

Reciprocity laws allow us to quickly determine $\left(\frac{-1}{p}\right)$ and $\left(\frac{2}{p}\right)$, and thus to determine whether $-1$ or $2$ is a square mod $p$. For more general Legendre symbols such as $\left(\frac{3}{p}\right)$, $\left(\frac{5}{p}\right)$, etc., we will use a *dynamic* interpretation of the Legendre symbol studied by Zolotarev. This requires a long digression[12] into the study of permutations.

[12] Or a short excursion, since the theory of permutations is a grand expanse of mathematics.

PERMUTATIONS are often introduced as reorderings. For example, we could reorder 123456 to obtain 362145. We can encode this reordering in a function $f$, by defining

$$f(1) = 3, \quad f(2) = 6, \quad f(3) = 2, \quad f(4) = 1, \quad f(5) = 4, \quad f(6) = 5.$$

In other words, $f(a) = b$ whenever the permutation puts $b$ in the place where $a$ had been. We can also envision this information by placing an arrow $a \to b$ whenever $f(a) = b$. The result is the **cycle diagram** displayed in the margin.

In this way, we can consider permutations as functions from a set to itself. In the example above, the permutation is a function from $\{1, 2, 3, 4, 5, 6\}$ to itself.

**Definition 8.11** A **permutation** of a set $S$ is a bijection[13] from $S$ to $S$.

If $f$ is a function from a finite set $S$ to itself, then we may draw a diagram with elements of $S$ as nodes, and arrows pointing from $x$ to $y$ whenever $f(x) = y$. For the function $f$ to be a permutation, i.e., for $f$ to be a bijection, the following condition must be satisfied: every node must be the target of exactly one arrow.

The diagram of a permutation of a finite set must be composed of cycles. For if we begin at a node $s$, and walk along arrows, we must fall into a repetition. If we never returned to $s$, we would find a $\rho$-shape as in the margin, violating the condition of bijectivity. Thus every node $s$ belongs to a cycle.

A **transposition** is a special sort of permutation – one which exchanges two elements of a set and fixes the rest. For example, consider the transposition of $\{1, 2, 3, 4, 5, 6, 7\}$ which switches $2$ and $5$. It is the function $f$ for which $f(1) = 1$, $f(3) = 3$, $f(4) = 4$, $f(6) = 6$, $f(7) = 7$, and for which $f(2) = 5$ and $f(5) = 2$. Its cycle diagram is displayed below.

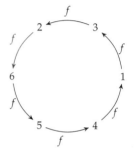

Figure 8.9: The red arrow from 5 to 4 displays the data $f(5) = 4$.

[13] In other words, a one-to-one and onto function.

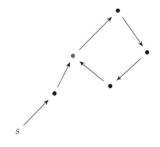

Figure 8.10: Observe that the red node is the target of two arrows. Thus such a diagram cannot occur in a permutation.

The **identity permutation** is the permutation which fixes every element. It is not so interesting, but it is a good starting point.

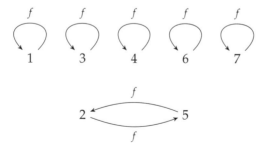

PERMUTATIONS CAN BE COMPOSED by composing functions, when the set remains fixed. If we consider permutations as actions performed on a finite set, then composition means the performance of one action followed by another. We demonstrate this with two permutations $f$ and $g$ of the set $\{1,2,3,4,5,6\}$. Define

$$f(1) = 4, \quad f(2) = 3, \quad f(3) = 1, \quad f(4) = 5, \quad f(5) = 6, \quad f(6) = 2;$$
$$g(1) = 1, \quad g(2) = 5, \quad g(3) = 3, \quad g(4) = 4, \quad g(5) = 2, \quad g(6) = 6.$$

The permutation "$f$ then $g$" is typically denoted $g \circ f$ when considering the composition of functions. To compute this composition, one nests the output of $f$ in the input of $g$. For example, $[g \circ f](1) = g(f(1)) = g(4) = 4$. In this way, we compute $h = f \circ g$,

$$h(1) = 4, \quad h(2) = 3, \quad h(3) = 1, \quad h(4) = 2, \quad h(5) = 6, \quad h(6) = 5.$$

The cycle diagram of $h = g \circ f$ (i.e., "$f$ then $g$") can be obtained by following the arrows in the cycle diagrams of $f$ and $g$.

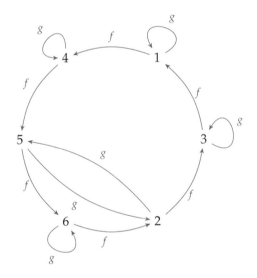

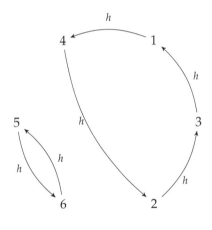

Figure 8.11: To see the permutation $h$ – the effect of $f$ then $g$ – begin at any number in the diagram. Travel along one red arrow (via $f$) then along one blue arrow (via $g$). The result of the two-step process is called $h$, and is displayed in the black arrows. Composition with $g$ shortcuts the 6-cycle, cleaving it into a 4-cycle and a 2-cycle.

**Proposition 8.12 (Permutations are generated by transpositions)**
*Every permutation may be constructed by composing transpositions.*

PROOF: An example should suffice to describe the construction. To permute 123456 to 362145, carry out five transpositions:

- Switch 3 and 1: 123456 becomes 321456.
- Switch 6 and 2: 321456 becomes 361452.
- Switch 2 and 1: 361452 becomes 362451.
- Switch 1 and 4: 362451 becomes 362154.
- Switch 4 and 5: 362154 becomes 362145.

Thus 5 transpositions suffice to put 5 numbers in their correct place. The last number is in the only remaining place, where it belongs. ■

Cycles of a permutation are the loops that arise in a cycle diagram. Suppose that $f$ is a permutation of a finite set $S$, and $g$ is a transposition of $S$ which switches two elements $a$ and $b$. We compare the cycle diagrams of $f$ and of $g \circ f$ below, in two cases.

**If $a$ and $b$ belong to the same cycle** of length $\ell$ in the cycle diagram of $f$, restrict attention to that cycle. All other cycles of $f$ are unchanged in the cycle diagram of $g \circ f$. The result of $h = g \circ f$ is displayed below.

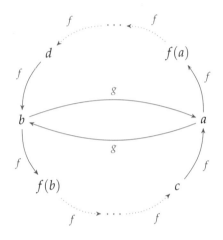 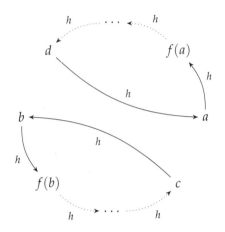

Figure 8.12: The cycle diagram of $f$ on the left, and the cycle diagram of $h = g \circ f$ on the right. We have given the name $d$ for the element satisfying $f(d) = b$, and the name $c$ for the element satisfying $f(c) = a$.

The single cycle containing $a$ and $b$, in the cycle diagram of $f$, is cleft into two cycles in the cycle diagram of $h$. The length of the single cycle in $f$ is the sum of the lengths of the two cycles of $h$.

**If $a$ and $b$ belong to distinct cycles** in the cycle diagram of $f$, we restrict our attention to the two cycles containing $a$ and $b$. All other cycles of $f$ are unchanged in the cycle diagram of $g \circ f$. The result of $h = g \circ f$ is displayed below.

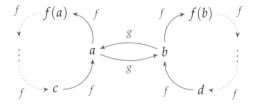 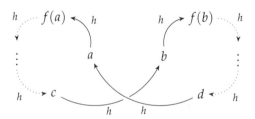

The distinct cycles containing $a$ and $b$, in the cycle diagram of $f$, are merged into one cycle in the cycle diagram of $h$. The length of the new cycle in $h$ is the sum of the lengths of the two cycles in the cycle diagram of $f$.

Figure 8.13: The cycle diagram of $f$ on the left, and the cycle diagram of $h = g \circ f$ on the right. We have given the name $d$ for the element satisfying $f(d) = b$, and the name $c$ for the element satisfying $f(c) = a$.

THE SIGN of a permutation is a number, 1 or −1, depending on the lengths of its cycles. It is defined by two properties.

- A cycle of odd length has **sign** 1, and a cycle of even length has sign −1. In other words, a cycle of length $\ell$ has sign $(-1)^{\ell+1}$.
- The **sign of a permutation** is the product of the signs of its cycles.

When we compute the sign of a permutation, we place a $+$ or a $-$ within each cycle, according to whether its sign is 1 or −1. We see quickly that the sign of a transposition is −1. The geometric results on the previous page yield the following lemma.

**Lemma 8.13** *Let $f$ be a permutation of a finite set $S$. Let $g$ be a transposition of the set $S$. Then*
$$\operatorname{sgn}(g \circ f) = -\operatorname{sgn}(f).$$

PROOF: The cycle diagrams of $g \circ f$ and $f$ are the same, except that two cycles in one permutation correspond to a single cycle in the other. Let $x$ and $y$ be the lengths of the two cycles. As two cycles, they have sign $(-1)^{x+1} \cdot (-1)^{y+1} = (-1)^{x+y+2}$. Merged into one cycle of length $x + y$, the sign is $(-1)^{x+y+1}$. Observe that
$$(-1)^{x+y+2} = -(-1)^{x+y+1}.$$

Thus the sign of $g \circ f$ is opposite the sign of $f$. ∎

**Theorem 8.14 (The transposition interpretation of sgn)** *Suppose that $f$ is a permutation of a finite set $S$. If $f$ can be constructed by composing $n$ transpositions, then*
$$\operatorname{sgn}(f) = (-1)^n.$$

PROOF: The identity permutation of $S$ has sign 1. Indeed, its cycle diagram has only cycles of length 1, each with sign 1. Composing the identity permutation of $S$ with $n$ transpositions leads to the permutation $f$, and so Lemma 8.13 gives
$$\operatorname{sgn}(f) = (-1)^n \cdot 1 = (-1)^n.$$
∎

**Corollary 8.15 (Composition formula for sgn)** *Let $f$ and $g$ be permutations of a finite set $S$. Then*
$$\operatorname{sgn}(g \circ f) = \operatorname{sgn}(g) \cdot \operatorname{sgn}(f).$$

PROOF: All permutations may be constructed from a series of transpositions. If $g$ can be constructed from $m$ transpositions and $f$ from $n$ transpositions, then $g \circ f$ can be constructed from $m + n$ transpositions. Hence
$$\operatorname{sgn}(g \circ f) = (-1)^{m+n} = (-1)^m \cdot (-1)^n = \operatorname{sgn}(g) \cdot \operatorname{sgn}(f).$$
∎

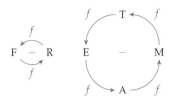

Figure 8.14: Here $f$ is a permutation of {F,E,R,M,A,T}. As a reordering, it sends FERMAT to RAFTME. Since both cycles have even length, its sign is
$$\operatorname{sgn}(f) = (-1) \cdot (-1) = 1.$$

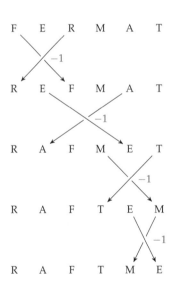

Figure 8.15: The permutation $f$ can be obtained by composing four transpositions. Therefore its sign is
$$\operatorname{sgn}(f) = (-1)^4 = 1.$$
This agrees with the result computed from cycle lengths above.

THREE INTERPRETATIONS of the sign of a permutation will be needed in what follows. We have seen two so far: one related to lengths of cycles and another related to transpositions. A third interpretation of the sign of a permutation relates to *inversions*. It applies only when permuting a set which is ordered from the beginning, e.g., a set of numbers $\{1, 2, \ldots, n\}$.

**Definition 8.16** Let $S$ be a set of numbers and $f$ a permutation of $S$. An **inversion** of $f$ is an ordered pair $(s, t)$ in $S$ such that $s < t$ and $f(s) > f(t)$.

Thus an inversion of a permutation is a pair of numbers which become "out-of-order" after applying the permutation. Write $\mathrm{Inv}(f)$ for the number of inversions of a permutation.

**Problem 8.17** Compute the number of inversions in the permutation $f$ of $\{1, 2, 3, 4, 5, 6\}$ given by

$$f(1) = 4, \quad f(2) = 3, \quad f(3) = 1, \quad f(4) = 5, \quad f(5) = 6, \quad f(6) = 2.$$

SOLUTION: We display all "$<$" relations among $\{1, 2, 3, 4, 5, 6\}$ on the left in the figure below. On the right, we apply the permutation $f$, and highlight pairs where the inequality reverses.

| | | | | | | | | | | |
|---|---|---|---|---|---|---|---|---|---|---|
| $1 < 6$ | $2 < 6$ | $3 < 6$ | $4 < 6$ | $5 < 6$ | | $4 > 2$ | $3 > 2$ | $1 < 2$ | $5 > 2$ | $6 > 2$ |
| $1 < 5$ | $2 < 5$ | $3 < 5$ | $4 < 5$ | | | $4 < 6$ | $3 < 6$ | $1 < 6$ | $5 < 6$ | |
| $1 < 4$ | $2 < 4$ | $3 < 4$ | | | | $4 < 5$ | $3 < 5$ | $1 < 5$ | | |
| $1 < 3$ | $2 < 3$ | | | | | $4 > 1$ | $3 > 1$ | | | |
| $1 < 2$ | | | | | | $4 > 3$ | | | | |

We count 7 inversions in the table above: $\mathrm{Inv}(f) = 7$. ✓

Figure 8.16: The original inequalities on the left, and the output of $f$ on the right. We color the output red if the inequality is reversed.

COUNTING INVERSIONS is a reasonable way of describing how "out-of-order" a permutation is. If $S$ consists of the numbers between $1$ and $n$, then there are $\frac{1}{2}n(n-1)$ inequalities among the numbers in $S$. This is the maximal number of inversions – the permutation which sends $1, 2, \ldots, n$ to $n, (n-1), \ldots, 1$ reverses every inequality.

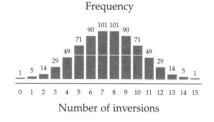

Figure 8.17: The distribution of $\mathrm{Inv}(f)$, as $f$ varies over the 720 permutations of $\{1, 2, 3, 4, 5, 6\}$, so $0 \leq \mathrm{Inv}(f) \leq 15$.

INVERSIONS can be connected to the sign of a permutation, by decomposing a permutation into transpositions again. More specifically, an **adjacent transposition** is a transposition which switches consecutive numbers, e.g. a transposition that switches 1 and 2, or 2 and 3, etc.. Every transposition can be constructed by a series of adjacent transpositions. For example, to switch 1 and 4, each row in the table in the margin is obtained by adjacent transposition from the row above it.

Inversions are increased or decreased by one, when an adjacent transposition is applied.

**Lemma 8.18** *Let $f$ be a permutation of $\{1, 2, \ldots, n\}$. Let $g$ be an adjacent transposition of $S$. Then*

$$\mathrm{Inv}(g \circ f) = \mathrm{Inv}(f) \pm 1.$$

PROOF: Let $a$ and $a + 1$ be the numbers transposed by $g$. Let $u$ be the number for which $f(u) = a$. Let $v$ be the number for which $f(v) = a + 1$.

Let $h = g \circ f$. Except for $u$ and $v$, the inversions of $f$ are identical to the inversions of $h = g \circ f$. But $f(u) < f(v)$ while $h(u) > h(v)$, as shown in the margin. If $u < v$ then $\mathrm{Inv}(h) = \mathrm{Inv}(f) + 1$, and if $u > v$ then $\mathrm{Inv}(h) = \mathrm{Inv}(f) - 1$. ∎

A third interpretation of the sign of a permutation arises.

**Theorem 8.19 (The inversion interpretation of sgn)** *Let $f$ be a permutation of a finite set $S$ of numbers. Then*

$$\mathrm{sgn}(f) = (-1)^{\mathrm{Inv}(f)}.$$

PROOF: The identity permutation of $S$ has sign 1. It is possible to construct $f$ by composing the identity permutation with some number $n$ of adjacent transpositions. By Theorem 8.14, $\mathrm{sgn}(f) = (-1)^n$. On the other hand, the number of inversions changes by 1 for each adjacent transposition. Therefore, throughout the construction of $f$, the number of adjacent transpositions is even or odd according to whether the number of inversions is even or odd. Hence $n$ and $\mathrm{Inv}(f)$ are both even or both odd. Thus

$$\mathrm{sgn}(f) = (-1)^n \cdot 1 = (-1)^{\mathrm{Inv}(f)}.$$
∎

RETURNING TO NUMBER THEORY, we will study the signs of permutations that arise in modular arithmetic. The reader can find all three interpretations of the sign used in the next few pages.

| 1 | 2 | 3 | 4 |
| 1 | 2 | 4 | 3 |
| 1 | 4 | 2 | 3 |
| 4 | 1 | 2 | 3 |
| 4 | 2 | 1 | 3 |
| 4 | 2 | 3 | 1 |

Table 8.1: Move 4 to the first place by a series of adjacent transpositions. Then move 2 to the second place, then 3 to the third place. The last number falls into the last remaining place.

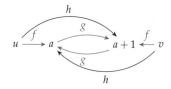

Figure 8.18: Since $a < a + 1$, we find that $f(u) < f(v)$. Since $a + 1 > a$, we find that $h(u) > h(v)$.

LET $p$ BE AN ODD PRIME. On this page and the next, we consider permutations of the set $S = \{0, 1, \ldots, p-1\}$. If $a$ is an integer, let add($a$ mod $p$) be the permutation of $S$ obtained by addition of $a$, mod $p$. Its sign is a straightforward computation.

**Proposition 8.20** *For every integer $a$, sgn(add($a$ mod $p$)) = 1.*

PROOF: If $a \equiv 0$ mod $p$, then add($a$ mod $p$) is the identity permutation, which has sign 1. Otherwise, Proposition 6.2 implies that addition by $a$ mod $p$ yields a full-length cycle (since GCD($a,p$) = 1). Since $p$ is odd, a cycle of length $p$ has sign 1. ∎

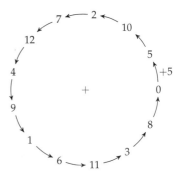

Figure 8.19: Addition of 5 mod 13 yields a full cycle of length 13, which is a permutation of sign 1.

Now let mult($a$ mod $p$) be the permutation of $S$ given by multiplication by $a$, mod $p$. In 1872, Zolotarev[14] linked this sign to the Legendre symbol.

[14] See "Nouvelle démonstration de la loi de réciprocité de Legendre," by Zolotareff, in *Nouvelles Annales de Mathématiques* 2$^e$ série, tome 11 (1872).

**Lemma 8.21 (Zolotarev's Lemma)** *Let $p$ be an odd prime number, and suppose GCD($a,p$) = 1. Then $\left(\frac{a}{p}\right)$ equals the sign of mult($a$ mod $p$).*

PROOF: Multiplication by $a$ sends 0 to 0, forming a cycle of length 1. This doesn't affect the sign, so we focus on the remaining permutation of $\Phi(p) = \{1, \ldots, p-1\}$ in what follows. Let $\ell$ be the cycle length, and $c$ the number of cycles, in the dynamics of multiplication by $a$, modulo $p$. From Lemma 6.9, all cycles have the same length, and so $\ell \cdot c = p - 1$. Therefore,

$$\text{sgn}(\text{mult}(a \bmod p)) = \left((-1)^{\ell+1}\right)^c. \tag{8.4}$$

**If $c$ is even**, the sign above is 1, regardless of $\ell$. Since the cycle length is $\ell$, we have $a^\ell \equiv 1$ mod $p$ and

$$\left(\frac{a}{p}\right) \equiv a^{\frac{p-1}{2}} = (a^\ell)^{c/2} \equiv 1^{c/2} = 1 = \text{sgn}(\text{mult}(a \bmod p)).$$

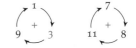

**If $c$ is odd**, then as $\ell c = p - 1$ is even, $\ell$ must be even.

$$\text{sgn}(\text{mult}(a \bmod p)) = \left((-1)^{\ell+1}\right)^c = (-1)^c = -1.$$

Next consider the number $b$ in $\Phi(p)$ satisfying

$$b \equiv a^{\ell/2} \bmod p.$$

Since $\ell$ is the cycle length and $\ell/2 < \ell$, we find that

$$b^2 \equiv 1 \text{ and } b \not\equiv 1 \bmod p.$$

By Proposition 6.21, $b \equiv -1$ mod $p$. Furthermore,

$$\left(\frac{a}{p}\right) \equiv a^{\frac{p-1}{2}} = (a^{\ell/2})^c = b^c = (-1)^c = -1.$$

So when $c$ is odd, we also find

$$\left(\frac{a}{p}\right) = \text{sgn}(\text{mult}(a \bmod p)).$$

∎

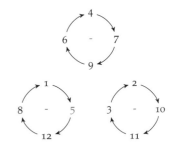

Figure 8.20: Illustration of Zolotarev's Lemma. Modulo 13, multiplication by 3 yields four cycles of length three, giving sign $(+1)^4 = 1$. Multiplication by 5 yields three cycles of length four, giving sign $(-1)^3 = -1$.

LET $p$ AND $q$ BE DISTINCT ODD PRIMES. Quadratic reciprocity relates two questions: is $p$ a square modulo $q$? is $q$ a square modulo $p$? It was first proven by Gauss, not once but six times in his published work and twice in his unpublished notes. Quadratic reciprocity is the first step in a path that leads to cubic and quartic reciprocity, to Hilbert's general reciprocity, to class field theory, to the Langlands[15] program today. Reciprocity laws are more than formulae – they have driven research for over two centuries.

Reciprocity laws are deep by nature – their proofs never come easy. The proof we give is no exception. It adapts an argument by Zolotarev, to uncover quadratic reciprocity in the signs of permutations. With no further ado,

[15] The Langlands program originates in the visionary ideas of Robert Langlands in the late 1960s. For a survey, see Stephen Gelbart's "An elementary introduction to the Langlands program" in the *Bulletin of the American Mathematical Society*, vol. 10 **2** (1984)

**Theorem 8.22 (Quadratic Reciprocity)** *If $p$ and $q$ are distinct odd primes, then*

$$\left(\frac{p}{q}\right)\left(\frac{q}{p}\right) = (-1)^{\frac{p-1}{2}\cdot\frac{q-1}{2}}.$$

PROOF: We lay out our plan of attack first. Define $S$ to be the set of numbers between $0$ and $pq - 1$. There will be three permutations of $S$, called $\alpha$, $\beta$, and $\gamma$. We will compute their signs to find

$$\operatorname{sgn}(\alpha) = \left(\frac{q}{p}\right) \text{ and } \operatorname{sgn}(\beta) = \left(\frac{p}{q}\right);$$

$$\operatorname{sgn}(\gamma) = (-1)^{\frac{p-1}{2}\cdot\frac{q-1}{2}}.$$

Finally, we will see that $\gamma \circ \alpha = \beta$. This implies $\operatorname{sgn}(\gamma) \cdot \operatorname{sgn}(\alpha) = \operatorname{sgn}(\beta)$. This gives[16] quadratic reciprocity, in the form

$$\operatorname{sgn}(\beta) \cdot \operatorname{sgn}(\alpha) = \operatorname{sgn}(\gamma).$$

To begin, place the numbers of $S$ onto $pq$ pigeons.

[16] If $u, v, w$ are all 1 or $-1$, then the equalities

$$uvw = 1, uv = w, uw = v, vw = u$$

are all equivalent to each other.

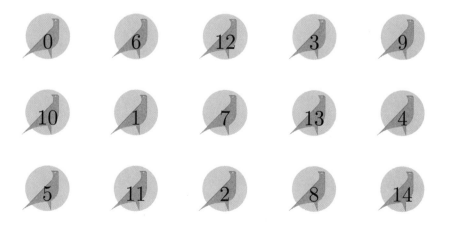

The elements of $S = \{0, 1, \ldots, pq - 1\}$ can be described using three "bracket notations".[17]

**Definition 8.23** Let $a$ be a number between 0 and $p - 1$. Let $b$ be a number between 0 and $q - 1$.

- Let $[a, b]$ be the unique element of $S$ congruent to $a \bmod p$ and $b \bmod q$.
- Let $\langle a, b]$ be the number $aq + b$.
- Let $[a, b\rangle$ be the number $a + bp$.

Every element of $S$ can be uniquely expressed in each bracket notation. We place the number labeled $[a, b]$ at column $a$ and row $b$, and we supply all three bracket notations for each below.

[17] The bracket notation $[a, b]$ is familiar from the Chinese remainder theorem. The other bracket notations are compatible, in the sense that $\langle a, b] = aq + b$ is congruent to $b$, modulo $q$. The right-bracket next to $b$ suggests the compatibility:

$$\langle a, b] = [a, b] \bmod q, \quad [a, b\rangle = [a, b] \bmod p.$$

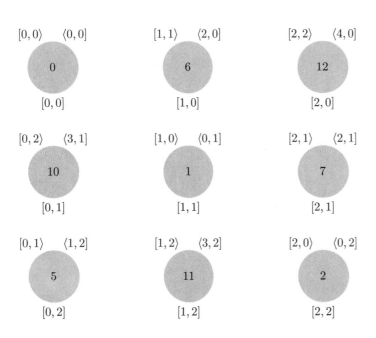

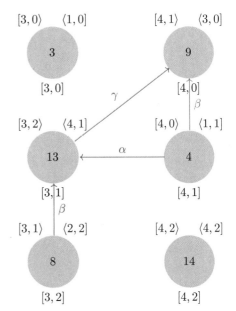

**Definition 8.24** Define three permutations $\alpha$, $\beta$, $\gamma$ of $S$:

- Let $\alpha$ be the permutation which sends the number labeled $[a, b]$ to that labeled $\langle a, b]$.
- Let $\beta$ be the permutation which sends the number labeled $[a, b]$ to that labeled $[a, b\rangle$.
- Let $\gamma$ be the permutation which sends the number labeled $\langle a, b]$ to that labeled $[a, b\rangle$.

As strange as these permutations might be, observe that

$$\gamma \circ \alpha = \beta.$$

Figure 8.21: In this figure, $p = 5$ and $q = 3$. The numbers have been arranged into $p$ columns and $q$ rows, as determined by the Chinese remainder theorem. Our permutations scramble the birds. For example, $\alpha(4) = 13$, because 4 is labeled $[4, 1]$ and 13 is labeled $\langle 4, 1]$.

**Lemma 8.25** *The signs of $\alpha$ and $\beta$ are given by*

$$\text{sgn}(\alpha) = \left(\frac{q}{p}\right) \text{ and } \text{sgn}(\beta) = \left(\frac{p}{q}\right).$$

PROOF: It suffices by symmetry[18] to prove the result for $\text{sgn}(\alpha)$. Arranging our numbers according to the Chinese remainder theorem, the row-number of $x$ is the representative[19] of $x$ modulo $q$, and the column-number of $x$ is the representative of $x$ modulo $p$.

Observe that $\alpha$ permutes each row independently. Indeed, $\alpha$ sends $\langle a, b\rangle$ to $[a, b]$, preserving the row-number $b$. The column-number of $[a, b]$ is $a$, by how we arrange our numbers. Since $\langle a, b\rangle$ labels $aq + b$, the column number of $\langle a, b\rangle$ is the representative of $aq + b$ mod $p$.

[18] Switch $p$ and $q$, rows and columns, etc., and the proof for $\alpha$ becomes the proof for $\beta$.

[19] We use natural representatives

$$0, \ldots, p - 1 \text{ and } 0, \ldots, q - 1.$$

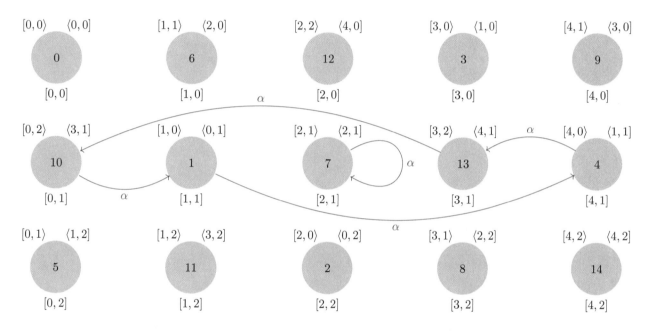

Figure 8.22: In this figure, $p = 5$ and $q = 3$. We highlight the effect of the permutation $\alpha$ on row number 1. Within this row, the cycle diagram of $\alpha$ consists of one cycle of length 4, and one cycle of length 1.

Within row $b$, the permutation $\alpha$ sends column $a$ to column $aq + b$ modulo $p$. The function sending $a$ to $aq + b \pmod{p}$ can be described as "multiply by $q$, then add $b$" modulo $p$. Proposition 8.20 and Zolotarev's Lemma 8.21 allow us to compute.

$$\text{sgn}\,(\text{add}(b \bmod p) \circ \text{mult}(q \bmod p)) = \text{sgn}(\text{add}(b \bmod p)) \cdot \text{sgn}(\text{mult}(q \bmod p)) = 1 \cdot \left(\frac{q}{p}\right) = \left(\frac{q}{p}\right).$$

The sign of $\alpha$ is the product of the signs obtained from each row. The number of rows equals $q$, an odd number. So $(\pm 1)^q = \pm 1$ and

$$\text{sgn}(\alpha) = \left(\frac{q}{p}\right)^q = \left(\frac{q}{p}\right).$$

∎

We are left to compute the sign of the third permutation $\gamma$. Recall that $\gamma$ is the permutation which sends the number labeled $\langle a,b]$ to the number labeled $[a,b\rangle$. Translating these bracket notations, we can compute the sign of $\gamma$.

**Lemma 8.26** *Let $\gamma$ be the permutation of $\{0,\ldots,pq-1\}$ given by*

$$\gamma(aq+b) = a+pb$$

*whenever $0 \le a \le p-1$ and $0 \le b \le q-1$. Then*

$$\mathrm{sgn}(\gamma) = (-1)^{\frac{p-1}{2} \cdot \frac{q-1}{2}}.$$

PROOF: We count the inversions of $\gamma$ to determine its sign. For this, consider two elements $x, x'$ in our set $\{0,\ldots,pq-1\}$. Dividing each by $q$ with remainder yields quotients $a, a'$ and remainders $b, b'$:

$$x = aq+b \text{ and } x' = a'q+b'.$$

Since $0 \le b, b' \le q-1$, we find that

$$x < x' \text{ if and only if } (a < a') \text{ or } (a = a' \text{ and } b < b'). \tag{8.5}$$

Now we compare $\gamma(x)$ and $\gamma(x')$. We have

$$\gamma(x) = a+bp \text{ and } \gamma(x') = a'+b'p.$$

Since $0 \le a, a' \le p-1$, we find that

$$\gamma(x) > \gamma(x') \text{ if and only if } (b > b') \text{ or } (b = b' \text{ and } a > a'). \tag{8.6}$$

An inversion occurs in $\gamma$ for each time that (8.5) and (8.6) hold simultaneously. A few cases are eliminated, giving the following.

$$(x < x' \text{ and } \gamma(x) > \gamma(x')) \text{ if and only if } (a < a' \text{ and } b > b'). \tag{8.7}$$

We are left to enumerate the possible quadruples $(a, a', b, b')$ satisfying $0 \le a < a' \le p-1$ and $0 \le b' < b \le q-1$. This gives $\frac{1}{2}p(p-1)$ possible pairs $(a, a')$ and $\frac{1}{2}q(q-1)$ pairs $(b, b')$, which can be chosen independently. Therefore,

$$\mathrm{Inv}(\gamma) = \frac{p(p-1)}{2} \cdot \frac{q(q-1)}{2}.$$

Since $p$ and $q$ are odd, we compute

$$\mathrm{sgn}(\gamma) = (-1)^{\mathrm{Inv}(\gamma)} = (-1)^{\frac{p(p-1)}{2} \cdot \frac{q(q-1)}{2}} = ((-1)^{pq})^{\frac{p-1}{2} \cdot \frac{q-1}{2}} = (-1)^{\frac{p-1}{2} \cdot \frac{q-1}{2}}.$$

∎

| $x$ | = | $aq+b$ |
|---|---|---|
| 0 | = | $0q+0$ |
| 1 | = | $0q+1$ |
| 2 | = | $0q+2$ |
| 3 | = | $1q+0$ |
| 4 | = | $1q+1$ |
| 5 | = | $1q+2$ |
| 6 | = | $2q+0$ |
| 7 | = | $2q+1$ |
| 8 | = | $2q+2$ |
| 9 | = | $3q+0$ |
| 10 | = | $3q+1$ |
| 11 | = | $3q+2$ |
| 12 | = | $4q+0$ |
| 13 | = | $4q+1$ |
| 14 | = | $4q+2$ |

Table 8.2: Here as before, $q=3$. If $x = aq+b$ and $x' = a'q+b'$, then $x < x'$ if (1) $a < a'$ or (2) $a = a'$ and $b < b'$.

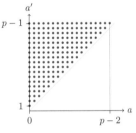

Figure 8.23: The pairs $(a, a')$ with $0 \le a < a' \le p-1$ correspond to dots in the triangle above (for $p = 19$). They can be enumerated by the techniques of Chapter 0.

Since $\gamma \circ \alpha = \beta$, we arrive at **quadratic reciprocity**.

If $p$ and $q$ are distinct odd primes, then

$$\left(\frac{p}{q}\right)\left(\frac{q}{p}\right) = (-1)^{\frac{p-1}{2} \cdot \frac{q-1}{2}}.$$

∎

The right side's bark is worse than its bite. As soon as $(p-1)/2$ or $(q-1)/2$ is even, the right side is 1. So we may rephrase the above formula as below.

$$\left(\frac{p}{q}\right)\left(\frac{q}{p}\right) = \begin{cases} -1 & \text{if } p \equiv 3 \text{ and } q \equiv 3 \bmod 4; \\ 1 & \text{otherwise.} \end{cases}$$

Consider two questions again: is $p$ a square modulo $q$? is $q$ a square modulo $p$? These questions about squares in different modular worlds have the same answer, unless both $p$ and $q$ are congruent to 3 modulo 4. When one question is easy, the other becomes easy.

**Problem 8.27** Is 5 a square modulo 101?

SOLUTION: By quadratic reciprocity, $\left(\frac{5}{101}\right) = \left(\frac{101}{5}\right)$, so we can ask instead whether 101 is a square modulo 5. But $101 \equiv 1 \bmod 5$, and 1 is as square as it gets. Therefore 5 is a square[20] modulo 101. ✓

**Problem 8.28** Is 7 a square modulo 43?

SOLUTION: By quadratic reciprocity, $\left(\frac{7}{43}\right) = -\left(\frac{43}{7}\right)$. But $42 \equiv 1 \bmod 7$, and so $\left(\frac{43}{7}\right) = \left(\frac{1}{7}\right) = 1$. Hence $\left(\frac{7}{43}\right) = -1$, and 7 is not a square modulo 43. ✓

[20] Unfortunately, quadratic reciprocity doesn't answer the question – the square of **what**? In this case,

$$5 \equiv 45^2 \equiv 56^2 \bmod 101.$$

But these were found by a different technique.

QUADRATIC RECIPROCITY allows us to answer **all** questions of the form "Is $a$ a square modulo $p$?" when $p$ is an odd prime. By lifting (Corollary 7.21), this knowledge suffices to answer all questions of the form "Is $a$ a square modulo $p^e$?" when $p$ is an odd prime, $e \geq 1$, and $\text{GCD}(a, p) = 1$. By lifting and shifting (Corollary 7.23), we can also answer all questions of the form "Is $a$ a square modulo $2^e$?" by examining squareness modulo 2, 4, or 8.

According to Corollary 7.6, the Chinese remainder theorem permits us to answer all questions of the form "Is $a$ a square modulo $m$?" when $\text{GCD}(a, m) = 1$. Just consider the prime decomposition $m = 2^{e_2} 3^{e_3} 5^{e_5} \cdots$, and answer the questions "Is $a$ a square modulo $2^{e_2}$?" and "Is $a$ a square modulo $3^{e_3}$?" etc. Squareness modulo $m$ is equivalent to squareness modulo all prime powers $p^{e_p}$.

**Problem 8.29** Is 77 a square modulo 101?

SOLUTION: We compute the Legendre symbol in a series of steps.

$$\left(\frac{77}{101}\right) = \left(\frac{7}{101}\right) \cdot \left(\frac{11}{101}\right),$$
$$= \left(\frac{101}{7}\right) \cdot \left(\frac{101}{11}\right) \text{ by quadratic reciprocity,}$$
$$= \left(\frac{3}{7}\right) \cdot \left(\frac{2}{11}\right) \text{ by reducing mod 7 and mod 11,}$$
$$= (-1) \cdot (-1) \text{ by observation and by reciprocity for 2,}$$
$$= 1.$$

Hence 77 is a square modulo 101. ✓

**Problem 8.30** For which prime numbers $p$ is 3 a square modulo $p$?

SOLUTION: 3 is a square modulo 2 and modulo 3. Beyond these special cases, quadratic reciprocity tells us that

$$\left(\frac{3}{p}\right) = \left(\frac{p}{3}\right) \cdot \begin{cases} 1 & \text{if } p \equiv 1 \bmod 4; \\ -1 & \text{if } p \equiv 3 \bmod 4. \end{cases}$$

Note that $\left(\frac{p}{3}\right)$ is 1 or $-1$ depending on whether $p \equiv 1 \bmod 3$ or $p \equiv 2 \bmod 3$. Putting these congruences together via the Chinese remainder theorem, we find that

$$\left(\frac{3}{p}\right) = \begin{cases} 1 & \text{if } p \equiv 1 \text{ or } 11 \bmod 12; \\ -1 & \text{if } p \equiv 5 \text{ or } 7 \bmod 12. \end{cases}$$

✓

The chance that 3 is a square modulo $p$ is asymptotically 50%, as $p$ ranges over all prime numbers. But Chebyshev's bias appears again – in large ranges of primes, we often find a slightly higher probability that 3 is nonsquare rather than square, mod $p$.

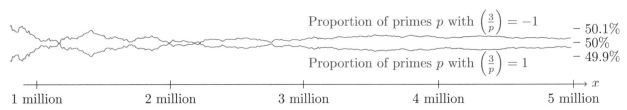

Figure 8.24: Proportion of primes $p$ for which $\left(\frac{3}{p}\right) = \pm 1$, among all primes $p$ with $1 < p \leq x$.

As a final application of quadratic reciprocity, we prove a series of infinite results about prime numbers.[21] We begin with a lemma, whose proof is a clever adaptation of Euclid's proof of the infinite of primes.

**Lemma 8.31** *Let $f(x) = a_0 + a_1 x + \cdots + a_d x^d$ be a nonconstant polynomial with integer coefficients. Then among the prime factors of $f(n)$ (for all integers $n$), there are infinitely many primes.*

PROOF: If $a_0 = 0$, then every prime divides $f(0) = 0$, and the lemma is proven. So assume that $a_0 \neq 0$ in what follows.

Let $p_1, \ldots, p_t$ be a list of primes, dividing $f(n_1), f(n_2), \ldots, f(n_t)$, respectively. Let $P = p_1 \cdots p_t$ be their product. Then

$$\frac{1}{a_0} \cdot f(ka_0 P) = \frac{1}{a_0} \cdot \left( a_0 + a_1 k a_0 P + a_2 k^2 a_0^2 P^2 + \cdots + a_d k^d a_0^d P^d \right),$$
$$= 1 + a_1 k P + a_2 k^2 a_0 P^2 + \cdots + a_d k^d a_0^{d-1} P^d.$$

We make a few observations about the right side. First, any prime divisor of the right side is not a divisor of $P$ (or else it would be a divisor of 1). Second, as a function of $k$, it is a nonconstant polynomial (since $d > 0$ and $a_d, a_0, P \neq 0$). Hence when $k$ is large, the right side does have a prime divisor (it cannot always be $-1$ or 1 for all values of $k$). Third, any divisor of the right side is a divisor of $f(ka_0 P)$. Putting these observations together, we find that there must exist a prime number $p$, not within the original list, which divides $f(n)$ for some integer $n$. ∎

**Proposition 8.32** *Let $a$ be a nonzero integer. There are infinitely many primes $p$ such that $\left( \frac{a}{p} \right) = 1$.*

PROOF: Let $f(x) = x^2 - a$. A prime $p$ is a factor of $f(n)$ if and only if $n^2 - a \equiv 0 \bmod p$, if and only if $n^2 \equiv a \bmod p$. Hence a prime $p$ divides $f(n)$ (for some $n$) if and only if $\left( \frac{a}{p} \right) = 1$ or $\left( \frac{a}{p} \right) = 0$. Since $\left( \frac{a}{p} \right) = 0$ for only finitely many primes $p$ (those dividing $a$), the lemma completes the proof. ∎

**Corollary 8.33** *There are infinitely many primes $p$ congruent to 1 mod 4.*

PROOF: Apply the proposition to $a = -1$ and use reciprocity. ∎

**Corollary 8.34** *There are infinitely many primes congruent to 1 mod 3.*

PROOF: Apply the proposition to $a = -3$ and use reciprocity. ∎

These corollaries demonstrate how quadratic reciprocity can be a transformative tool; by transforming a statement modulo $p$ (e.g., $f(n) \equiv 0 \bmod p$) to a statement about $p$ modulo something else (e.g., $p \equiv 1 \bmod 4$), one may quickly prove infinitely many statements about the infinite of primes.

[21] I learned the following approach from a post by Qiaochu Yuan at the Mathematics Stack Exchange. See "Uses of quadratic reciprocity theorem," URL (version: 2010-11-17): http://math.stackexchange.com/q/10716.

For example, let $f(x) = x^2 + 1$. Below we tabulate the prime factors of $f(n)$ for integers $n$; it suffices to consider positive $n$ since $f(x) = f(-x)$ and $f(0) = 1$.

| $n$ | $n^2 + 1$ | Prime factors |
|---|---|---|
| 1 | 2 | 2 |
| 2 | 5 | 5 |
| 3 | 10 | 2, 5 |
| 4 | 17 | 17 |
| 5 | 26 | 2, 13 |
| 6 | 37 | 37 |
| 7 | 50 | 2, 5 |
| 8 | 65 | 5, 13 |
| 9 | 82 | 41, 2 |
| 10 | 101 | 101 |

According to the lemma, infinitely many primes will appear in the rightmost column. In this case, all odd primes appearing in the rightmost column will be congruent to 1 mod 4, proving the infinitude of such primes.

| $a$ | $\left( \frac{a}{p} \right) = 1$ iff |
|---|---|
| $-3$ | $p \equiv 1 \bmod 3$ |
| $-2$ | $p \equiv 1$ or $3 \bmod 8$ |
| $-1$ | $p \equiv 1 \bmod 4$ |
| 1 | All $p$ |
| 2 | $p \equiv 1$ or $7 \bmod 8$ |
| 3 | $p \equiv 1$ or $11 \bmod 12$ |

Table 8.3: Each row is an application of quadratic reciprocity. For example, $\left( \frac{-3}{p} \right) = 1$ if and only if $p \equiv 1 \bmod 3$. Our method here is not strong enough to isolate a single residue in each row; for example, we cannot prove that there are infinitely many primes congruent to 1 mod 8. Such results require Dirichlet's Theorem, whose proof is beyond the scope of this text.

## Historical notes

The history of reciprocity laws is a thick thread in the history of number theory, at least from Fermat to the present day. Franz Lemmermeyer has assembled a bibliography of 1099 articles related to reciprocity laws,[22] and his book "Reciprocity Laws: from Euler to Eisenstein"[23] covers the time period most relevant here.

By 1751, Euler was directly studying quadratic residues. His table below separates the (nonzero) residues into an equal number of squares and nonsquares.[24]

$$3 \left\{ \begin{matrix} 1 \\ 2 \end{matrix} \right\}; \quad 5 \left\{ \begin{matrix} 1, 4 \\ 2, 3 \end{matrix} \right\}; \quad 7 \left\{ \begin{matrix} 1, 4, 2 \\ 3, 5, 6 \end{matrix} \right\}; \quad 11 \left\{ \begin{matrix} 1, 4, 9, 5, 3 \\ 2, 6, 7, 8, 10 \end{matrix} \right\};$$

$$13 \left\{ \begin{matrix} 1, 4, 9, 3, 12, 10 \\ 5, 6, 7, 8, 11 \end{matrix} \right\}; \quad 17 \left\{ \begin{matrix} 1, 4, 9, 16, 8, 2, 15, 13 \\ 3, 5, 6, 7, 10, 11, 12, 14 \end{matrix} \right\};$$

$$19 \left\{ \begin{matrix} 1,4,9,16,6,17,11,7,5 \\ 2,3,8,10,12,13,14,15,18 \end{matrix} \right\}; \quad 23 \left\{ \begin{matrix} 1,4,9,16,2,13,3,18,12,8,6 \\ 5,7,10,11,14,15,17,19,20,21,22 \end{matrix} \right\};$$

$$29 \left\{ \begin{matrix} 1,4,9,16,25,7,20,6,23,13,5,28,24,27 \\ 2,3,8,10,11,12,14,15,17,18,19,21,26,27 \end{matrix} \right\};$$

In Theorem 4 of the same paper, Euler proves that exactly half of the numbers between 1 and $p - 1$ are quadratic residues. Later in the paper (see §35. Scholion), Euler observes that a quadratic residue $a$ mod $p$ satisfies $a^{(p-1)/2} \equiv 1 \bmod p$ (in Gauss's notation), and proves this criterion for squareness. By 1772, Euler presents the theory of quadratic residues as review,[25] including reciprocity for $-1$. At the end of this work, we find the table below.

| diuisor numerus primus formae | tum est |
|---|---|
| $4ns + a$ | $+s$. refiduum et $-s$ refiduum |
| $4ns - a$ | $+s$ refiduum et $-s$ non-refiduum |
| $4ns + \mathfrak{A}$ | $+s$ non-refiduum et $-s$ non-refiduum |
| $4ns - \mathfrak{A}$ | $+s$ non-refiduum et $-s$ refiduum. |

While organized differently, the results of this table are equivalent to those of quadratic reciprocity. But this was empirical truth – Euler did not prove quadratic reciprocity.

In his "Essai sur la théorie des nombres" (1797), Legendre introduces the symbol that now bears his name.

> Comme les quantités analogues à $N^{\frac{c-1}{2}}$ se rencontreront fréquemment dans le cours de nos recherches, nous emploierons le caractère abrégé $\left(\frac{N}{c}\right)$ pour exprimer le reste que donne $N^{\frac{c-1}{2}}$ divisé par $c$; reste qui, suivant ce qu'on vient de voir, ne peut être que $+1$ ou $-1$.[26]

On p.196, Legendre finds the formula for $\left(\frac{2}{p}\right)$, and on p.214, we find the first statement of quadratic reciprocity in the form we present here. After a case-by-case treatment, Legendre writes,

---

[22] As of 2016, see http://www.rzuser.uni-heidelberg.de/~hb3/frecbib.html

[23] Published in Springer Monographs in Mathematics, 2000.

[24] From "Demonstratio theorematis Fermatiani omnem numerum sive integrum sive fractum esse summam quatuor paucioremve quadratorum," presented in 1751, and published in *Novi Commentarii academiae scientiarum Petropolitanae* 5 (1760). Available as E242 at the Euler Archive, http://eulerarchive.maa.org/.

[25] See "Observationes circa divisionem quadratorum per numeros primos", presented in 1772, published in *Opuscula varii argumenti* 1 (1783). Available as E552 at the Euler Archive, http://eulerarchive.maa.org/. Theorem IV and V contain the results about $-1$.

Figure 8.25: The notation is the following: $s$ is an odd prime, and an auxiliary (unnamed odd) prime we will call $p$. The left column describes four cases: (1) $p$ is a square mod $s$, and $p \equiv 1 \bmod 4$; (2) $-p$ is a square mod $s$ and $p \equiv 3 \bmod 4$; (3) $p$ is a nonsquare mod $s$ and $p \equiv 1 \bmod 4$; (4) $-p$ is a nonsquare mod $s$ and $p \equiv 3 \bmod 4$. The right column describes whether $\pm s$ is a square mod $p$ according to the four cases above.

[26] See Part II of Legendre's "Essai," §1, p.186, for his "Explication du caractère abrégé $\left(\frac{N}{c}\right)$".

Ces deux cas généraux sont compris dans la formule

$$\left(\frac{n}{m}\right) = (-1)^{\frac{m-1}{2} \cdot \frac{n-1}{2}} \cdot \left(\frac{m}{n}\right).$$

Legendre proves quadratic reciprocity, but assumes a fact not yet proven in his time – the infinitude of primes in arithmetic progressions.[27] With Dirichlet's proof of this fact in 1837, Legendre's proof was completed.

Reciprocity drove Gauss to write the *Disquisitiones*. He writes,

> What happened was this. Engaged in other work I chanced on an extraordinary arithmetic truth (if I am not mistaken, it was the theorem of Art. 108). Since I considered it so beautiful in itself and since I suspected its connection with even more profound results, I concentrated on it all my efforts in order to understand the principles on which it depended and to obtain a rigorous proof.[28]

Gauss refers to his Art. 108, which states

> $-1$ is a quadratic residue of all prime numbers of the form $4n+1$ and a nonresidue of all prime numbers of the form $4n+3$.[29]

Reciprocity for $-1$, in Art. 108, is followed by reciprocity for $\pm 2$ in Art. 112–116, for $\pm 3$ in Art. 117–120, for $\pm 5$ in Art. 121–123, for $\pm 7$ in Art. 124. A general treatment of quadratic reciprocity begins in Art. 130. Gauss proves the theorem in Art. 135. He calls quadratic reciprocity the *fundamental theorem*, when writing the following,

> The fundamental theorem must certainly be regarded as one of the most elegant of its type.

Gauss, recognizing the incompleteness of Legendre's proof at the time, was the first to prove quadratic reciprocity. After his first proof (Art. 135), Gauss went on to give a second proof in the *Disquisitiones* (Art. 262), and published four more proofs later in life.[30]

We have followed Zolotarev's proof of quadratic reciprocity (1872),[31] as it fits well with the *dynamics* of modular arithmetic. A disadvantage is that Zolotarev's proof seems not to generalize beyond the quadratic case.

Other proofs (e.g., using *Gauss sums*) generalize to cubic (1844, Eisenstein) and quartic (1828,1832, Gauss) reciprocity. For example, if $p,q$ are Eisenstein primes, and $p,q \equiv \pm 2 \bmod 3$, then

$$\left(\frac{p}{q}\right)_3 = \left(\frac{q}{p}\right)_3,$$

where $\left(\frac{a}{b}\right)_3$ denotes a cubic analogue of the Legendre symbol.[32] These reciprocity laws are the first steps in two centuries of research, leading through Hilbert's general *power reciprocity* (1897, in Hilbert's *Zahlbericht*), Artin's reciprocity (1929) and class field theory, and continuing today in the Langlands program.

---

[27] Legendre requires an auxiliary prime $A$ satisfying congruence conditions; he writes: "Or il est facile de s'assurer qu'il y aura toujours une infinité de nombres premiers qui satisferont à ces conditions." (p.219).

[28] From the Author's Preface to the *Disquisitiones*. Translation by Arthur A. Clarke.

[29] Art. 108, loc. cit.

[30] Lemmemeyer has compiled a chronology of 246 proofs of quadratic reciprocity at http://www.rzuser.uni-heidelberg.de/~hb3/fchrono.html (in 2016); eight are attributed directly to Gauss, though only six were published during his life.

[31] See G. Zolotarev, "Nouvelle démonstration de la loi de réciprocité de Legendre," in *Nouv. Ann. Math* vol. 2 **11** (1872).

[32] Cubic reciprocity describes when one prime is a cube modulo another, and quartic reciprocity when one prime is a fourth power modulo another. The cubic symbol can be defined by

$$\left(\frac{p}{q}\right)_3 = p^{\frac{qq-1}{3}} \bmod q.$$

## Exercises

1. Compute the following Legendre symbols, and interpret them in terms of squareness.
$$\left(\frac{3}{7}\right), \left(\frac{2}{37}\right), \left(\frac{-5}{29}\right), \left(\frac{105}{101}\right), \left(\frac{23}{37}\right), \left(\frac{62}{71}\right).$$

2. Is 41 a square modulo 1 000 000?

3. Sketch the cycle diagram and compute the sign for the permutation which sends $1, 2, 3, 4, 5, 6, 7, 8, 9, 10$ to $1, 10, 2, 9, 3, 8, 4, 7, 5, 6$.

4. Find a number $N$ such that $0 \leq N \leq 76$, and $N \equiv [1, 6] \bmod [7, 11]$. Describe $N$ in the other two bracket notations, in the form $\langle a, b]$ and as $[c, d\rangle$ using the same primes 7 and 11.

5. Challenge: Let $a$ be a positive integer, $p$ an odd prime, $a \not\equiv 0 \bmod p$, and consider the natural representatives of
$$a, 2a, 3a, \ldots, \frac{p-1}{2} \cdot a \bmod p.$$
Let $e$ be the number of these natural representatives which are greater than $(p-1)/2$. Prove that $\left(\frac{a}{p}\right) = (-1)^e$. Hint: adapt the treatment of $\left(\frac{2}{p}\right)$, studying the half-product and negations.

6. Let $p$ be a **Sophie Germain prime** – a prime for which $2p + 1$ is also prime. Suppose moreover that $p \equiv 3 \bmod 4$. Prove that $2p + 1$ is a divisor of $2^p - 1$. Hint: show that 2 is a square mod $2p + 1$ along the way.[33]

7. Let $p$ be an odd prime number. Recall that a primitive root, mod $p$, is an integer $g$ such that $g^{p-1} \equiv 1 \bmod p$, and no smaller power of $g$ is congruent to 1 mod $p$. Some results in this chapter can be proved via the existence of a primitive root (Theorem 6.26).

    (a) Use a primitive root $g$ to demonstrate that $-1$ is a square mod $p$ if and only if $p \equiv 1 \bmod 4$.

    (b) Use a primitive root $g$ to prove Wilson's Theorem. Hint: show that
    $$(p-1)! \equiv g^{1+2+\cdots+(p-2)} \bmod p.$$

    (c) Given a primitive root $g$, and an integer $a$ such that $a \not\equiv 0 \bmod p$, prove that $a$ is a square modulo $p$ if and only if $a = g^e$ for an even number $e$. Use this to prove Euler's criterion: $a$ is a square mod $p$ if and only if $a^{(p-1)/2} \equiv 1 \bmod p$.

8. The following exercises lead to a proof[34] of

[33] Most of this exercise appeared in Chapter 6. But the additional hypothesis $p \equiv 3 \bmod 4$ allows one to conclude that $2p + 1$ is a divisor of $2^p - 1$ and not of $2^p + 1$.

[34] The treatment in this exercise is influenced by notes of Peter L. Clark.

**Theorem 8.35 (Lagrange's Four-Square Theorem)** *If $n$ is a natural number, then $n$ can be expressed as the sum of four squares.*

A lattice $\Lambda$ in 4-space is a set of the form
$$\{(x,y,z,w) \cdot M : x,y,z,w \in \mathbb{Z}\}$$
where $M$ is a 4-by-4 matrix of nonzero determinant. The *covolume* $V$ of $\Lambda$ is defined to be the absolute value of Det $M$.

Let $\Lambda$ be such a lattice, of covolume $V$, and let $S$ be the 4-sphere $x^2 + y^2 + z^2 + w^2 \leq r^2$ of radius $r$.[35] Minkowski's theorem in 4 dimensions states that if
$$\frac{1}{2}\pi^2 r^4 \geq 2^4 \cdot V,$$
then the sphere contains a non-origin point of the lattice.

[35] The volume of a 4-sphere of radius $r$ is $\frac{1}{2}\pi^2 r^4$.

(a) Let $p$ be an odd prime. Prove that there exist integers $r, s$ such that $r^2 + s^2 \equiv -1 \bmod p$. Hint: consider the equivalent congruence $r^2 + 1 \equiv -s^2 \bmod p$. As $r$ and $s$ vary, why must the value on the left be congruent to the value on the right, at least once?

(b) Suppose that $r^2 + s^2 \equiv -1 \bmod p$. Consider the lattice
$$\Lambda = \{(px + rz + sw, py + sz - rw, z, w) : x,y,z,w \in \mathbb{Z}\}.$$
Prove that if $(a,b,c,d) \in \Lambda$, then $a^2 + b^2 + c^2 + d^2 \equiv 0 \bmod p$.

(c) Use Minkowski's theorem to prove that there exist $a, b, c, d$ such that $a^2 + b^2 + c^2 + d^2 = p$.

(d) In a 1748 letter to Goldbach, Euler wrote the text in the margin. Now prove Lagrange's Four-Square Theorem.

"Si $m = aa + bb + cc + dd$ et $n = pp + qq + rr + ss$ erit $mn = A^2 + B^2 + C^2 + D^2$ existente

$A = ap + bq + cr + ds$
$B = aq - bp - cs + dr$
$C = ar + bs - cp - dq$
$D = as - br + cq - dp$."

(From Euler's May 4, 1748 letter to Goldbach. This letter can be found as OO841 at the Euler Archive, http://eulerarchive.maa.org/correspondence/letters/000841.pdf.)

9. If $n = 3^{e_3} 5^{e_5} 7^{e_7} \cdots$ is an **odd** positive integer, and $a$ is an integer, the **Jacobi symbol** $\left(\frac{a}{n}\right)$ is defined by
$$\left(\frac{a}{n}\right) = \left(\frac{a}{3}\right)^{e_3} \cdot \left(\frac{a}{5}\right)^{e_5} \cdot \left(\frac{a}{7}\right)^{e_7} \cdots.$$

Prove the following properties.

Here, the convention is that $0^0 = 1$ whenever it may appear.

(a) If $a \equiv b \bmod n$ then $\left(\frac{a}{n}\right) = \left(\frac{b}{n}\right)$.

(b) If $a, b$ are integers, then $\left(\frac{a}{n}\right)\left(\frac{b}{n}\right) = \left(\frac{ab}{n}\right)$.

(c) If $m, n$ are coprime odd positive integers, then
$$\left(\frac{m}{n}\right) \cdot \left(\frac{n}{m}\right) = (-1)^{\frac{m-1}{2} \cdot \frac{n-1}{2}}.$$

(d) Prove that, if $\gcd(a,n) = 1$, then $\left(\frac{a}{n}\right)$ is the sign of the multiplication-by-$a$-mod-$n$ permutation on the set $\{0, \ldots, n-1\}$.

10. Challenge: suppose that $x$ is an integer, and $x$ is a square modulo every prime. Prove that $x$ is a square integer.

# Part III

# Quadratic Forms

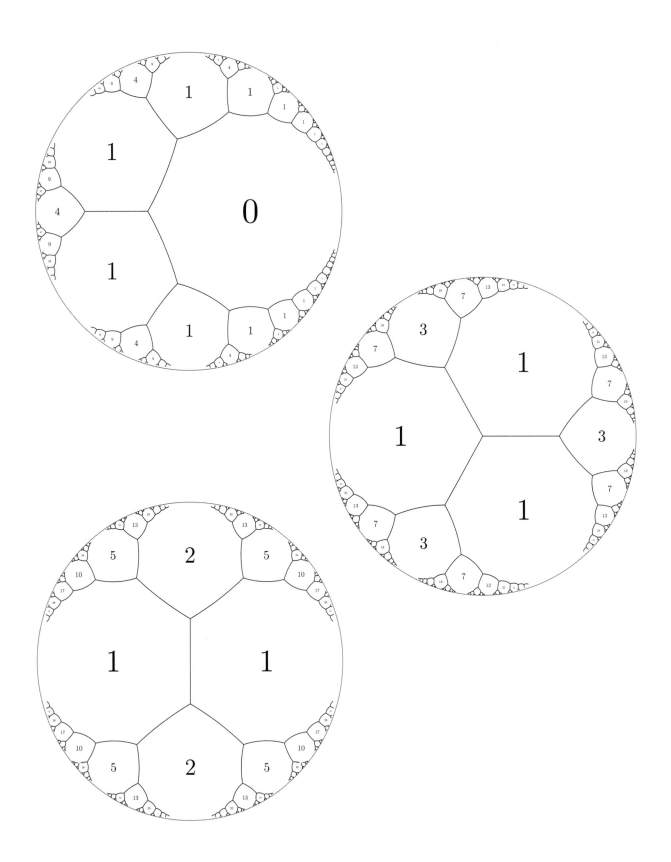

# 9
# The Topograph

In 1733, Leonhard Euler presented his solution to a problem inspired by Diophantus.[1] Euler gave a solution $(x, y)$, both positive integers, for the equation
$$31x^2 + 1 = y^2.$$
Euler's solution, in which $x$ is positive but as small as possible, is
$$x = 273, \quad y = 1520.$$

In the 1990s, John H. Conway discovered a remarkable visual approach to solving such equations. His invention, the *topograph*, is the subject of this and the next two chapters.[2]

To introduce Conway's topograph, we present a two-dimensional Hop and Skip problem.

**Problem 9.1** Imagine you are standing in the Cartesian plane (the usual $xy$-plane). You can hop, moving northeast along the vector $(23, 5)$ or southwest along the vector $(-23, -5)$. You can skip, moving northeast along the vector $(14, 3)$ or southwest along the vector $(-14, -3)$.

[1] See "De solutione problematum diophanteorum per numeros integros," originally in *Commentarii academiae scientarum Petropolitanae* **6** (1738), pp. 175–188, now translated by D. Otero as "On the solution of a problem of Diophantus," in the Euler Archive (http://www.eulerarchive.org) as E29.

[2] John H. Conway first described his topograph in "The (sensual) quadratic form," by J.H. Conway with assistance by F. Fung, Carus Mathematical Monographs **26** (1997). Almost all of the material in this chapter follows from the first chapter of Conway's book.

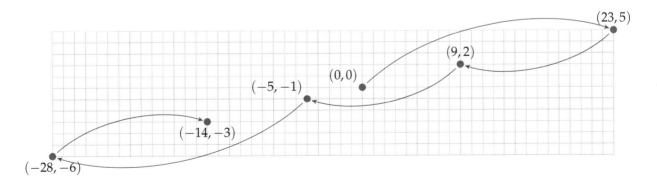

Using only hops and skips, and beginning at $(0, 0)$, where in the plane can you travel?

Figure 9.1: Beginning at zero, you can hop northeast, skip southwest twice, hop southwest, then skip northeast to arrive at $(-14, -3)$.

SOLUTION: If one can hop $\pm(23,5)$ and skip $\pm(14,3)$, then one can combine these to jump $\pm(9,2)$ by hopping northeast and skipping southwest. The jump is our first compound move.

If we can skip $\pm(14,3)$ and jump $\pm(9,2)$, then we can leap $\pm(5,1)$.
If we can jump $\pm(9,2)$ and leap $\pm(5,1)$ then we can bound $\pm(4,1)$.

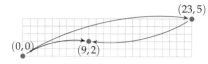

Figure 9.2: A jump is built out of a hop and a skip in the opposite direction.

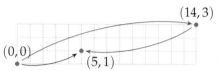
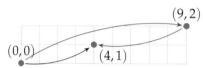

Figure 9.3: A leap is a skip minus a jump. A bound is a jump minus a leap.

If we can leap $\pm(5,1)$ and bound $\pm(4,1)$ then we can step to the right $(1,0)$, and step left $(-1,0)$.

If we can bound $\pm(4,1)$ and step left $(-1,0)$, then we can step up $(0,1)$ and step down $(0,-1)$.

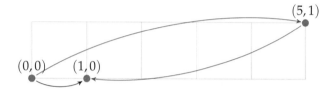

Figure 9.4: A step right is a leap minus a bound. A step up is a bound, and four steps left.

Since we can step right or left by one unit and step up or down by one unit, then we can walk to any point with integer coordinates. ✓

To SOLVE two-dimensional Hop and Skip problems, perform the Euclidean algorithm on the vectors,[3] guided first by the $x$-coordinate, then by the $y$-coordinate if necessary. Beginning with a hop of $(23,5)$ and skip of $(14,3)$, the Euclidean algorithm continues:

$$(23,5) = 1(14,3) + (9,2),$$
$$(14,3) = 1(9,2) + (5,1),$$
$$(9,2) = 1(5,1) + (4,1),$$
$$(5,1) = 1(4,1) + (1,0),$$
$$(4,1) = 4(1,0) + (0,1).$$

The quotients here are dictated by the Euclidean algorithm applied the $x$-coordinates. For example, the first line is dictated by division-with-remainder on the $x$-coordinates: $23 = 1(14) + 9$. Using the quotient 1, the first line reads

$$(23,5) = 1(14,3) + (9,?).$$

The $y$-coordinate is filled in as it must be; $5 = 1(3) + 2$, and so:

$$(23,5) = 1(14,3) + (9,2).$$

[3] We are assuming some familiarity with vectors. To add vectors, simply add $x$- and $y$-coordinates, as in

$$(1,2) + (3,5) = (4,7).$$

To multiply a number and a vector, the result is a vector obtained by distributing the number to each coordinate, as in

$$2(3,5) = (6,10).$$

The steps in the two-dimensional Euclidean algorithm are guided by the following steps in the one-dimensional Euclidean algorithm.

$$23 = 1(14) + 9,$$
$$14 = 1(9) + 5,$$
$$9 = 1(5) + 4,$$
$$5 = 1(4) + 1,$$
$$4 = 4(1) + 0.$$

For vectors, the Euclidean algorithm does not always end well.

**Problem 9.2** Perform the Euclidean algorithm on $(97, 14)$ and $(28, 37)$

SOLUTION: Guided by the $x$-coordinates, we proceed:
$$(97, 14) = 3(28, 37) + (13, -97)$$
$$(28, 37) = 2(13, -97) + (2, 231)$$
$$(13, -97) = 6(2, 231) + (1, -1483)$$
$$(2, 231) = 2(1, -1483) + (0, 3197).$$

The Euclidean algorithm on the $x$-coordinates proceeds below:
$$97 = 3(28) + 13,$$
$$28 = 2(13) + 2,$$
$$13 = 6(2) + 1,$$
$$2 = 2(1) + 0.$$

The final moves are $(1, -1483)$ and $(0, 3197)$. Wherever one can travel with the moves $(97, 14)$ and $(28, 37)$, one can travel with the moves $(1, -1483)$ and $(0, 3197)$. ✓

In the above problem, note that $\text{GCD}(97, 28) = 1$ and $\text{GCD}(14, 37) = 1$. However, one cannot[4] use these moves to travel everywhere in the plane!

[4] Why not? Using the moves $(1, -1483)$ and $(0, 3197)$, one can only travel to $(0, y)$ if $y$ is a multiple of 3197.

In one dimension, the crucial *invariant* throughout the Euclidean algorithm was the greatest common divisor: if $a = q(b) + r$, then
$$\text{GCD}(a, b) = \text{GCD}(b, r).$$

In two dimensions, the crucial invariant is the determinant.

**Definition 9.3** Let $(a, b)$ and $(c, d)$ be vectors. The **determinant** of this pair[5] is the number $ad - bc$. It is written with the notation[6] Det, as in
$$\text{Det}\begin{pmatrix} a & b \\ c & d \end{pmatrix} = ad - bc.$$

[5] The determinant switches sign if the order of the vectors is switched. Notice that
$$\text{Det}\begin{pmatrix} c & d \\ a & b \end{pmatrix} = cb - da$$
$$= -(ad - bc)$$
$$= -\text{Det}\begin{pmatrix} a & b \\ c & d \end{pmatrix}.$$

**Proposition 9.4** *Let $(a, b)$ and $(c, d)$ be vectors of integers. Suppose that $q$ is an integer, and $(r, s)$ is a vector, satisfying*
$$(a, b) = q(c, d) + (r, s).$$
*Then*
$$\text{Det}\begin{pmatrix} a & b \\ c & d \end{pmatrix} = -\text{Det}\begin{pmatrix} c & d \\ r & s \end{pmatrix}.$$

[6] Other texts use a lowercase *det*, and another common notation is a pair of vertical lines, as in
$$\text{Det}\begin{pmatrix} a & b \\ c & d \end{pmatrix} = \begin{vmatrix} a & b \\ c & d \end{vmatrix}.$$
But we think that vertical lines are overloaded enough already.

PROOF: Since $a = qc + r$ and $b = qd + s$, we have
$$ad - bc = (qc + r)d - (qd + s)c = qcd + rd - qdc - sc = -(cs - rd).$$
∎

In the previous problem, the determinant of the pair $(97, 14)$ and $(28, 37)$ is
$$97 \cdot 37 - 14 \cdot 28 = 3197.$$

In accordance with the previous result, the determinant of the "final moves" $(1, -1483)$ and $(0, 3197)$ is the same:
$$1 \cdot 3197 - (-1483) \cdot 0 = 3197.$$

THE DETERMINANT has a geometric interpretation.

**Theorem 9.5 (Area interpretation of the determinant)** *Let $(a,b)$ and $(c,d)$ be nonzero vectors[7]. Suppose that the angle[8] swept out, counterclockwise from $(a,b)$ to $(c,d)$, is less than $180°$. The determinant $ad - bc$ equals the area of the parallelogram with sides given by the vectors $(a,b)$ and $(c,d)$.*

PROOF: We dissect the rectangle of width $a + c$ and height $b + d$.

[7] In this theorem, we may take $(a,b)$ and $(c,d)$ to be vectors of real numbers; by the next page, $a$, $b$, $c$, and $d$ must be integers.

[8] We give a proof which is valid for angles between 0 and $90°$; for angles between $90°$ and $180°$, see the exercises.

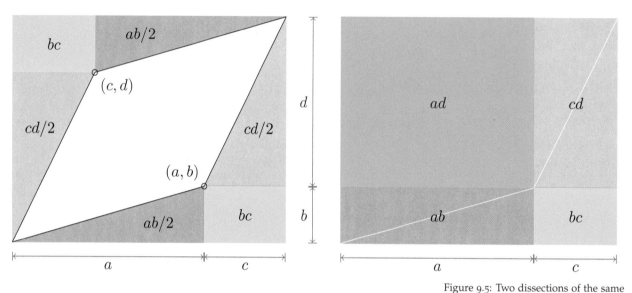

Figure 9.5: Two dissections of the same rectangle, with width $a + c$ and height $b + d$.

The blue triangles on the left add up to the blue rectangle on the right. The green triangles on the left add up to the green rectangle on the right. The yellow rectangles at the southeast corners are equal. Thus the remaining areas, a parallelogram and rectangle on the left, and a rectangle on the right, must be equal.

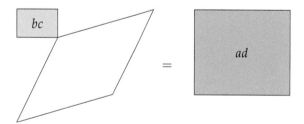

Subtracting $bc$ from both sides demonstrates that the area of the parallelogram equals $ad - bc$. ∎

What happens if $(a,b)$ and $(c,d)$ are vectors lying on the same line?[9] The parallelogram degenerates to a line segment, a figure with area zero. If $(a,b)$ and $(c,d)$ lie on the same line, $\mathrm{Det}\begin{pmatrix} a & b \\ c & d \end{pmatrix} = 0$.

The pictured parallelogram lies in the counterclockwise sweep from $(a,b)$ to $(c,d)$, and our area interpretation of the determinant is valid in this setting. If instead, the parallelogram lies in the clockwise sweep from $(a,b)$ to $(c,d)$, then swap the vectors $(a,b)$ and $(c,d)$, or equivalently, swap the sign of the determinant.

In any case, the *absolute value* of the determinant equals the area of the parallelogram.

[9] Nonzero vectors $(a,b)$ and $(c,d)$ lie on a line if $(a,b) = k \cdot (c,d)$ for some number $k$. In this case,

$$\mathrm{Det}\begin{pmatrix} a & b \\ ka & kb \end{pmatrix} = akb - bka = 0.$$

IMPOSING A GRID, we can compute area by counting dots.

**Theorem 9.6 (Pick's theorem for parallelograms)** *On a unit grid, draw a parallelogram with corners at grid-points. Then the area of the parallelogram equals $I + 1/2E + 1$, where $I$ is the number of interior grid-points, and $E$ the number of grid-points crossed by edges (not including corners).*

To illustrate, we count $I$ and $E$ for the parallelogram shown here. There are 137 grid points in the interior of the parallelogram, so $I = 137$. There is one grid-point each on the top and bottom edges, and five each on the left and right edges. Hence $E = 1 + 1 + 5 + 5 = 12$. The area equals
$$137 + \frac{1}{2} \cdot 12 + 1 = 144.$$

PROOF: We dissect the rectangle of width $a + c$ and height $b + d$, taking care to dissect thickened dots along the way.

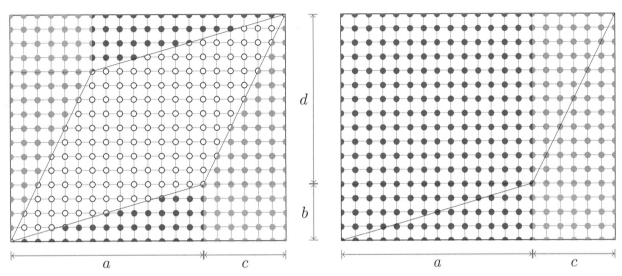

Figure 9.6: The same dissection as the opposite page, now with dots.

The blue and green dots on the left, whole or dissected, can be arranged and stitched[10] to form the blue and green dots on the right. The yellow dots at the southeast are the same, left and right.

The remaining dots – white and yellow on left, and pink on right – have the same number[11]. The number of yellow dots on top-left is:[12]

$$(c-1)(b-1) + (c-1) + (b-1) + 1 = bc.$$

For the same reason, the number of pink dots on right is:

$$(a-1)(d-1) + (a-1) + (d-1) + 1 = ad.$$

Hence the number of white dots equals the number of pink on right minus the number of yellow on left; this number is $ad - bc$, which equals the area of the parallelogram.

The number of white dots equals $I + 1/2E + 1$. Indeed, each edge-point contributes a half white dot, and the corners together contribute one whole white dot.

Thus $I + 1/2E + 1 = ad - bc$, the area of the parallelogram. ∎

[10] By "stitching" a dot, we mean that two half-dots can be assembled to a full dot:

[11] Even if the number of dots were fractional, they are the same number on either side.

[12] To count the dots in a rectangle,

there are $(b-1) \times (c-1)$ full dots in the interior. The top and bottom half-dots stitch to form $c - 1$ full-dots. The left and right half-dots stitch to form $b - 1$ full-dots. The four corners stitch to form one full dot. The total number of dots is

$$(b-1)(c-1) + (b-1) + (c-1) + 1.$$

A BASIS is a pair of moves – a hop and a skip – which suffice for traveling to every grid-point in the plane.

**Definition 9.7** Let $(a,b)$ and $(c,d)$ be vectors of integers. We say that they form a **basis** when hops of $\pm(a,b)$ and skips of $\pm(c,d)$ suffice to travel to every grid-point in the plane.

**Theorem 9.8 (Determinant criterion for bases)** *Let $(a,b)$ and $(c,d)$ be vectors of integers. They form a basis*[13] *if and only if*

$$\mathrm{Det}\begin{pmatrix} a & b \\ c & d \end{pmatrix} = 1 \text{ or } -1.$$

[13] Recall that
$$\mathrm{Det}\begin{pmatrix} a & b \\ c & d \end{pmatrix} = -\mathrm{Det}\begin{pmatrix} c & d \\ a & b \end{pmatrix}.$$
Hence, in this theorem, it does not matter whether we consider $(a,b)$ then $(c,d)$, or the other way around.

PROOF: The grid-points to which one may hop and skip form a grid of parallelograms in the plane. The $I$ points on the interior and $E$ points on the edges of each parallelogram are **inaccessible**. So $(a,b)$ and $(b,c)$ form a basis if and only if $I = 0$ and $E = 0$.

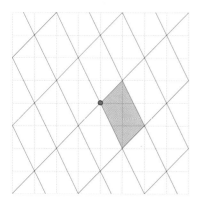
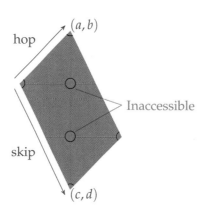

Figure 9.7: A hop of $(1,1)$ and skip of $(-1,2)$ allows one to travel along the red grid of parallelograms. But two grid-points are inaccessible within each parallelogram.

The area of the parallelogram is

$$\text{Area} = \mathrm{Det}\begin{pmatrix} 1 & 1 \\ -1 & 2 \end{pmatrix} = 3.$$

The fact that there are inaccessible points reflects the fact that the area of the parallelogram is greater than 1.

There are two ways to express the area of the parallelogram.

$$\left| \mathrm{Det}\begin{pmatrix} a & b \\ c & d \end{pmatrix} \right| = \text{Area} = I + 1/2\,E + 1.$$

The vectors $(a,b)$ and $(c,d)$ form a basis if and only if $I = 0$ and $E = 0$, if and only if $Area = 1$, if and only if the determinant is $\pm 1$. ∎

A flexible mathematician[14] should comfortably use whichever characterization of "basis" is convenient, when talking about a pair of vectors $(a,b)$ and $(c,d)$. The following are equivalent.

[14] This book favors the geometric and dynamic, while many others fall back onto the algebraic.

*Algebraic:* $ad - bc = \pm 1$.

*Geometric:* The area of the parallelogram swept out by $(a,b)$ and $(c,d)$ is equal to 1.

*Dynamic:* Using hops of $\pm(a,b)$ and skips of $\pm(c,d)$, one can hop and skip to every grid-point.

Not all vectors participate[15] in a basis. For example, $(2,4)$ never participates[16] in a basis, and neither does $(-7,14)$.

**Proposition 9.9** $(a,b)$ *participates in a basis if and only if* $GCD(a,b) = 1$, *i.e., if and only if* $(a,b)$ *is a **primitive vector**.*

PROOF: Given a vector $(a,b)$, finding a collaborator $(c,d)$ for a basis is the same as solving the equation

$$ad - bc = \pm 1.$$

But, letting $x = d$ and $y = -c$, this is the same as solving the linear Diophantine equation

$$ax + by = \pm 1.$$

According to Theorem 1.14, this equation has a solution if and only if $GCD(a,b) = 1$. ∎

Given a vector $(a,b)$, finding a collaborator to form a basis is precisely as difficult as solving a linear Diophantine equation.

**Problem 9.10** Find a basis containing the vector $(17, 12)$.

SOLUTION: We begin by solving the linear Diophantine equation $17x + 12y = 1$; a solution exists, since $GCD(17, 12) = 1$. Carry out the Euclidean algorithm on 17 and 12.

$$17 = 1(12) + 5$$
$$12 = 2(5) + 2$$
$$5 = 2(2) + 1.$$

Solve for the "step" 1 in terms of hops of 17 and skips of 12:

$$1 = 5 - 2(2),$$
$$= 5 - 2(12 - 2(5)) = 5(5) - 2(12),$$
$$= 5(17 - 1(12)) - 2(12) = 5(17) - 7(12).$$

The equality $1 = 5(17) - 7(12)$ allows us to cook up a matrix, whose first row is the vector $(17, 12)$ and whose determinant is 1.

$$\mathrm{Det}\begin{pmatrix} 17 & 12 \\ 7 & 5 \end{pmatrix} = (17)(5) - (12)(7) = 1,$$

and so $(17, 12)$ forms a basis with $(7, 5)$. ✓

Using hops of $(17, 12)$ and skips of $(7, 5)$, one can hop and skip to every point with integer coordinates.

[15] We say that a vector $(a,b)$ participates in a basis if there exists a vector $(c,d)$ such that the pair $(a,b)$ and $(c,d)$ form a basis.

[16] This can be seen geometrically as well. For if one had a parallelogram with one side connecting $(0,0)$ to $(2,4)$, then such a parallelogram would have an inaccessible grid-point on its edge at $(1,2)$.

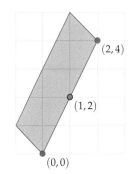

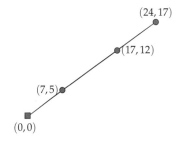

Figure 9.8: The thin parallelogram swept out by $(17, 12)$ and $(7, 5)$ has area 1, the same area as the unit square drawn at $(0,0)$.

To denote a variable vector, we use notation like $\vec{v}$. The negative of a vector is found by negating both coordinates. If $\vec{v} = (2, -3)$ then $-\vec{v} = (-2, 3)$. To add or subtract vectors, add or subtract their $x$- and $y$-coordinates. If $\vec{v} = (1, 2)$ and $\vec{w} = (1, 3)$ then

$$\vec{v} + \vec{w} = (2, 5), \quad \vec{v} - \vec{w} = (0, -1).$$

A nonzero vector $\vec{v}$ can be seen as an arrow pointing from the origin $(0, 0)$ in some direction; likewise, the pair $\{\vec{v}, -\vec{v}\}$ denoted compactly $\pm\vec{v}$, can be seen as two arrows of the same lengths, pointing from the origin $(0, 0)$ in two opposite directions. Such a pair is called a **lax vector**.[17] A lax vector is like a hop; if we can hop $(2, 1)$ then we can hop backwards, along the vector $-(2, 1) = (-2, -1)$. A lax vector is a *bidirectional* vector.

How should one write a lax vector? Should one write $\pm(2, -1)$ or should one write $\pm(-2, 1)$? They are the same as lax vectors. We prefer to make the first entry positive or zero. So we prefer $\pm(2, -1)$ over $\pm(-2, 1)$. When the first entry is zero, we prefer the second entry to be positive. So we prefer $(0, 1)$ over $(0, -1)$.

A **lax basis** is an unordered pair $\{\pm\vec{v}, \pm\vec{w}\}$ of lax vectors, for which any choice of signs yields a basis. The previous problem demonstrates that $\{\pm(17, 12), \pm(7, 5)\}$ is a lax basis.

THE **domain topograph** is a way to visualize all primitive lax vectors, all lax bases, and the connections they share. In the domain topograph, every lax basis is represented by a line segment separating its two lax vectors. For example, the lax basis with lax vectors $(17, 12)$ and $(7, 5)$ is visualized by

$$\frac{\pm(17, 12)}{\pm(7, 5)}$$

Some lax bases, like $\{\pm(17, 12), \pm(7, 5)\}$ are not so easy to discover. But one particular lax basis is so familiar that it deserves a name. **Home basis** is defined to be $\{\pm(1, 0), \pm(0, 1)\}$. Algebraically, we compute $\mathrm{Det}\begin{pmatrix} 1 & 0 \\ 0 & 1 \end{pmatrix} = 1$. Geometrically, we observe that the area of the unit square 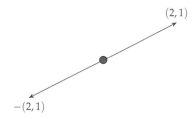 is equal to 1. Dynamically, by hopping left and right one unit, and skipping up and down one unit, one can travel to every grid-point. In the domain topograph, home basis appears as below.

$$\text{Home basis:} \quad \frac{\pm(1, 0)}{\pm(0, 1)}$$

[17] The "lax" terminology is due to Conway. The lax vector $\pm(2, 1)$ is the bidirectional vector including $(2, 1)$ and $(-2, -1)$, displayed below.

In this way, the lax vector $\pm(2, 1)$ is a pair of vectors, pointed in opposite directions. Of course, the zero vector $(0, 0)$ equals its negative; the lax vector $\pm(0, 0)$ does not point in any particular direction.

NEW BASES ARISE FROM OLD, by addition and subtraction.

**Proposition 9.11** *Suppose that $\vec{v}$ and $\vec{w}$ form a basis. Then*

$$\{\vec{v},\vec{v}+\vec{w}\} \text{ is a basis}; \quad \{\vec{v},\vec{v}-\vec{w}\} \text{ is a basis};$$
$$\{\vec{v}+\vec{w},\vec{w}\} \text{ is a basis}; \quad \{\vec{v}-\vec{w},\vec{w}\} \text{ is a basis}.$$

PROOF: If $\vec{v}$ and $\vec{w}$ form a basis, then hops of $\vec{v}$ and skips of $\vec{w}$ suffice to travel everywhere. Thus hops of $\vec{v}$ and leaps of $\vec{v}+\vec{w}$ suffice to travel everywhere (since a skip equals a leap minus a hop). Hence $\{\vec{v},\vec{v}+\vec{w}\}$ forms a basis. The other cases are similar. ∎

Here the dynamic approach is most useful for proving the theorem. The algebraic approach is not difficult but not enlightening. One would have to check that the algebraic condition

$$\text{Det}\begin{pmatrix} a & b \\ c & d \end{pmatrix} = ad - bc = \pm 1$$

implies the algebraic condition

$$\text{Det}\begin{pmatrix} a & b \\ a+c & b+d \end{pmatrix} = \pm 1.$$

This is left as an exercise. The geometric approach, using areas of parallelograms, is more interesting and more difficult.

The domain topograph displays the consequence of the theorem. Every lax basis yields four **adjacent lax bases**, using two new vectors.

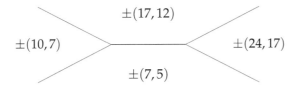

Figure 9.9: In the middle are the original two lax vectors $\pm(17,12)$ and $\pm(7,5)$ which form a lax basis. On the left and right are the lax vectors obtained by adding or subtracting $\pm(17,12)$ and $\pm(7,5)$. Four new lax bases are displayed by green lines.

THE DOMAIN TOPOGRAPH contains an infinitude of lax bases. We apply the previous theorem to branch out and find more lax bases by addition and subtraction. Starting at home basis we expand the domain topograph three times, recursively, to produce no less than 29 lax bases among 16 lax vectors.

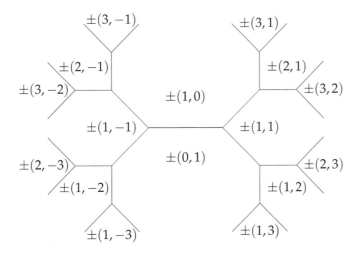

Figure 9.10: Begin at home basis with $\pm(1,0)$ and $\pm(0,1)$. Adding and subtracting yields two new lax vectors, $\pm(1,1)$ and $\pm(1,-1)$ and four new lax bases, in green. But adding and subtracting along the four new bases yields four new lax vectors and eight new lax bases, in blue. One more step yields eight new lax vectors, and sixteen new lax bases, in red. Notice that the determinant is $\pm 1$, looking at any two lax vectors across an edge.

THE DOMAIN TOPOGRAPH contains *all* lax bases. Beginning at home basis, branching, and branching further, one will eventually find *every* lax basis. To see why this is the case, we illustrate how to **walk home**. We begin at[18] the lax basis $\{\pm(17,12), \pm(7,5)\}$ and walk to home basis $\{\pm(1,0), \pm(0,1)\}$. Reversing the path gives directions from home basis to our chosen lax basis.

[18] We use the preposition "at" because now we think of a lax basis as a location – an edge – in the domain topograph.

**Problem 9.12** Find a path along edges in the domain topograph, beginning at the lax basis $\{\pm(17,12), \pm(7,5)\}$, and ending at home basis $\{\pm(1,0), \pm(0,1)\}$.

SOLUTION: Begin by drawing line segments for the lax bases adjacent to $\pm(17,12)$ and $\pm(7,5)$. The numbers seem to get smaller, as one travels southwest. So walk southwest, to the lax basis $\{\pm(10,7), \pm(7,5)\}$ and add edges for the adjacent lax bases.

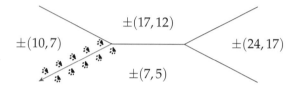

The numbers get smaller as one travels south, so walk south to the lax basis $\{\pm(3,2), \pm(7,5)\}$, and add edges for the adjacent lax bases.

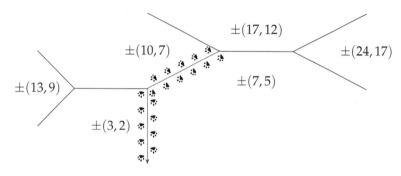

The numbers get smaller as one travels southwest, so walk southwest to $\{\pm(3,2), \pm(4,3)\}$, and add edges for the adjacent lax bases.

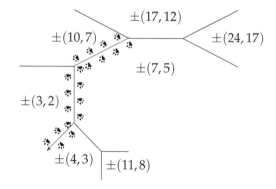

Walk west, and add edges for lax bases adjacent to $\{\pm(3,2), \pm(1,1)\}$.

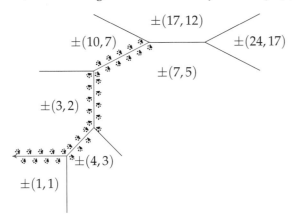

Walk southwest, and add edges for lax bases adjacent to $\{\pm(2,1), \pm(1,1)\}$.

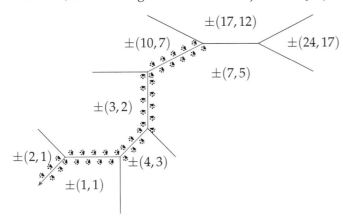

Walk south, and arrive at home basis.

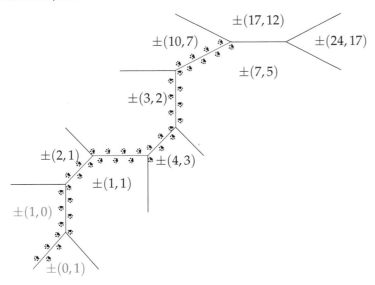

We have walked from $\{\pm(17,12), \pm(7,5)\}$ to home basis. ✓

This process is called *following your nose*. In other words, the goal is to find *home basis*, a basis whose vectors have coordinates as small as possible. Follow your nose by proceeding towards vectors with coordinates as small as possible at all times.

Our ability to walk from any basis to home basis exhibits the connectedness of the domain topograph.

**Lemma 9.13** *If $\{\vec{v}, \vec{w}\}$ is a basis, then the Euclidean algorithm applied to $\vec{v}$ and $\vec{w}$ leads to a basis of the form $\{\pm(1,s), \pm(0,1)\}$.*

PROOF: Suppose that $\vec{v} = (a, b)$ and $\vec{w} = (c, d)$. Since $\{\vec{v}, \vec{w}\}$ is a basis, $ad - bc$ equals 1 or $-1$. Hence $x = d, y = -b$ is a solution to the Diophantine equation $ax + cy = \pm 1$. Thus $\text{GCD}(a, c) = 1$.

Now perform the Euclidean algorithm on $\pm \vec{v}$ and $\pm \vec{w}$, guided by the $x$-coordinates. Since $\text{GCD}(a, c) = 1$, the last two vectors in the Euclidean algorithm will be $(1, s)$ and $(0, t)$ for some integers $s$ and $t$. But the absolute value of the determinant is unchanged throughout the Euclidean algorithm, and so we find

$$\pm 1 = \text{Det} \begin{pmatrix} a & b \\ c & d \end{pmatrix} = \text{Det} \begin{pmatrix} 1 & s \\ 0 & t \end{pmatrix} = t.$$

Hence $t = \pm 1$ and the result is proven. ■

The Euclidean algorithm is guided by the Euclidean algorithm on $a$ and $c$. It begins

$$(a, b) = q \cdot (c, d) + (e, f)$$
$$(c, d) = q' \cdot (e, f) + (g, h)$$
$$\vdots$$

Guided by the $x$-coordinates, the last $x$-coordinates are 1 then 0.

$$\vdots$$
$$(p, q) = p \cdot (1, s) + (0, t).$$

**Theorem 9.14 (Connectedness of the topograph)** *The topograph is connected; one may walk from any lax basis to home basis.*

PROOF: Let $\{\pm \vec{v}, \pm \vec{w}\}$ be a lax basis. Apply the Euclidean algorithm to $\vec{v}$ and $\vec{w}$. Each step in the Euclidean algorithm corresponds to an arc in the domain topograph; a step like $\vec{v} = q \cdot \vec{w} + \vec{r}$ corresponds to the arc below, from $\{\pm v, \pm w\}$ to $\{\pm \vec{w}, \pm \vec{r}\}$.

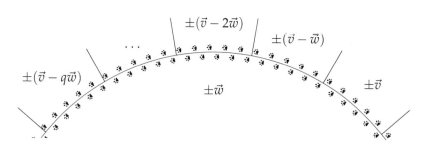

Figure 9.11: One can walk from the lax basis $\{\pm v, \pm w\}$ to the lax basis $\{\pm w, \pm r\}$. Thus, one can walk along an arc for each division-with-remainder in the Euclidean algorithm.

The length of the arc is $q + 1$ segments, where $q$ is the quotient occurring in the division-with-remainder.

By Lemma 9.13, one may walk from $\{\pm v, \pm w\}$ to $\{\pm(1, s), \pm(0, 1)\}$ along a series of such arcs. Observe that

$$(1, s) = s \cdot (0, 1) + (1, 0).$$

Thus one more arc in the domain topograph leads to home basis. ■

THE INFINITE DOMAIN TOPOGRAPH does not fit on the page. It is hard to see, all at once, so we focus on small elements: vertices, edges, and regions.

If we focus on a vertex,[19] we see a **triad** ⊢. Around each triad are three lax vectors[20], any two of which form a lax basis.

[19] A vertex is a point where edges meet

[20] A triple $\{\pm\vec{u}, \pm\vec{v}, \pm\vec{w}\}$, any two of which form a lax basis, is what J.H. Conway calls a **lax superbasis**.

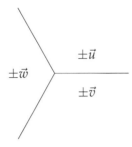

Figure 9.12: A triad with lax vectors $\pm\vec{u}$, $\pm\vec{v}$, $\pm\vec{w}$, and three lax bases.

If we focus on an edge, we see a **cell** ⊢⊣. At each cell, we see four lax vectors: $\pm\vec{v}, \pm\vec{w}$, their difference, and their sum.

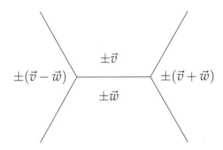

Figure 9.13: A cell with lax vectors $\pm\vec{v}$, $\pm\vec{w}$, their sum and difference. The cell contains five lax bases, and two triads.

If we focus on a region, we see an **infinity-gon**[21]. We see the lax vector $\pm\vec{w}$ in the region, and an infinite sequence of lax vectors $\pm(\vec{v} + n\vec{w})$ as one travels around the region.

[21] Of course, we cannot draw all of an infinity-gon, so we draw a small number of edges and leave the rest to the imagination.

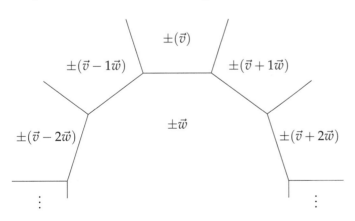

Figure 9.14: The central region has lax vector $\pm\vec{w}$. Surrounding $\pm\vec{w}$ are lax vectors of the form $\pm(\vec{v} + n\vec{w})$ for all integers $n$.

SUPERBASES are inventions of J.H. Conway which we will use to study the intersections between bases, and to track orientations throughout the domain topograph. Just as we distinguish vectors $(a,b)$ from lax vectors $\pm(a,b)$, there are "strict" and "lax" definitions of a superbasis.

**Definition 9.15** A **strict superbasis** is an (unordered) triple $\{\vec{u},\vec{v},\vec{w}\}$ of vectors such that $\vec{u} + \vec{v} + \vec{w} = 0$ and from which any two vectors form a basis.

By the previous Proposition 9.11, the conditions that $\{\vec{u},\vec{v}\}$ forms a basis and that $\vec{u} + \vec{v} + \vec{w} = 0$ together imply that $\{\vec{u},\vec{w}\}$ and $\{\vec{v},\vec{w}\}$ form a bases. In this way, a superbasis is attached to three bases.

**Lemma 9.16** *If $\{\pm\vec{u}, \pm\vec{v}, \pm\vec{w}\}$ is a set of three lax vectors, any two of which form a lax basis, then $\pm\vec{u} \pm \vec{v} \pm \vec{w} = 0$ for some choice of signs.*

PROOF: Consider a basis $\{\vec{v},\vec{w}\}$ with $\vec{v} = (a,b)$ and $\vec{w} = (c,d)$. If $\vec{u} = (x,y)$ forms a basis with both $\vec{v}$ and $\vec{w}$, then

$$bx - ay = \pm 1 \text{ and } dx - cy = \pm 1.$$

These four equations (two equations with two sign choices) describe four lines; two of slope $b/a$ and two of slope $d/c$. These four lines intersect at four corners of a parallelogram; these four points yield exactly four possible vectors $\vec{u}$ (or two pairs of lax vectors).

On the other hand, the four vectors $\vec{u} = \pm\vec{v} \pm \vec{w}$ (arising from four sign-choices) all satisfy the condition of being a basis with both $\vec{v}$ and $\vec{w}$. Hence, if $\{\pm\vec{u}, \pm\vec{v}, \pm\vec{w}\}$ satisfies the hypothesis of the Lemma, then $\vec{u} = \pm\vec{v} \pm \vec{w}$ for some choice of signs, and the result follows. ∎

**Definition 9.17** A **lax superbasis** is an unordered triple $\{\pm\vec{u}, \pm\vec{v}, \pm\vec{w}\}$ of lax vectors, from which any two form a lax basis.

The previous lemma illustrates that every lax superbasis arises from a strict superbasis, by relaxing about signs. In fact, every lax superbasis arises from *exactly two* strict superbases – if $\{\vec{u},\vec{v},\vec{w}\}$ forms a strict superbasis, then among the eight possible sign-choices in $\{\pm\vec{u}, \pm\vec{v}, \pm\vec{w}\}$, only the choices with all + or all − form a strict superbasis.

In the domain topograph, the vertices correspond to lax superbases. Each such vertex has three protruding edges – the three lax bases found within a given lax superbasis – and three surrounding regions – the three lax vectors within the lax superbasis. Every edge corresponds to a lax basis; as expected, every edge is incident with two vertices (its endpoints) – corresponding to the two lax superbases containing the given lax basis.

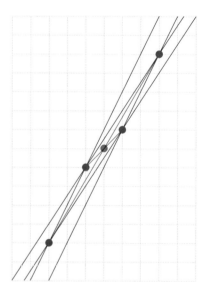

Figure 9.15: Lax vectors $\pm(1,2)$ and $\pm(2,3)$ correspond to four lines $2x - y = \pm 1$ and $3x - 2y = \pm 1$. Their intersection points give two lax vectors $\pm(1,1)$ and $\pm(3,5)$. The resulting triples $\{\pm(1,2), \pm(2,3), \pm(1,1)\}$ and $\{\pm(1,2), \pm(2,3), \pm(3,5)\}$ form lax superbases.

Figure 9.16: A lax superbasis: any two of the three lax vectors forms a lax basis.

THE ORIENTATION of the domain topograph depends on how one begins. But if one begins at home basis,[22] then places the vector $\pm(1,1)$ to its right, the **orientation** of every lax superbasis is determined. In other words, for every lax superbasis $\{\pm\vec{u}, \pm\vec{v}, \pm\vec{w}\}$, the three lax vectors occur in either a clockwise or a counterclockwise fashion around the vertex. But how do we tell the orientation of a lax superbasis without expanding the topograph from home basis?

**Lemma 9.18** *Suppose that $\{\vec{u}, \vec{v}, \vec{w}\}$ is a (strict) superbasis, with $\vec{u} = (a,b)$ and $\vec{v} = (c,d)$ and $\vec{w} = (e,f)$. Then all three determinants are equal:*

$$\mathrm{Det}\begin{pmatrix} a & b \\ c & d \end{pmatrix} = \mathrm{Det}\begin{pmatrix} c & d \\ e & f \end{pmatrix} = \mathrm{Det}\begin{pmatrix} e & f \\ a & b \end{pmatrix}.$$

PROOF: The superbasis condition implies that $e = -a - c$ and $f = -b - d$. Therefore

$$\mathrm{Det}\begin{pmatrix} c & d \\ e & f \end{pmatrix} = cf - de = c(-b-d) - d(-a-c) = ad - bc.$$

The first two determinants are equal; the other equality is similar. ∎

**Theorem 9.19 (Orientation of the topograph by determinants)** *At every vertex in the domain topograph, choose signs to form a strict superbasis. Then all counterclockwise determinants equal 1.*

PROOF: Note that there are only two sign-choices to form a strict superbasis from a lax superbasis. The resulting determinants are the same. At home basis, a strict superbasis is $(1,0)$, $(0,1)$, and $(-1,-1)$. Their counterclockwise determinants are equal to 1.

Now consider a cell in the domain topograph, with $\vec{v} = (a,b)$ and $\vec{w} = (c,d)$.

To form strict superbases, we choose the signs as below.

Directly computing, observe that

$$\mathrm{Det}\begin{pmatrix} c & d \\ -a & -b \end{pmatrix} = \mathrm{Det}\begin{pmatrix} a & b \\ c & d \end{pmatrix}.$$

Hence the counterclockwise determinants at both triads are the same. Since one can walk from triad to triad along cells, all counterclockwise determinants will equal 1 throughout the topograph. ∎

---

[22] The standard home basis is extended to a **home triad**.

In counterclockwise progression, this triad proceeds from $\pm(1,0)$ to $\pm(0,1)$ to $\pm(1,1)$ and back to $\pm(1,0)$ again.

Choose signs to make a strict superbasis near home basis.

The counterclockwise determinants are

$$\mathrm{Det}\begin{pmatrix} 1 & 0 \\ 0 & 1 \end{pmatrix}, \mathrm{Det}\begin{pmatrix} 0 & 1 \\ -1 & -1 \end{pmatrix},$$

and $\mathrm{Det}\begin{pmatrix} -1 & -1 \\ 1 & 0 \end{pmatrix}$,

which all equal 1.

So how is the lax superbasis

$$\{\pm(17,12), \pm(7,5), \pm(10,7)\}$$

oriented in the topograph? Choose signs to make a strict superbasis

$$(-17,-12), (7,5), (10,7).$$

The determinant

$$\mathrm{Det}\begin{pmatrix} 7 & 5 \\ 10 & 7 \end{pmatrix} = 49 - 50 = -1,$$

and so the travelling from $\pm(7,5)$ to $\pm(10,7)$ must be clockwise around the vertex.

A MATRIX $M = \begin{pmatrix} a & b \\ c & d \end{pmatrix}$ can be viewed as a *transformation* of vectors. The transformation of a vector $(x,y)$ by $M$ is defined by the rule

$$\begin{pmatrix} a & b \\ c & d \end{pmatrix} \textbf{ sends } (x,y) \textbf{ to } (ax+cy, bx+dy).$$

In particular,

$$\begin{pmatrix} a & b \\ c & d \end{pmatrix} \text{ sends } (1,0) \text{ to } (a,b), \text{ and sends } (0,1) \text{ to } (c,d).$$

**Lemma 9.20** *Let $M$ be a matrix and let $\vec{v}, \vec{w}$ be vectors. Suppose that $M$ sends $\vec{v}$ to $\vec{v}'$ and $\vec{w}$ to $\vec{w}'$. Then $M$ sends $\vec{v} \pm \vec{w}$ to $\vec{v}' \pm \vec{w}'$.*

PROOF: Let $M = \begin{pmatrix} a & b \\ c & d \end{pmatrix}$, and $\vec{v} = (x,y)$ and $\vec{w} = (s,t)$. Then $M$ sends $\vec{v}$ to $(ax+cy, bx+dy)$, and sends $\vec{w}$ to $(as+ct, bs+dt)$. Also, $M$ sends $(x \pm s, y \pm t)$ to $(a(x \pm s) + c(y \pm t), b(x \pm s) + d(y \pm t))$. The result follows from addition or subtraction of vectors:

$$(ax+cy, bx+dy) \pm (as+ct, bs+dt)$$
$$= (a(x \pm s) + c(y \pm t), b(x \pm s) + d(y \pm t)). \blacksquare$$

An **automorphism** is a matrix $M$ with integer entries whose determinant is $\pm 1$. If $M$ is an automorphism, then $M$ sends home basis to another basis (one whose vectors are the rows of $M$). Such matrices transform the domain topograph according to the following.

**Proposition 9.21** *Let $M$ be an automorphism and let $\{\vec{v}, \vec{w}\}$ be a basis. Suppose that $M$ sends $\vec{v}$ to $\vec{v}'$ and $\vec{w}$ to $\vec{w}'$. Then $\{\vec{v}', \vec{w}'\}$ is a basis.*

PROOF: Let $M = \begin{pmatrix} a & b \\ c & d \end{pmatrix}$ be an automorphism and $\{\vec{v}, \vec{w}\}$ be a basis. Then one can hop via $\vec{v}$ and skip via $\vec{w}$ to find the vector $(1,0)$, and *then* transform $(1,0)$ via $M$ to arrive at $(a,b)$. The lemma implies that we may equivalently transform first and then hop via $\vec{v}'$ and skip via $\vec{w}'$ to arrive at $(a,b)$. Since one can hop via $\vec{v}$ and skip via $\vec{w}$ to find $(0,1)$ and then transform via $M$ to find $(c,d)$, one can hop via $\vec{v}'$ and skip via $\vec{w}'$ to arrive at $(c,d)$.

Since $\text{Det } M = \pm 1$, the pair $\{(a,b), (c,d)\}$ is a basis. Since hops of $\vec{v}'$ and skips of $\vec{w}'$ suffice to travel to both vectors $(a,b)$ and $(c,d)$ in a basis, they suffice to travel everywhere in the grid. Hence $\{\vec{v}', \vec{w}'\}$ is a basis. $\blacksquare$

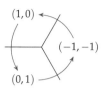

Figure 9.17: The matrix $\begin{pmatrix} 0 & 1 \\ -1 & -1 \end{pmatrix}$ rotates the topograph. In particular, it sends $(1,0)$ to $(0,1)$ and sends $(0,1)$ to $(-1,-1)$ and sends $(-1,-1)$ to $(1,0)$.

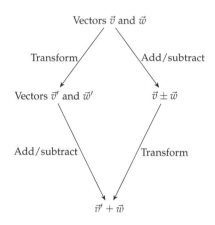

Figure 9.18: Both processes yield the same result.

**Corollary 9.22** *Let M be an automorphism, and let $\vec{v}$ be a primitive vector. Then M sends $\vec{v}$ to a primitive vector.*

PROOF: A primitive vector $\vec{v}$ participates in a basis with some vector $\vec{w}$. If $M$ sends $\vec{v}$ to $\vec{v}'$ and $\vec{w}$ to $\vec{w}'$, then $\{\vec{v}', \vec{w}'\}$ is a basis too. Hence $\vec{v}'$ is a primitive vector. ∎

**Corollary 9.23** *If $\{\vec{u}, \vec{v}, \vec{w}\}$ is a superbasis, and M sends $\vec{u}$ to $\vec{u}'$, $\vec{v}$ to $\vec{v}'$, $\vec{w}$ to $\vec{w}'$, then $\{\vec{u}', \vec{v}', \vec{w}'\}$ is a superbasis.*

PROOF: Proposition 9.21 implies that the triple $\{\pm\vec{u}', \pm\vec{v}', \pm\vec{w}'\}$ is a lax superbasis. Moreover, the sign choices yield a strict superbasis: since $\vec{u} + \vec{v} + \vec{w} = (0,0)$, and $M$ sends $(0,0)$ to $(0,0)$, we find that

$$M \text{ sends } \vec{u} + \vec{v} + \vec{w} \text{ to } (0,0).$$

Thus $\vec{u}' + \vec{v}' + \vec{w}' = (0,0)$ by the previous lemma. ∎

Automorphisms can be composed and can be undone. If $A$ and $B$ are automorphisms, then there is a **composite automorphism** $C$: if $A$ sends a vector $\vec{v}$ to $\vec{v}'$ and $B$ sends that $\vec{v}'$ to $\vec{v}''$, then $C$ sends $\vec{v}$ directly to $\vec{v}''$. If $M$ is an automorphism, then there exists an automorphism $N$ called the **inverse automorphism** which has the following effect: if $M$ sends a vector $\vec{v}$ to a vector $\vec{v}'$, then $N$ sends $\vec{v}'$ back to $\vec{v}$. Details can be found in the exercises.

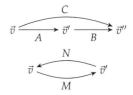

AUTOMORPHISMS send regions to regions, edges to edges, triads to triads, cells to cells. The domain topograph is uniform – it looks the same throughout – because any region, edge, or triad can be sent to any other by an automorphism.

**Theorem 9.24 (Uniformity of the topograph)** *Let $\begin{array}{c}\vec{u}\\ \diagup\\ \longrightarrow \!\!\!\!\!\!< \vec{w}\\ \diagdown\\ \vec{v}\end{array}$ and $\begin{array}{c}\vec{u}'\\ \diagup\\ \longrightarrow \!\!\!\!\!\!< \vec{w}'\\ \diagdown\\ \vec{v}'\end{array}$ be two strict superbases in the domain topograph. Then there exists an automorphism which sends $\vec{u}$ to $\vec{u}'$, $\vec{v}$ to $\vec{v}'$, and $\vec{w}$ to $\vec{w}'$.*

PROOF: Let $\vec{u} = (a,b)$ and $\vec{v} = (c,d)$. Then the matrix $M = \begin{pmatrix} a & b \\ c & d \end{pmatrix}$ sends home basis to $\{\vec{u}, \vec{v}\}$. It follows that $M$ sends $(-1, -1)$ to $\vec{w} = (-a - c, -b - d)$ as well. Hence there is an automorphism $M$ sending the home triad $\begin{array}{c}(1,0)\\ \diagup\\ \longrightarrow \!\!\!\!\!\!< (-1,-1)\\ \diagdown\\ (0,1)\end{array}$ to $\begin{array}{c}\vec{u}\\ \diagup\\ \longrightarrow \!\!\!\!\!\!< \vec{w}\\ \diagdown\\ \vec{v}\end{array}$. In the same way, there is an automorphism $M'$ sending the home triad to $\begin{array}{c}\vec{u}'\\ \diagup\\ \longrightarrow \!\!\!\!\!\!< \vec{w}'\\ \diagdown\\ \vec{v}'\end{array}$. Let $N$ be the automorphism inverse to $M$. The composite automorphism, $N$ followed by $M'$, is the automorphism we seek. ∎

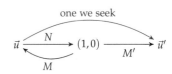

To see the entire domain topograph, one must imagine an infinitude of regions, each surrounded by an infinity-gon, or a cell branching into four more cells, each branching into new cells, or triads glued at their ends to more triads to more triads. Near home basis the domain topograph is pictured below.

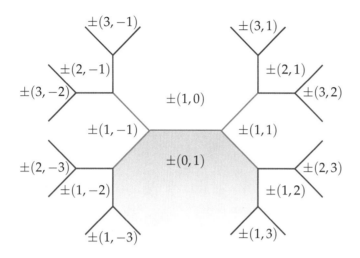

Figure 9.19: The topograph, near home basis. The region around $\pm(0,1)$ is shaded, and edges are colored, for comparison with the next page.

Cells can be pulled into the familiar shape ⟩–⟨.

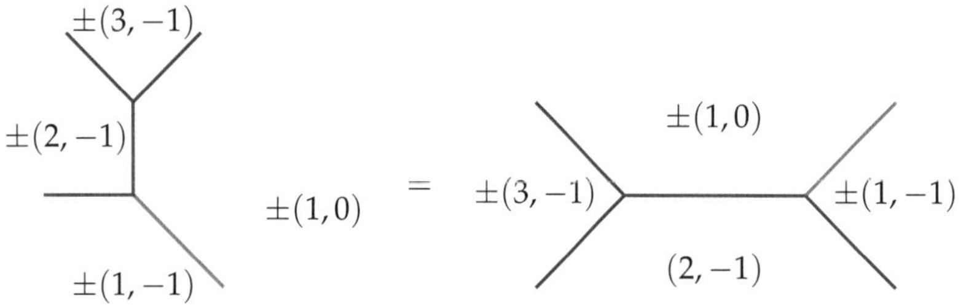

Figure 9.20: On the left, the cell as it occurs in the domain topograph at the top of the page. On the right, the same cell pulled into the familiar cell shape.

Regions can be pulled into the familiar shape ⟩⟅.

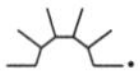

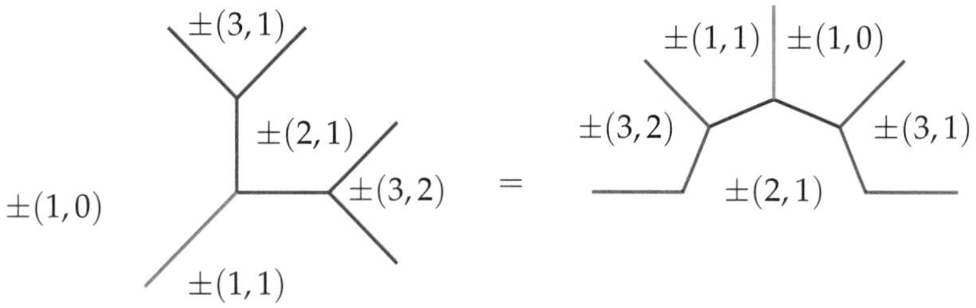

Figure 9.21: On the left, the region as it occurs in the domain topograph at the top of the page. On the right, the same region pulled into the familiar infinity-gon shape.

One may bend and stretch and shrink the domain topograph as much as desired, as long as edges never cross, points never touch each other, and regions are demarcated by infinity-gons.

Exploiting the flexibility of the domain topograph, we can pull the infinity-gon surrounding $\pm(0,1)$ until it forms a nearly straight horizontal line, with descending forking branches. In this way, we can see the domain topograph in the *negative space* surrounding Ford's circles. The edges and regions below are attached to each other in the same way on this page as on the previous page.

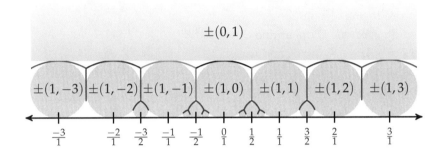

Figure 9.22: To each lax vector $\pm(x,y)$ corresponds the rational number $y/x$. On top of each rational number $y/x$ lies a Ford circle labelled by the lax vector $\pm(x,y)$. The lax vector $\pm(0,1)$ corresponds to $1/0$, sometimes called $\infty$. The endless gray region on top is the Ford circle at $\infty$.

The connection between the domain topograph and Ford's circles is no mere coincidence. Rather, the concepts of this chapter (so far) are the same as those of Chapter 3 in disguise.

In this chapter, we discussed *primitive lax vectors* $\pm(a,b)$. But, with the exception of $\pm(0,1)$, each primitive lax vector corresponds to a *rational number* $b/a$. Observe that $(a,b)$ and $(-a,-b)$ correspond to the same rational number, since $b/a = (-b)/(-a)$. Since every rational number can be expressed as a reduced fraction,[23] there is a one-to-one correspondence:

[23] By convention, the reduced fraction for $\infty$ is $1/0$.

Coprime lax vectors $\leftrightarrow$ Rational numbers, and $\infty$.

In this chapter, we discussed the *determinant* and *lax bases*. Given two vectors $(a,b)$ and $(c,d)$, the determinant of the resulting matrix is $ad - bc$ (if $(a,b)$ is the first row and $(c,d)$ the second). But recall that $b/a \heartsuit d/c$ when $bc - ad = \pm 1$. Thus we find

$$\text{Det}\begin{pmatrix} a & b \\ c & d \end{pmatrix} = \pm 1 \text{ if and only if } \frac{b}{a} \heartsuit \frac{d}{c}.$$

In this way, there is a one-to-one correspondence[24]:

Lax bases $\leftrightarrow$ Kissing pairs of fractions.

[24] To be precise, the set of "Kissing pairs of fractions" refers to the set of unordered pairs $\{u,v\}$ where $u,v$ are rational numbers or $\infty$, such that the reduced fraction of $u$ kisses the reduced fraction of $v$.

The *domain topograph* contains precisely the same information as a table of all rational numbers (and $\infty$) and all kissings between their reduced fractions.

THE RANGE TOPOGRAPH is a tool, invented by J.H. Conway, for visualizing binary quadratic forms. A **binary quadratic form** is a function of two variables, of the general[25] form

$$Q(x,y) = ax^2 + bxy + cy^2,$$

in which (at least in this text) $a$, $b$, and $c$ are integers.

In previous chapters, we studied *linear* Diophantine equations; we approached an equation like $2x + 3y = 7$, we found a criterion for the existence of a solution (does $GCD(2,3)$ divide 7?), we found a method of producing a solution (apply the Euclidean algorithm and "solve for 1"), and we found a formula producing all solutions.

In this and following chapters, we study *quadratic* Diophantine equations; we approach equations like $2x^2 + 3xy + 4y^2 = 7$, and use Conway's topograph to visualize its solutions.

One says that a binary quadratic form $Q(x,y) = ax^2 + bxy + cy^2$ **represents**[26] a number $m$ if there exists a vector $(x,y) \neq (0,0)$ such that $Q(x,y) = m$. To solve a quadratic Diophantine equation, we study the values represented by binary quadratic forms.

**Proposition 9.25** *If $Q$ is a binary quadratic form, then for every integer $n$,*

$$Q(nx, ny) = n^2 \cdot Q(x,y).$$

*In particular,*

$$Q(-x,-y) = Q(x,y), \quad Q(0,0) = 0.$$

PROOF: If $Q(x,y) = ax^2 + bxy + cy^2$, then we find

$$Q(nx, ny) = a(nx)^2 + b(nx)(ny) + c(ny)^2$$
$$= n^2 \cdot (ax^2 + bxy + cy^2) = n^2 \cdot Q(x,y). \blacksquare$$

For any binary quadratic form $Q(x,y)$, its **range topograph** is produced by drawing the domain topograph, and labelling regions not by the primitive lax vectors $\pm \vec{v}$, but rather by their outputs $Q(\pm \vec{v})$.

[25] Specific examples of binary quadratic forms include:

$$Q(x,y) = 2x^2 + 3xy - y^2,$$
$$Q(x,y) = -x^2 + xy,$$
$$Q(x,y) = x^2 + y^2.$$

Each is a function of two variables. For example, if

$$Q(x,y) = 2x^2 + 3xy - y^2,$$

then

$$Q(4,5) = 2 \cdot 4^2 + 3(4)(5) - 5^2$$
$$= 32 + 60 - 25 = 67.$$

[26] The condition that $(x,y) \neq (0,0)$ is convenient. Others would say "nontrivially represents" to exclude $(0,0)$.

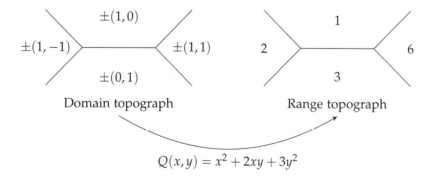

Figure 9.23: The range topograph of $Q(x,y) = x^2 + 2xy + 3y^2$, near home basis. The drawing reflects the facts:

$$Q(1,0) = 1, \quad Q(0,1) = 3.$$
$$Q(1,1) = 6, \quad Q(1,-1) = 2.$$

THE DOMAIN TOPOGRAPH sets down a geometry of points, edges, and regions. The range topograph labels the regions by integers to illustrate the values represented by a quadratic form $Q$. Below is the range topograph of $Q(x,y) = x^2 - y^2$ (values in red), laid over the domain topograph.

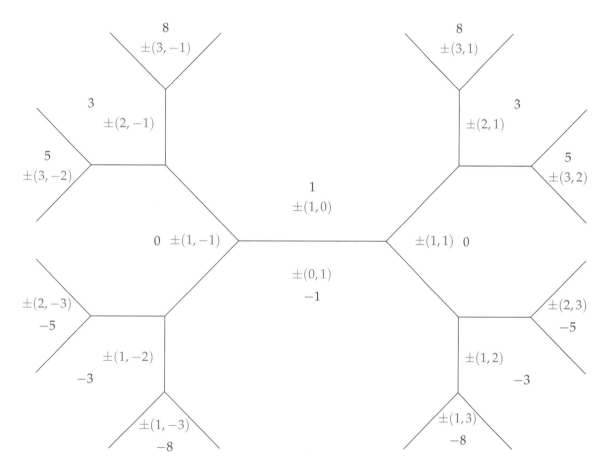

Figure 9.24: Sixteen regions in the domain topograph, near home basis, and the corresponding values in the range topograph of $Q(x,y) = x^2 - y^2$.

This range topograph exhibits two types of symmetry. A left-right flip leaves the range topograph unchanged. A top-down flip exchanges the numbers in the range topograph with their negatives. These symmetries of the range topograph reflect two facts about the quadratic form $Q(x,y) = x^2 - y^2$:

1. Switching the sign of either $x$ or $y$ leaves $Q(x,y)$ unchanged. In other words, $Q(x,y) = Q(x,-y)$.

2. Switching $x$ and $y$ switches the sign of $Q(x,y)$. In other words, $Q(x,y) = -Q(y,x)$.

Finding other patterns within the range topograph requires us to look at its elements: the triads, cells, and infinity-gons.

A CELL in the range topograph contains four integers. But any three of these integers determine the fourth, using Conway's *Arithmetic Progression Rule*. To prove this rule, we begin with an ancient lemma.

**Lemma 9.26** *If $x_1, x_2, y_1, y_2$ are real numbers, then the sequence*

$$(x_1 - x_2)(y_1 - y_2), \quad x_1 y_1 + x_2 y_2, \quad (x_1 + x_2)(y_1 + y_2)$$

*is an arithmetic progression with common difference $x_1 y_2 + x_2 y_1$.*

PROOF: The terms of the sequence are expanded in the left column; their common differences are in the right column.

$$(x_1 y_1 - x_1 y_2 - x_2 y_1 + x_2 y_2)$$
$$(x_1 y_2 + x_2 y_1)$$
$$(x_1 y_1 + x_2 y_2)$$
$$(x_1 y_2 + x_2 y_1)$$
$$(x_1 y_1 + x_1 y_2 + x_2 y_1 + x_2 y_2)$$

∎

**Theorem 9.27 (Arithmetic progression rule)** *In every cell in the range topograph of a binary quadratic form, as pictured below, the sequence $e, (u + v), f$ forms an arithmetic progression.*

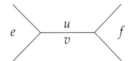

PROOF: Let $Q(x, y) = ax^2 + bxy + cy^2$ be a binary quadratic form. Let $(p, q)$ and $(r, s)$ be a basis at which the cell is located:

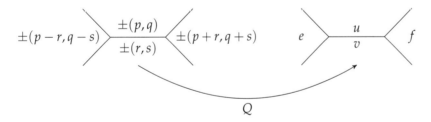

The values of $e, u + v$, and $f$ are expressed in the following table.

| $Q(x,y) =$ | $ax^2$ | + | $bxy$ | + | $cy^2$ |
|---|---|---|---|---|---|
| $e =$ | $a(p-r)^2$ | + | $b(p-r)(q-s)$ | + | $c(q-s)^2$ |
| $u+v =$ | $a(p^2+r^2)$ | + | $b(pq+rs)$ | + | $c(q^2+s^2)$ |
| $f =$ | $a(p+r)^2$ | + | $b(p+r)(q+s)$ | + | $c(q+s)^2$ |

Lemma 9.26 implies[27] that the three columns, blue, red, and green, form arithmetic progressions. The sum of three arithmetic progressions is an arithmetic progression, so $e, u + v, f$ form an arithmetic progression. ∎

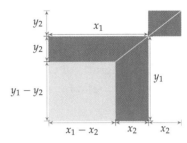

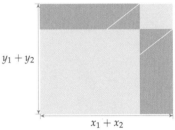

Figure 9.25: When $0 < x_2 < x_1$ and $0 < y_2 < y_1$, a geometric proof of the Lemma is given above. The area of the red region is the difference $(x_1 y_1 + x_2 y_2) - (x_1 - x_2)(y_1 - y_2)$. The area of the green region is the difference $(x_1 + x_2)(y_1 + y_2) - (x_1 y_1 + x_2 y_2)$. The red region and green region have the same area, by the dissection given above.

[27] For the first column, set
$$x_1 = y_1 = p, \quad x_2 = y_2 = r.$$
For the second column, set
$$x_1 = p, y_1 = q, \quad x_2 = r, y_2 = s.$$
For the third column, set
$$x_1 = y_1 = q, \quad x_2 = y_2 = s.$$

THE ARITHMETIC PROGRESSION PROPERTY allows one to quickly sketch a range topograph, beginning with only three values. Every time one sees a cell with three values, one can compute the fourth.

**Problem 9.28** Identify the missing value in the cell, according to the arithmetic progression rule.

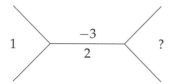

SOLUTION: We must have an arithmetic progression: $1, (-3+2), ?$. The value that completes the arithmetic progression $1, -1, ?$ is $-3$. ✓

**Problem 9.29** Sketch a range topograph containing the triad $\genfrac{}{}{0pt}{}{-1}{3}\!\!\bigg\langle 1$.

SOLUTION: We may extend this topograph, by noting that the three unknown values below must fit into arithmetic progressions.

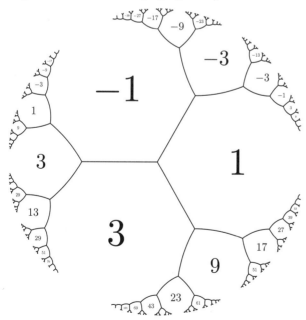

We may extend the topograph indefinitely in this fashion. In the **hyperbolic plane**,[28] the entire topograph may be drawn.

[28] We have used the Poincaré disc model of the hyperbolic plane here. The entire hyperbolic plane fits inside a circle. All angles at triads are 120°. "Line segments" in the hyperbolic plane appear as arcs of circles which intersect the boundary circle at right angles.

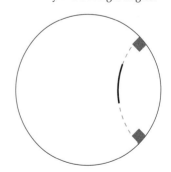

"Length" is measured differently in the hyperbolic plane – the "lengths" of all edges in the diagram on the left are equal.

EVERY TRIPLE of integers occurs in a triad of a range topograph, for *some* binary quadratic form. Consider the triad near home basis, in the range topograph of $Q(x,y) = ax^2 + bxy + cy^2$.[29]

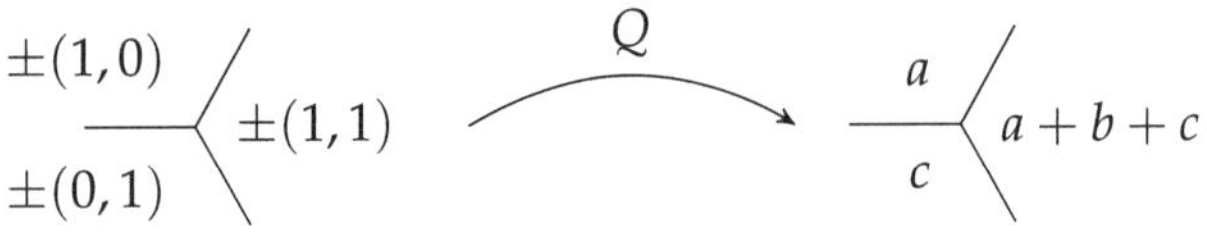

[29] This triad reflects the following facts:
$$Q(1,0) = a,$$
$$Q(1,1) = a + b + c,$$
$$Q(0,1) = c.$$

If we know the three numbers $a, a+b+c, c$, the we can determine the numbers $a, b, c$. This implies the following basic result.

**Proposition 9.30** *A quadratic form is determined by its values at $(1,0)$, $(0,1)$, $(1,1)$.*[30]

[30] If $Q(1,0) = p$ and $Q(0,1) = q$ and $Q(1,1) = r$, then
$$Q(x,y) = px^2 + (r - p - q)xy + qy^2.$$

**Problem 9.31** Find a binary quadratic form whose range topograph includes the triad $\begin{smallmatrix} & 1 \\ -5 & \!\!\!\!\!\big\langle\, 2 \end{smallmatrix}$ near home basis.

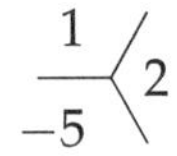

SOLUTION: We seek a quadratic form $Q(x,y) = ax^2 + bxy + cy^2$, in which $a = 1$, $c = -5$, and $a + b + c = 2$. Substituting yields $1 + b - 5 = 2$ and so $b = 6$. So $Q(x,y) = x^2 + 6xy - 5y^2$. ✓

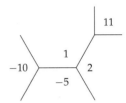

Figure 9.26: A larger piece of the range topograph of $Q(x,y) = x^2 + 6xy - 5y^2$. Three triads are displayed.

DIFFERENT QUADRATIC FORMS can sometimes have *essentially* the same range topograph. Below are three different quadratic forms – their coefficients bear no obvious relationship to one another.

$$Q_1(x,y) = x^2 + 6xy - 5y^2 \qquad Q_2(x,y) = 2x^2 + 8xy + y^2 \qquad Q_3(x,y) = -10x^2 + 16xy - 5y^2$$

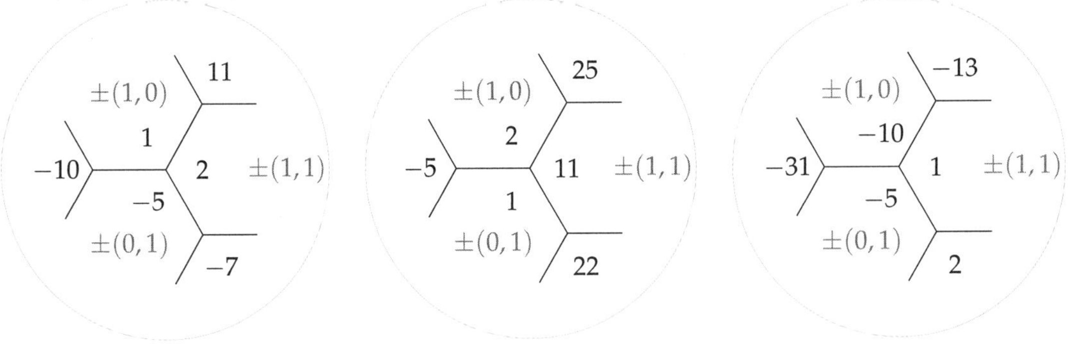

Figure 9.27: The topographs of $Q_1$, $Q_2$, and $Q_3$, near home basis. These three binary quadratic forms are equivalent to each other: $Q_1$ is properly equivalent to $Q_3$ (no mirror reflection is needed), while the equivalences between $Q_1$ and $Q_2$ and between $Q_2$ and $Q_3$ are improper.

But the triad $\begin{smallmatrix} & 1 \\ -5 & \!\!\!\!\!\big\langle\, 2 \end{smallmatrix}$ occurs in all three topographs – after rotation (on the right) and rotation and reflection (on the left). Since the values at a triad determine *all* values on a range topograph via the arithmetic progression rule, the three quadratic forms represent the same set of values.

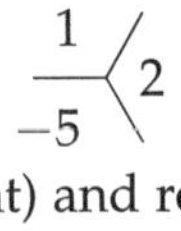

AUTOMORPHISMS allow us transform binary quadratic forms. Let $M = \begin{pmatrix} a & b \\ c & d \end{pmatrix}$ be an automorphism.[31] If $Q(x,y)$ is a binary quadratic form, then we can define a new form by[32]

$$Q'(x,y) = Q(ax + cy, bx + dy).$$

In other words, if $M$ sends $\vec{v}$ to $\vec{v}'$, then $Q'(\vec{v}) = Q(\vec{v}')$.

When $Q$ and $Q'$ are related in this way, we say that $M$ is an **equivalence** from $Q$ to $Q'$; if moreover, $\det M = 1$, we say that $M$ gives a **proper equivalence** from $Q$ to $Q'$. Accordingly, we call two quadratic forms $Q$ and $Q'$ (properly) equivalent if there exists a (proper) equivalence $M$ relating them.

When two quadratic forms are equivalent, their range topographs are essentially the same. Every triad which occurs in the range topograph of $Q$ must occur *somewhere* in the range topograph of an equivalent form $Q'$.

**Proposition 9.32** *If $p, q, r$ are three integers, and $\{\vec{u}, \vec{v}, \vec{w}\}$ is a strict superbasis, then there exists a unique binary quadratic form $Q$ such that*

$$Q(\vec{u}) = p \text{ and } Q(\vec{v}) = q \text{ and } Q(\vec{w}) = r.$$

PROOF: Let $Q'$ be the binary quadratic form taking values $p, q, r$ at $(1, 0)$, $(0, 1)$, and $(-1, -1)$ respectively. Let $M$ be an automorphism which sends $(1, 0), (0, 1), (-1, -1)$ to the strict superbasis $\vec{u}, \vec{v}, \vec{w}$ (by Theorem 9.24). Let $Q$ be the binary quadratic form sent to $Q'$ via the automorphism $M$. Then

$$Q'(1,0) = Q(\vec{u}), \quad Q'(0,1) = Q(\vec{v}), \quad Q'(-1,-1) = Q(\vec{w}).$$

Hence $Q$ is the binary quadratic form we seek.

For uniqueness, the values of $Q$ at $\vec{u}, \vec{v}, \vec{w}$ determine the values of $Q$ everywhere by the arithmetic progression rule and the connectedness of the domain topograph. Hence those values determine the values of $Q$ at $(1,0), (0,1), (-1,-1)$. These values determine the coefficients of $Q$ uniquely by Proposition 9.30. ■

**Theorem 9.33 (Triad criterion for equivalence)** *If a triad $\begin{smallmatrix} p \\ q \end{smallmatrix}\!\!\!-\!\!\!<\!r$ occurs in the range topographs of two quadratic forms $Q_1$ and $Q_2$, then $Q_1$ is equivalent to $Q_2$. If the triad occurs in the same orientations in both $Q_1, Q_2$, then the equivalence is proper.*

PROOF: Let $Q$ be the binary quadratic form in which the triad $\begin{smallmatrix} p \\ q \end{smallmatrix}\!\!\!-\!\!\!<\!r$ occurs at the home superbasis. The previous proof demonstrates that $Q$ is equivalent to $Q_1$ and to $Q_2$. Hence[33] $Q_1$ is equivalent to $Q_2$. ■

[31] Recall, this means that $M$ is a matrix of integers of determinant $\pm 1$.

[32] Indeed, this is a binary quadratic form; if $Q(x,y) = ux^2 + vxy + wy^2$ then expanding yields

$$\begin{aligned} Q'(x,y) &= Q(ax+cy, bx+dy) \\ &= u(ax+cy)^2 \\ &\quad + v(ax+cy)(bx+dy) \\ &\quad + w(bx+dy)^2 \\ &= (ua^2 + vab + wb^2)x^2 \\ &\quad + (2uac + vad + vbc + 2wbd)xy \\ &\quad + (uc^2 + vcd + wd^2)y^2. \end{aligned}$$

So $Q'$ is a binary quadratic form, albeit one with complicated coefficients.

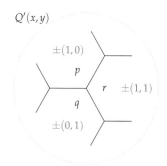

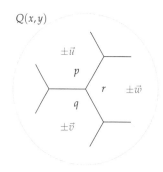

Figure 9.28: The triad which occurs at $\pm(1,0), \pm(0,1), \pm(1,1)$ in $Q'$ occurs at $\pm\vec{u}, \pm\vec{v}, \pm\vec{w}$ in $Q$.

[33] We're using the fact that equivalence of forms is an "equivalence relation" (a reflexive, symmetric, and transitive relation). See the exercises for more.

THE DISCRIMINANT of a quadratic form is a single integer which captures some, but not all, of its important features. Binary quadratic forms with different discriminants are very different, as we will see. We give at least three perspectives on the discriminant in this chapter, beginning with the most fundamental.

**Definition 9.34** Let $Q(x,y) = ax^2 + bxy + cy^2$ be a binary quadratic form. Its **discriminant** is the integer

$$\Delta(Q) = b^2 - 4ac.$$

**Proposition 9.35** *For every $Q$, $\Delta(Q) \equiv 0$ or $1$ mod $4$.*

PROOF: If $Q(x,y) = ax^2 + bxy + cy^2$, then

$$\Delta(Q) = b^2 - 4ac \equiv b^2 \bmod 4.$$

Zero and one are the only squares modulo 4. ∎

Therefore, the only possible discriminants of binary quadratic forms are integers in the following sequence.

$$\ldots, -12, -11, -8, -7, -4, -3, 0, 1, 4, 5, 8, 9, 12, 13, \ldots.$$

As the following two examples[34] demonstrate, every such integer **does** occur as the discriminant of a binary quadratic form.

[34] The reader can verify that the principal forms have the claimed discriminants.

**Definition 9.36** If $\Delta$ is a multiple of 4, then the **principal form** of discriminant $\Delta$ is defined to be

$$Q(x,y) = x^2 - \frac{\Delta}{4}y^2.$$

If $\Delta$ is one more than a multiple of 4, then the **principal form** of discriminant $\Delta$ is defined to be

$$Q(x,y) = x^2 + xy - \frac{\Delta - 1}{4}y^2.$$

The discriminant $\Delta(Q)$ is easy enough to compute. But for what follows it will be helpful to relate $\Delta(Q)$ to the range topograph in two ways. The first is the discriminant of a triad.

**Definition 9.37** Let $\begin{array}{c} e \\ \diagup\!\!\!\diagdown\, f \\ g \end{array}$ be a triad. Its **discriminant** is

$$\Delta\!\left(\begin{array}{c} e \\ \diagup\!\!\!\diagdown\, f \\ g \end{array}\right) = e^2 + f^2 + g^2 - 2(ef + fg + ge).$$

Note that the quantity $e^2 + f^2 + g^2 - 2(ef + fg + ge)$ does not change after permuting $e, f, g$ in any way, e.g., after rotating or reflecting the triad.

A CELL too has an intrinsic discriminant.

**Definition 9.38** Let $e \succ\!\!\frac{u}{v}\!\!\prec f$ be a cell. Its **discriminant** is

$$\Delta\left(e \succ\!\!\frac{u}{v}\!\!\prec f\right) = (u-v)^2 - ef.$$

Note that the quantity $(u-v)^2 - ef$ does not change after swapping $u \leftrightarrow v$ or after swapping $e \leftrightarrow f$.

The arithmetic progression rule implies that in any cell of a range topograph, $e \succ\!\!\frac{u}{v}\!\!\prec f$, the values satisfy $e = 2(u+v) - f$.[35] We deduce the compatibility of the previous two definitions.

[35] Since $e, u+v, f$ form an arithmetic progression, they have a common difference:
$$(u+v) - e = f - (u+v).$$
The identity $e = 2(u+v) - f$ follows directly.

**Lemma 9.39** *The discriminant of a cell in the range topograph equals the discriminant of either triad it contains.*

$$\Delta\left(e \succ\!\!\frac{u}{v}\right) = \Delta\left(e \succ\!\!\frac{u}{v}\!\!\prec f\right) = \Delta\left(\frac{u}{v}\!\!\prec f\right).$$

PROOF: Using the identity $e = 2(u+v) - f$, we simplify the discriminant of the cell.

$$\Delta\left(e \succ\!\!\frac{u}{v}\!\!\prec f\right) = (u-v)^2 - ef$$

$$= u^2 - 2uv + v^2 - (2(u+v) - f)f$$

$$= u^2 + v^2 + f^2 - 2uv - 2uf - 2vf = \Delta\left(\frac{u}{v}\!\!\prec f\right).$$

By symmetry in swapping $e, f$, the lemma follows. ∎

**Theorem 9.40 (Discriminants are equal at all cells and triads)** *In the range topograph of any binary quadratic form $Q(x, y) = ax^2 + bxy + cy^2$, the discriminants of all cells and all triads are equal to $\Delta(Q)$.*

PROOF: At home basis, we compute the discriminant of the triad directly[36] to find $b^2 - 4ac$, which equals $\Delta(Q)$

Hence the discriminants of all three cells adjacent to this triad are equal to $\Delta(Q)$ by the previous lemma. Hence the discriminants of all triads adjacent to these cells are equal to $\Delta(Q)$. As all triads and cells are linked through each other to home basis, the discriminants of every triad and every cell must be equal: all are equal to $\Delta(Q)$. ∎

The discriminant of a quadratic form can be seen *everywhere* in its range topograph.

[36] The triad at home basis is
$$\frac{a}{c}\!\!\prec a+b+c\;.$$ Its discriminant is
$$a^2 + (a+b+c)^2 + c^2$$
$$-2a(a+b+c) - 2c(a+b+c) - 2ac.$$
This simplifies miraculously to
$$\Delta\left(\frac{a}{c}\!\!\prec a+b+c\right) = b^2 - 4ac.$$

**Corollary 9.41 (Equivalent forms have equal discriminants)** *If two binary quadratic forms are equivalent, then they have the same discriminant.*

PROOF: If two binary quadratic forms are equivalent, then they have a triad in common. The discriminant of this triad then equals the discriminant of both quadratic forms. ∎

As we classify binary quadratic forms, the discriminant is our first tool although it is a bit coarse. For example, there are 405 quadratic forms $Q(x,y) = ax^2 + bxy + cy^2$ in which

$$-4 \leq a \leq 4, \quad -4 \leq b \leq 4, \quad 0 \leq c \leq 4.$$

We could organize them by coefficients $a, b, c$, in a 9x9x5 cube, as in the margin. But it is better to sort them by discriminant, and then place properly equivalent quadratic forms together. In Gauss's biological terminology, we say that two quadratic forms are in the same **class** if they are properly equivalent. After filtering, sorting, and clustering, the dots in the margin fall into the diagram below.

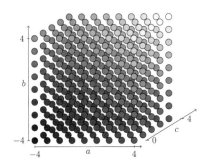

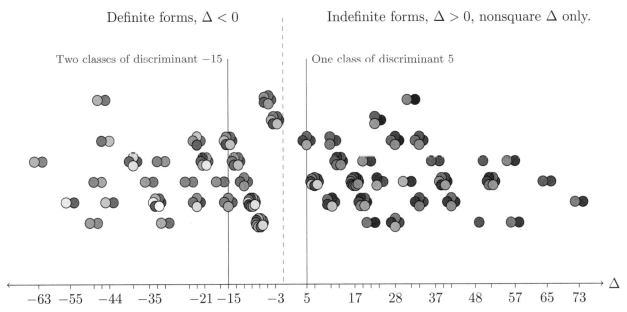

Figure 9.29: We included only quadratic forms $ax^2 + bxy + cy^2$ with $\gcd(a,b,c) = 1$ – the **primitive** forms, for technical reasons. We also excluded those with square discriminants. Left over, 199 of the 405 quadratic forms are plotted here.

The converse to Corollary 9.41 is false – there are binary quadratic forms of the same discriminant but in different classes. The *number* of primitive classes (only positive-definite forms are typically counted if $\Delta < 0$) of a given discriminant $\Delta$ is called the **class number** of $\Delta$ and denoted $h(\Delta)$. For example $h(5) = 1$ and $h(-15) = 2$. The class number is subtle and deep and turns up in surprising places around number theory. We will study it in the following two chapters.

EVERY binary quadratic form is equivalent to itself. This is trivial. But a binary quadratic form can be equivalent to itself in nontrivial ways – this is the phenomenon of symmetry.

**Definition 9.42** Let $Q$ be a binary quadratic form. A **proper isometry**[37] of $Q$ is a proper equivalence $\begin{pmatrix} a & b \\ c & d \end{pmatrix}$ from $Q$ to $Q$. In other words, $a, b, c, d$ are integers, $ad - bc = 1$, and for all integers $x, y$,

$$Q(x, y) = Q(ax + cy, bx + dy).$$

[37] The set of proper isometries is called the **special orthogonal group** of $Q$, written $SO(Q)$.

Proper isometries can often be seen in the range topograph as rotational and translational symmetries.

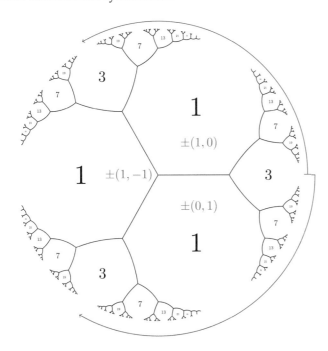

Figure 9.30: The topograph of the quadratic form $Q(x, y) = x^2 + xy + y^2$, as seen near home basis. By drawing the topograph in the hyperbolic plane, the nontrivial proper isometries become visible as a 120 degree rotation clockwise and counterclockwise.

The clockwise rotation sends $\pm(1, 0)$ to $\pm(0, 1)$ and $\pm(0, 1)$ to $\pm(1, -1)$. This corresponds to the fact that

$$Q(-y, x + y) = Q(x, y)$$

for all integers $x, y$.

We have studied the geometry of the domain topograph – lax vectors, bases, superbases, their attachments to each other. Automorphisms allow us to see that the topograph looks structurally the same, wherever one looks. When one imposes a quadratic form to place values on a range topograph, then we find geometric arrangements of integers governed throughout by the arithmetic progression rule. The discriminant is ubiquitous.

The next two chapters will examine the range topograph more closely, using it to study equivalence of quadratic forms and their isometries. This will lead to algorithms for solving all quadratic Diophantine equations in two variables. We divide the study into two chapters – one on definite forms (those with negative discriminant) and another on indefinite forms (those with positive discriminant).

## Historical notes

This and the following chapters treat the theory of binary quadratic forms systematically. For the origins of this theory, we trace a tradition from Diophantus to Fermat, Euler, Lagrange, Legendre and Gauss. Developments in India will be discussed in the last chapter.

Diophantus was a mathematician who probably lived between the 1st and 4th centuries of the common era. His *Arithmetica* is a complicated text, not only for its mathematical content, but for its manuscript history. Greek texts circulated in Byzantium between the 11th and 13th century, while Arabic texts[38] can be traced to Baghdad in the 9th century; the Greek and Arabic texts cover distinct sections of the *Arithmetica* and commentaries.

The history of the *Arithmetica* is foremost the history of the reception and transmission of the *Arithmetica*; two recent sources (and the sources we use here) are Norbert Schappacher's "Diophantus of Alexandria: A text and its history"[39] and Jean Christianidis' "The way of Diophantus: Some clarifications on Diophantus' method of solution."[40]

Diophantus introduces the *Arithmetica* with terminology for unknown numbers, their squares (τετραγώνοι), their cubes (κύβοι), and higher powers. His unknowns would today be called positive rational numbers; this contrasts with the modern notion of "Diophantine equation" which requires integer solutions. The body of the *Arithmetica* consists of solved problems. More precisely, Christianidis identifies a common structure to these problems – a general formulation, an instantiated formulation, invention and disposition, computation of sought numbers, and a test proof (verification).[41] For example, Problem II.8 begins with a general formulation, "Partition a given square into two squares."[42] For an instantiated formulation, Diophantus partitions 16 into two (positive rational) squares, making auxiliary choices along the way. By the end, he verifies a solution: 256/25 and 144/25.

For the history of quadratic Diophantine equations, it is better to begin with the Latin *Arithmetica* of Bachet de Méziriac (1621), a translation and adaptation of earlier manuscripts of Diophantus. In Proposition III.7 of Bachet's porisms (and not in earlier manuscripts of Diophantus), we find the identity sometimes attributed to Diophantus,[43]

$$(a^2 + b^2) \cdot (c^2 + d^2) = (ac - bd)^2 + (ad + bc)^2. \quad (9.1)$$

It follows that, if $x$ and $y$ can be expressed as the sum of two squares, then the product $xy$ can also be expressed as the sum of two squares.

It was this *Arithmetica* that Fermat read, and in which Fermat

[38] Arabic texts were discovered in 1971. See for example R. Rashed's "Les travaux perdus de Diophante, II" in *Revue d'Histoire des Sciences* **18** (1975) and J. Sesiano's "Books IV to VII of Diophantus's Arithmetica in the Arabic translation attributed to Qusṭā ibn Luqā" (1982).

[39] Schappacher's work can be found at http://www-irma.u-strasbg.fr/~schappa/NSch/Publications_files/1998cBis_Dioph.pdf, and is based on his earlier article "Wer war Diophant?" in *Mathematische Semesterberichte* **45/2** (1998). Schappacher notes that "we have no original parts of the text, nor any direct knowledge of what happened to it between its first writing and the 9th century AD."

[40] "The way of Diophantus: Some clarifications on Diophantus' method of solution" can be found in *Historia Mathematica* **34** (2007).

[41] See Christianidis, Section 5.

[42] Translation from Schappacher.

[43] Weil notes that such an identity occurs in Leonardo Pisano's (a.k.a. Fibonacci's) *Liber Quadratorum* in 1225. See "Number theory: An approach through history from Hammurapi to Legendre" by André Weil (Modern Birkhaüser Classics, 2001), Chapter 1, *Protohistory*, for more.

wrote his famous marginal note.[44] Fermat considered those numbers which could be expressed as the sum of two squares ($x^2 + y^2$), those expressible as the sum of a square and twice a square ($x^2 + 2y^2$), and those expressible as the sum of a square and thrice a square ($x^2 + 3y^2$).

When $D = 1, 2, 3$, it happens that if $N = x^2 + Dy^2$, and $p$ is a prime factor of $N$, then $p$ too can be expressed in the form $x^2 + Dy^2$. In 1748,[45] Euler studies the divisors of numbers of the form $x^2 + Dy^2$ for many integers $D$ – a prototypical Theorem 10 considers the case $D = 5$:

> All the prime divisors of numbers contained in the form $aa + 5bb$ are either 2, or 5, or are contained in one of four forms $20m + 1$, $20m + 3$, $20m + 7$, $20m + 9$.[46]

But Theorem 11 demonstrates that only those primes of the form $2m + 1$ and $20m + 9$ can be expressed again in the form $x^2 + 5y^2$.

An explanation came almost 30 years later, when Lagrange carried out a systematic study of quadratic Diophantine equations. His *Recherches D'Arithmétique* (1773,1775) begins

> Ces Recherches ont pour objet les nombres qui peuvent être représentés par la formule
> $$Bt^2 + Ctu + Du^2,$$
> où $B, C, D$ sont supposés des nombres entiers donnés, et $t, u$ des nombres aussi entier, mais indéterminés.[47]

Having read Lagrange, Euler returns to the divisors of $x^2 + 5y^2$ in 1775.[48]. To understand the prime divisors of $x^2 + 5y^2$, he must consider not only those of the form $x^2 + 5y^2$, but also those of the form $2x^2 + 2xy + 3y^2$. (Note their common discriminant $-20$). And so the study of binary quadratic forms of the special sort $x^2 + Dy^2$ requires the study of a broad class of binary quadratic forms.

Legendre (1797)[49] synthesizes the work of Lagrange, and gives a complete treatment of equations of the form $ax^2 + bxy + cy^2 = N$ (for integer $a, b, c, N$). While Lagrange's and Legendre's approaches are systematic, they focus on Diophantine equations – the solution of *equations* of the form $A = Bt^2 + Ctu + Du^2$, rather than on the *forms* $Bt^2 + Ctu + Du^2$ per se.

An important and intentional shift, from equations to forms, occurs in the *Disquisitiones* of Gauss (1801). Gauss carries out a "careful inquiry into the nature of the [binary quadratic] forms" – even abbreviating $ax^2 + 2bxy + cy^2$ as a triple $(a, b, c)$ "when we are not concerned with the unknowns $x, y$."[50] He studies the discriminant (he calls it the determinant), equivalence and the class number. Far from concluding the study of quadratic Diophantine equations, Gauss produces a fruitful theory of binary quadratic forms.

---

[44] See the historical notes of Chapters 1 and 3 for more on Fermat's Last Theorem.

[45] See "Theoremata circa divisores numerorum in hac forma $paa \pm qbb$ contentorum", published in *Commentarii academiae scientiarum Petropolitanae* 14 (1751). Listed as E164 in the Euler Archive http://eulerarchive.maa.org/ Euler treats values of $D$, first positive then negative, treating prime values of $|D|$ first. The values of $D$ studied by Euler are (in order):

$$1, 2, 3, 5, 7, 11, 13, 17, 19,$$
$$6, 10, 14, 15, 21, 35, 30,$$
$$-1, -2, -3, -5, -7, -11, -13, -17, -19,$$
$$-6, -10, -14, -22, -15, -21, -33, -35,$$
$$-30, -105.$$

[46] Translation by Jordan Bell.

[47] From *Recherches D'Arithmétique* (1773,1775), first published in N. Mém. Berlin, collected in Lagrange's *Oeuvres* 3, 695–795. Literally translated, "This research has for its object the numbers which can be represented by the formula

$$Bt^2 + Ctu + Du^2,$$

where $B, C, D$ are given integers, and $t$ and $u$ are also integers, but indeterminate."

[48] See "De insigni promotione scientiae numerorum", published in *Opuscula Analytica* 2 (1785). Listed as E598 in the Euler Archive http://eulerarchive.maa.org/ Euler's citation of Lagrange is explicit.

[49] Legendre first published the "Essai sur la Théorie des nombres" in 1797; a second edition in 1808 contains a preface remarking on the excellent work of the newly famous Gauss.

[50] See Art. 153 of the *Disquisitiones*, English translation by Arthur A. Clarke. Gauss limits his study to binary quadratic forms with *even* middle coefficient.

## Exercises

1. Walk from $\pm(20, 17)$ to home basis in the domain topograph.

2. Consider the Fibonacci sequence, with $F_0 = 0$, $F_1 = 1$, and $F_n = F_{n-1} + F_{n-2}$ for all $n \geq 2$. How long is the path from the lax vector $(F_n, F_{n+1})$ to home basis?

3. To prove that the area of a parallelogram is $|ad - bc|$, we used a dissection assuming all coordinates positive. Extend the theorem to cases where coordinates are positive or negative.[51]

   [51] Use symmetry to reduce the number of cases as much as possible.

4. Demonstrate that there cannot be a set of *four* lax vectors, any two of which form a basis.

5. This sequence of exercises leads to Pick's Theorem, relating the area of a grid-polygon to its interior and edge grid-points.

   (a) Consider a triangle in the plane whose vertices lie on grid-points. Let $I$ be the number of grid-points in the interior of the triangle, and $B$ the number of grid-points on the edges of the triangle (**including** the corners!). Prove that the area of the triangle equals $I + 1/2 B - 1$. Hint: Double the triangle to make a parallelogram.

   (b) A grid-polygon is a shape obtained by drawing edges from one grid-point to another, to form a closed path without self-intersections (edges may not cross). A grid-polygon is called convex if for any two points $p_1, p_2$ on the polygon, the entire line segment from $p_1$ to $p_2$ lies inside the polygon. Use induction on the number of vertices to prove that the area of a convex grid-polygon equals $I + 1/2 B - 1$, where $I$ and $B$ are as above.

   (c) (Challenge!) Extend this to nonconvex polygons.

Figure 9.31: To relate a convex polygon with $n$ vertices to a polygon with $n - 1$ vertices, one may slice off a triangle.

6. Sketch range topographs[52] containing the following triads:

   $$\begin{array}{c} 1 \\ \diagdown \\ 3 \end{array}\!\!\!\langle 2, \quad \begin{array}{c} 1 \\ \diagdown \\ 0 \end{array}\!\!\!\langle -1, \quad \begin{array}{c} 0 \\ \diagdown \\ 3 \end{array}\!\!\!\langle 0, \quad \begin{array}{c} 3 \\ \diagdown \\ 3 \end{array}\!\!\!\langle 3.$$

   Write down a binary quadratic form whose range topograph contains each triad.

   [52] 12 values comprise a good minimum for a range topograph.

7. Consider a topograph with values $1, 7, -2$ as in the margin.

   Fill in the ?s to satisfy the arithmetic progression rule?

8. If $Q(x, y) = ax^2 + bxy + cy^2$, then the **opposite** of $Q$ is defined to be $ax^2 - bxy + cy^2$ and the **associate** of $Q$ is defined to be $cx^2 + bxy + ay^2$. Demonstrate that $Q$ is improperly equivalent to both its opposite and its associate.

9. For those familiar with matrix algebra, the following exercises translate the ideas of the chapter into traditional algebraic form.

   If $\vec{v} = (x,y)$ is a vector, we may think of $\vec{v}$ as a row matrix $\vec{v} = (x\ y)$. If $M = \begin{pmatrix} a & b \\ c & d \end{pmatrix}$, then $M$ sends $(x,y)$ to $(ax+cy, bx+dy)$; this can be realized as matrix multiplication:
   $$\vec{v} \cdot M = (x\ y) \cdot \begin{pmatrix} a & b \\ c & d \end{pmatrix} = (ax+cy\ \ bx+dy).$$

   (a) Let $M$ and $N$ be automorphisms, and $C$ the composite automorphism "$M$ then $N$". Demonstrate that $C$ is represented by the matrix product $M \cdot N$.
   (b) Demonstrate that the inverse automorphism of $M$ is represented by the matrix inverse of $M$. Why is this an automorphism?[53]
   (c) If $Q(x,y) = ax^2 + bxy + cy^2$, then the matrix of $Q$ is defined by
   $$Q = \begin{pmatrix} a & \frac{1}{2}b \\ \frac{1}{2}b & c \end{pmatrix}.$$
   Check that if $\vec{v} = (x\ y)$, then $Q(x,y) = \vec{v} \cdot Q \cdot {}^t\vec{v}$.[54] Also check that $\Delta(Q) = -4\operatorname{Det}(Q)$.
   (d) Use matrix algebra to prove the arithmetic progression rule.
   (e) Demonstrate that an automorphism $M$ is an equivalence from $Q$ to $Q'$ if and only if $Q' = M \cdot Q \cdot {}^tM$. Use this to prove that equivalent binary quadratic forms have the same discriminant.
   (f) Characterize the proper isometries of a quadratic form $Q$ in terms of matrix algebra.

10. Let $Q(x,y) = ax^2 + bxy + cy^2$ and $Q'(x,y) = a'x^2 + b'xy + c'y^2$ be two binary quadratic forms. The form $Q'$ is called a (right) **neighbor** of $Q$ if $a' = c$, $b' + b$ is a multiple of $2c$, and $\Delta(Q) = \Delta(Q')$.

    (a) Prove that if $Q'$ is a right neighbor of $Q$, then $Q$ is properly equivalent to $Q'$. Hint: if $b' + b = 2kc$, then walk $k$ steps from home basis around the region with value $c$.
    (b) Suppose that $Q(x,y) = ax^2 + bxy + cy^2$ and $a \mid b$. (Gauss calls such forms **ambiguous**.) Prove that $Q$ is improperly equivalent to itself. Hint: Switch $a$ and $c$.

11. Considering Figure 9.10, expanding the topograph three times (from home basis) yields 29 lax bases in total. How many lax bases will be displayed after expanding the topograph $n$ times?

12. Considering the opening figure of the chapter, what proper isometries can you identify from the three range topographs?

[53] Automorphisms in this context must be matrices with integer entries and determinant $\pm 1$.

[54] Here ${}^t\vec{v}$ denotes the transpose of $\vec{v}$, the column vector
$${}^t\vec{v} = \begin{pmatrix} x \\ y \end{pmatrix}.$$
Transposition swaps the order of multiplication: if $A$ and $B$ are matrices, then
$${}^t(A \cdot B) = {}^tB \cdot {}^tA.$$

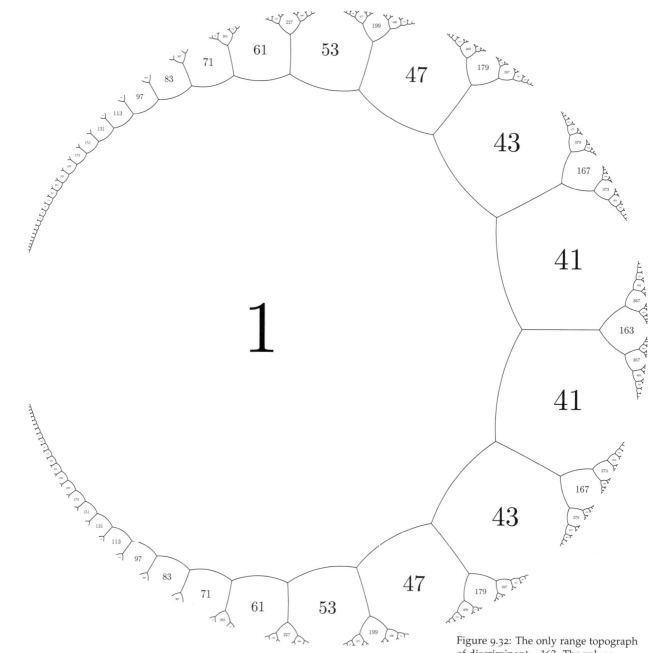

Figure 9.32: The only range topograph of discriminant −163. The values surrounding 1 are the values of the polynomial $x^2 + x + 41$. These values are prime when $0 \leq x \leq 39$. This extraordinary example of a prime-producing polynomial is due to Euler, in "Nouveaux Mémoires de l'Académie royale des Sciences", 1772.

# 10
# *Definite Forms*

To find a number in a range topograph, you can follow your nose. If the numbers start growing too large, you can usually turn around and travel in a different direction. The reason this works is:

**Theorem 10.1 (Conway's climbing principle)** *Consider a region in the range topograph, with entries $e, f, u, v, g, h$ as below,*

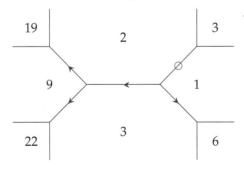

*If the arithmetic progression $e, u+v, f$ increases and $u$ and $v$ are positive, then the arithmetic progressions $v, u+f, g$ and $u, v+f, h$ increase too.*

PROOF: Suppose that the eastward progression increases: $e < u+v < f$. The inequalities $0 < u$ and $u+v < f$ imply

$$v < u+u+v < u+f.$$

Thus the northeast progression $v, u+f, g$ increases. The proof for the southeast progression is the same. ∎

To track our climbing, we decorate each line segment with an arrow, pointing from smaller numbers to larger numbers, in every nonconstant arithmetic progression. In a constant arithmetic progression, we draw a small circle on the line segment.

On the left, there are arrows pointing from 2 to 6, from 1 to 9, from 2 to 22, and from 3 to 19. These arrows illustrate the direction of increase. The small circle ○ illustrates the equality between 3 and 3 – no increase and no decrease takes place as you travel northeast.

THE CLIMBING PRINCIPLE tells us that when the numbers are positive, then the arrows maintain a flow of constant increase. If the numbers start growing, then they keep on growing!

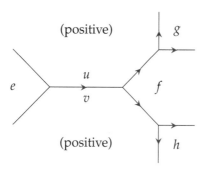

Figure 10.1: A single climbing arrow propogates. Since the arithmetic progression $e, u+v, f$ increases, so to do the arithmetic progressions $v, u+f, g$ and $u, v+f, h$. As $u, v, f$ are positive, this increase continues.

The climbing principle allows us to find *all* solutions to some quadratic Diophantine equations.

**Problem 10.2** Solve the Diophantine equation $x^2 - xy + y^2 = 7$.

SOLUTION: We look for 7 in the range topograph of $x^2 - xy + y^2$.

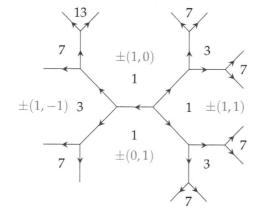

We find 7 in six locations in the range topograph. Moreover, by the climbing principle, the numbers in the range topograph will only grow larger if we branch further. As one travels outwards, past 7, one will find larger and larger numbers (like 13 at the northwest) – thus one will never find 7 again!

We find 7 at six locations, corresponding to six lax vectors in the domain topograph:

$$\pm(1,-2), \quad \pm(2,-1), \quad \pm(3,1), \quad \pm(3,2), \quad \pm(2,3), \quad \pm(1,3).$$

These give *twelve* solutions to the Diophantine equation $x^2 - xy + y^2 = 7$.

There is a subtle point now. The range topograph only tells us the outputs of a quadratic form when a *primitive* vector is input. What about nonprimitive vectors like $(0,0)$ or $(2,-4)$ or $(3,12)$?

The 12 solutions are

$(1,-2), (-1,2), \quad (2,-1), (-2,1),$

$(3,1), (-3,-1), \quad (3,2), (-3,-2),$

$(2,3), (-2,-3), \quad (1,3), (-1,-3).$

In this case we need not worry. If $m \neq \pm 1$ and $(ma, mb)$ is a non-primitive vector, then we have $Q(ma, mb) = m^2 Q(a, b)$. But 7 is **square-free**[1] and so $m^2 Q(a, b)$ cannot equal 7. Hence the twelve solutions we have found are all the solutions to $x^2 - xy + y^2 = 7$. ✓

[1] A square-free number is a number which is *not* a multiple of a square, except for 1.

In the same way, with the climbing principle, we can deduce that the equation $x^2 - xy + y^2 = 2$ has *no solutions*. Nor are there solutions to the equations $x^2 - xy + y^2 = 5$ and $x^2 - xy + y^2 = 6$

**Problem 10.3** Solve the Diophantine equation $x^2 - xy + y^2 = 4$.

SOLUTION: In the range topograph on the opposite page, we never see the number 4. Indeed, the numbers keep growing as we expand outwards, so the number 4 will never be found.

However, we find the number 1 at three locations:

$$\pm(1,0) \text{ and } \pm(0,1) \text{ and } \pm(1,1).$$

The scaling formula $Q(ma, mb) = m^2 Q(a, b)$ implies that $x^2 - xy + y^2 = 4$ when $\pm(x, y)$ is one of the following nonprimitive vectors:

$$\pm(2,0) \text{ and } \pm(0,2) \text{ and } \pm(2,2).$$

In this way, we find *six solutions*.[2]

[2] The six solutions are
$$(2,0), (0,2), (2,2),$$
$$(-2,0), (0,-2), (-2,-2).$$

We find no *primitive* solutions. But we find solutions which are not primitive by this method of **square-scaling**. No other solutions exist, since 4 can only be obtained by square-scaling in one way (scaling 1 by a factor of 4). ✓

**Problem 10.4** Solve the Diophantine equation $2x^2 + 3xy + 13y^2 = 18$.

SOLUTION: In the range topograph of $Q(x, y) = 2x^2 + 3xy + 13y^2$, we find 2 at $\pm(1, 0)$ and 18 at $\pm(1, 1)$. Note that $2 \times 3^2 = 18$. Thus square-scaling yields exactly four solutions:

$$(3,0), \quad (-3,0), \quad (1,1), \quad (-1,-1).$$

The climbing principle shows that there are no other solutions. ✓

DEFINITE binary quadratic forms include those which represent[3] only positive values (called **positive definite**) and those which represent only negative values (called **negative definite**). When $Q(x,y)$ is negative definite, i.e., $Q(x,y) < 0$ for every coprime vector $(x,y)$, we find that $-Q(x,y)$ is positive definite, i.e., $-Q(x,y) > 0$ for every coprime vector $(x,y)$. Hence the study of negative definite forms is reduced to the study of positive definite forms.

[3] Recall that $Q$ "represents" $n$ if there is a nonzero vector $\vec{v}$ such that $Q(\vec{v}) = n$.

Consider a positive definite binary quadratic form $Q(x,y) = ax^2 + bxy + cy^2$. This means that $Q(x,y) > 0$ for every primitive vector $(x,y)$; like any set of positive integers, there exists a *smallest*[4] number occurring in the range topograph. Let $u$ denote this minimum.

[4] When we talk about "smallest" here, we allow that ties might occur. There might be two or more regions in the range topogaph where the same smallest value $u$ occurs. The same non-uniqueness may occur for $v$ and $w$.

Along the infinity-gon surrounding $u$, we find values of $Q(x,y)$ in the range topograph. As these values are also positive, there exists a value $v$ which is *smallest* among those adjacent to $u$. Note that $u \leq v$.

There are two values adjacent to both $u$ and $v$. Let $w$ be the smallest of these two values.

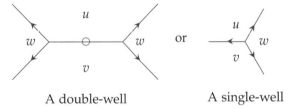

Since $v$ is the smallest value adjacent to $u$, we find that $v \leq w$. Since $w$ is the smallest of the two values adjacent to $u$ and $v$, the arithmetic progression $w, u+v, ?$ must be constant or increasing. We find that $w \leq u + v$. Depending on whether $w = u+v$ or $w < u+v$, we find one of the following patterns of climbing.[5]

[5] The climbing arrows require a bit of thought. For example, the northeast arrow states that $u + w \geq v$; this follows from the fact that $w \geq v$ and $u$ is positive.

A double-well           A single-well

**Definition 10.5** A **single-well** in a range topograph is a triad at which all climbing arrows are oriented outwards. A **double-well** in a range topograph is a cell at which the four outward arrows are oriented outwards, and the central edge is neutral. A **well** is a single- or double-well.

Here are examples of a single-well and double-well.

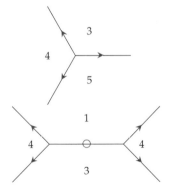

Values elsewhere in a range topograph must be greater than those found at a well, by the climbing principle. All climbing arrows point away from a well. In particular, no other well can occur as other flow-arrows must have the one-in and two-out form of Theorem 10.1.

**Theorem 10.6 (Existence and uniqueness of wells)** *Every positive definite binary quadratic form has exactly one well in its range topograph.*

The climbing principle can be cleverly[6] applied to prove an important fact about the *domain* topograph. We have taken for granted – so far – that the points and edges branch out and out further without ever rejoining, i.e., that there are no loops in the domain topograph. We make this precise below.

[6] The cleverness belongs to J.H. Conway, who presented the following argument in his book *The Sensual Quadratic Form*.

A **simple path** in the domain topograph, from a point $P$ to a point $Q$, is a path along edges which never touches the same point or the same edge twice. A **simple loop** in the domain topograph is a simple path which begin and ends at the same point.

**Theorem 10.7 (The topograph contains no loops)** *For any two points $P, Q$ in the domain topograph, there is only one simple path in the domain topograph from $P$ to $Q$. In particular, there are no simple loops in the domain topograph.*

PROOF: Suppose there were two simple paths from $P$ to $Q$. Follow these paths until they diverge, and call this point of divergence $P'$. From $P'$, follow each path until the first time they converge again; call this point of convergence $Q'$. It is possible, of course, that $P = P'$ or $Q = Q'$ or both. In this way, we construct a simple loop in the domain topograph, traveling along the first path from $P'$ to $Q'$ then back along the second path from $Q'$ to $P'$.

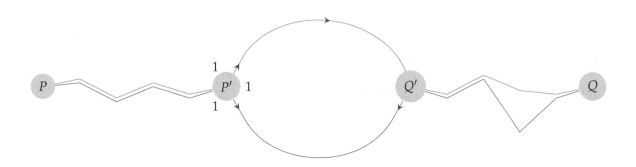

Consider the quadratic form whose values at the triad around $P'$ are $1, 1, 1$, using Proposition 9.32. We work on the range topograph of this quadratic form hereafter. The climbing principle implies that as one travels the simple loop from $P'$ to $Q'$ and back to $P'$, the values one encounters (on the right or left) must climb ever higher. But one cannot climb at every step and return to where one started. ∎

This result completes the proof that the domain topograph forms a **tree**. It is a network of points and edges, with no simple loops, in which one can walk from any point to any other point.

**Proposition 10.8** *If $Q$ is a definite binary quadratic form, then $\Delta(Q) < 0$.*

PROOF: If $Q$ is negative definite, then $-Q$ is positive definite[7], and $\Delta(Q) = \Delta(-Q)$. So it suffices to assume that $Q$ is positive definite. We focus on the well to compute the discriminant.

[7] If $Q(x,y) = ax^2 + bxy + cy^2$ then $-Q$ refers to the binary quadratic form $-Q(x,y) = -ax^2 - bxy - cy^2$. Its discriminant is the same:
$$\Delta(-Q) = (-b)^2 - 4(-a)(-c)$$
$$= b^2 - 4ac = \Delta(Q).$$

We label the westward arrow ambiguously, to indicate that $u + v \geq w$; the well may be single or double. The inequalities $u + v \geq w$ and $v \leq w$ imply that
$$0 \leq w - v \leq u.$$

Squaring yields the inequalities
$$0 \leq w^2 - 2vw + v^2 \leq u^2.$$

The discriminant of $Q$ equals the discriminant of the triad at the well.
$$\Delta(Q) = u^2 + v^2 + w^2 - 2uv - 2vw - 2wu.$$

Using the previous inequality, we find that
$$\Delta(Q) \leq u^2 + u^2 - 2uv - 2wu = 2u(u - (v + w)).$$

Since $v + w > u$, we find that $u - (v + w)$ is negative. Therefore
$$\Delta(Q) \leq 2u(u - (v + w)) < 0. \qquad \blacksquare$$

For single-wells, a geometric proof is possible by the **Hadwiger-Finsler inequality**.[8] This inequality states that if one is given a triangle with side lengths $u, v, w$, then its area can be bounded as below.

$$4\sqrt{3} \cdot \text{Area} \leq 2uv + 2vw + 2wu - u^2 - v^2 - w^2.$$

The single-well conditions are the triangle inequalities: $u + v > w$ and $v + w > u$ and $w + u > v$. Hence there exists a triangle with side-lengths $u, v,$ and $w$ and positive area. The Hadwiger-Finsler inequality implies that
$$4\sqrt{3} \cdot \text{Area} \leq -\Delta.$$

It follows that
$$\Delta \leq -4\sqrt{3} \cdot \text{Area},$$

and in particular the discriminant $\Delta$ is negative.

[8] The original statement is found in "Einige Relationem im Dreieck," by Paul Finsler and H. Hadwiger. *Commentarii Mathematici Helvetici* **10** (1937–1938). A lovely geometric proof is given by Claudi Alsina and Roger B. Nelsen in "Geometric Proofs of the Weitzenböck and Hadwiger-Finsler Inequalities", in *Mathematics Magazine*, Vol. 81, No. 3 (June 2008), pp.216-219.

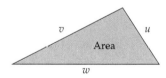

THE DISCRIMINANT of a positive-definite quadratic form contains a great deal of information about the range topograph. For example, we can use the discriminant to bound the minimum (nonzero) value of a positive-definite quadratic form.[9] Using this, we develop an algorithm for finding all positive definite binary quadratic forms with a given discriminant.

[9] Nonzero values of $Q$ obtained by square-scaling can only be larger than values of $Q$ occurring on the range topograph. Hence the minimum value of a positive-definite quadratic form can be found as a value on the range topograph.

**Theorem 10.9 (Minimum value bound for definite forms)** *Let $Q$ be a positive-definite binary quadratic form, with discriminant $\Delta$. Then the smallest nonzero value taken by $Q$ is no greater than $\sqrt{|\Delta|/3}$.*

PROOF: Begin at the well of $Q$, with $u \leq v \leq w$ as before. In particular, $u$ is the smallest nonzero value taken by $Q$. Let $h = (u+v) - w$, so the cell containing the well has the form

$$u+v+h \quad \overset{u}{\underset{v}{\diamond}} \quad u+v-h = w$$

The discriminant of $Q$ can be computed in terms of $u, v, h$.[10]

$$\Delta = h^2 - 4uv < 0.$$

[10] Observe that the cell here occurs too in the range topograph of $ux^2 - hxy + vy^2$ at home basis. This gives $\Delta = h^2 - 4uv$ quickly.

The inequality $u + v \geq w$ implies $h \geq 0$. The inequalities $0 \leq u \leq v \leq w$ imply the inequalities $0 \leq u \leq v \leq u+v-h$. Since $v \leq u+v-h$, we find $0 \leq u - h$ and so $h \leq u$. Summarizing, we find that

$$0 \leq h \leq u \leq v.$$

Therefore,

$$|\Delta| = 4uv - h^2 \geq 4u^2 - h^2 \geq 4u^2 - u^2 = 3u^2.$$

Therefore $u \leq \sqrt{|\Delta|/3}$. ∎

The triad $\overset{1}{\underset{1}{\diamond}} 1$ has discriminant $-3$ and demonstrates that this bound is sharp.

Four facts now allow us to enumerate all wells of a given (negative) discriminant.

1. $1 \leq u \leq \sqrt{|\Delta|/3}$. (The previous Theorem)

2. $0 \leq h \leq u$. (In the proof of the previous Theorem)

3. $h^2 - 4uv = \Delta$. (The definition of the discriminant)

4. $h$ is even or odd according to whether $\Delta$ is even or odd.

The last fact holds, since when $h$ is even/odd, $h^2$ is even/odd, and so $h^2 - 4uv = \Delta$ is even/odd, respectively.

**Problem 10.10** Find all wells of discriminant $-100$.

SOLUTION: Consider a well of discrimant $-100$, with $u \leq v \leq w$ and $h = (u+v) - w$ as before

Since $\sqrt{|-100|/3} = \sqrt{33.333\ldots} < 6$, we find that $1 \leq u \leq 5$. For each such $u$, $0 \leq h \leq u$. Since $-100$ is even, so too must $h$ be even. And, for each $u$ and $h$, the formula $h^2 - 4uv = -100$ determines $v$. We tabulate the possibilities below.

Solving for $v$ explicitly, we find
$$v = \frac{h^2 + 100}{4u}.$$
Of couse, to give a well, all three numbers $u, v, h$ must be integers!

| $u$ | $h$ | $v$ | |
|---|---|---|---|
| 1 | 0 | 25 | ✓ |
| 2 | 0 | 100/8 | ✠ |
| 2 | 2 | 13 | ✓ |
| 3 | 0 | 100/12 | ✠ |
| 3 | 2 | 104/12 | ✠ |
| 4 | 0 | 100/16 | ✠ |
| 4 | 2 | 104/16 | ✠ |
| 4 | 4 | 116/16 | ✠ |
| 5 | 0 | 5 | ✓ |
| 5 | 2 | 104/20 | ✠ |
| 5 | 4 | 116/20 | ✠ |

The three possible wells are displayed below.

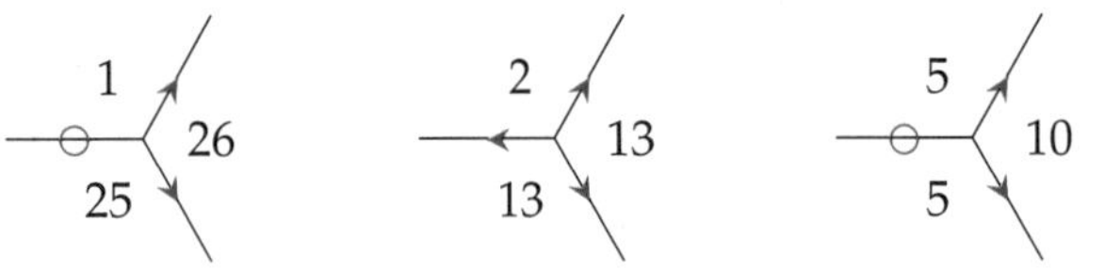

[Diagram showing three wells: (1, 25, 26), (2, 13, 13), (5, 5, 10)]

This solution, together with the existence of a unique well for each positive definite binary quadratic form, gives a powerful result. If $Q$ is a positive-definite binary quadratic form of discriminant $-100$, then $Q$ is equivalent to one of the following three forms:

$$x^2 + 25y^2 \quad \text{or} \quad 2x^2 - 2xy + 13y^2 \quad \text{or} \quad 5x^2 + 5y^2.$$

In this way, there are essentially only three positive-definite binary quadratic forms of discriminant $-100$; every such binary quadratic form has a well, and that well must be one of the three given above.

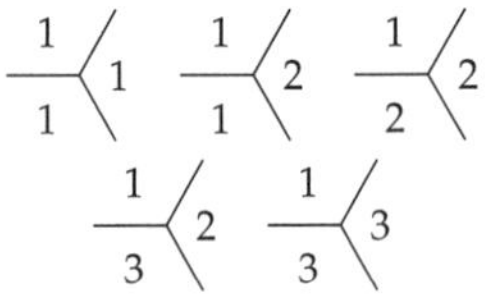

**Lemma 10.11** *If* $\Delta = -3, -4, -7, -8, -11$ *then there is only one well of discriminant* $\Delta$.

PROOF: We apply the tabulation method above. In all five cases $|\Delta| < 12$, so $\sqrt{|\Delta|/3} < 2$, and so $u = 1$. But this forces $h = 0$ or $h = 1$; the value of $h$ is then determined by whether $\Delta$ is even or odd. The values of $u$ and $h$ then determine $v$, since $h^2 - 4uv = \Delta$. ∎

Figure 10.2: There do exist wells of discriminants $-3, -4, -7, -8, -11$. They are pictured above.

**Theorem 10.12** *If Q is a positive-definite*[11] *binary quadratic form of discriminant* $\Delta = -3, -4, -7, -8,$ *or* $-11,$ *then*

$\Delta = -3$: *Q is properly equivalent to* $x^2 + xy + y^2$.

$\Delta = -4$: *Q is properly equivalent to* $x^2 + y^2$.

$\Delta = -7$: *Q is properly equivalent to* $x^2 + xy + 2y^2$.

$\Delta = -8$: *Q is properly equivalent to* $x^2 + 2y^2$.

$\Delta = -11$: *Q is properly equivalent to* $x^2 + xy + 3y^2$.

PROOF: The previous lemma states that there is a unique well for each of these discriminants. Each of the five wells occurs in the five quadratic forms above, respectively. The only subtle point is in the word "properly" – that each well occurs in the same orientation in $Q$ as in the five quadratic forms above. But observe that among the five wells, all are symmetric (the same in clockwise or counterclockwise orientation), except for the well $\frac{1}{2}\!\!\!\diagdown\!\!3$ of discriminant $-8$. But this well occurs in the cell $3\!\diagup\!\!\!\!\frac{1}{2}\!\!\!\!\diagdown\!3$ in both clockwise and counterclockwise orientations. Hence every positive-definite quadratic form $Q$ of discriminant $-8$ contains this well in both orientations, and is properly equivalent to $x^2 + 2y^2$ as desired. ∎

While uniqueness of wells occurs only for small discriminants,[12] finiteness is guaranteed for all.

**Theorem 10.13** *If* $\Delta$ *is a negative integer, then there are finitely many wells of discriminant* $\Delta$. *Their number is bounded by* $|\Delta|/3$.

PROOF: Each well of discriminant $\Delta$ arises from a triple of integers $u, v, h$ such that $h^2 - 4uv = \Delta$ and

$$0 \leq h \leq u \leq v.$$

As $u$ must be bounded by $\sqrt{|\Delta|/3}$, we find that the number of possible pairs $(h, u)$ is bounded[13] by $|\Delta|/3$. For each such pair, there exists at most one integer $v$ satisfying $h^2 - 4uv = \Delta$. Hence the number of wells of discriminant $\Delta$ is bounded by $|\Delta|/3$. ∎

Since proper equivalence of positive-definite binary quadratic forms can be seen from wells,[14] the finiteness theorem above yields the finiteness corollary below.

**Corollary 10.14 (Finiteness of class number for definite forms)**
*Given a negative integer* $\Delta$, *there exists a finite list of binary quadratic forms* $\{Q_1, \ldots, Q_t\}$, *such that every positive-definite binary quadratic form Q of discriminant* $\Delta$ *is properly equivalent to one member of the list.*

---

[11] Proposition 10.22 and Proposition 11.11 imply that every binary quadratic form of negative discriminant is positive-definite or negative-definite. But here is an algebraic argument. Suppose

$$Q(x, y) = ax^2 + bxy + cy^2$$

and $\Delta = b^2 - 4ac < 0$. This implies $a \neq 0$.

Define $z = x + \frac{b}{2a}y$ and observe

$$az^2 - \frac{\Delta}{4a}y^2,$$
$$= a(x + \frac{b}{2a}y)^2 + \frac{4ac - b^2}{4a}y^2,$$
$$= ax^2 + bxy + \frac{b^2}{4a}y^2 + \frac{4ac - b^2}{4a}y^2,$$
$$= ax^2 + bxy + cy^2.$$

So $Q(x, y) = az^2 - \frac{\Delta}{4a}y^2$.

Hence if $a > 0$ and $\Delta < 0$, then $Q(x, y) > 0$ for all real $x, y$ except $(x, y) = (0, 0)$. If $a < 0$ and $\Delta < 0$, then $Q(x, y) < 0$ for all real $x, y$ except $(x, y) = (0, 0)$.

[12] A complete list will be given in the following pages.

[13] Much better bounds are possible, even by elementary means. See the exercises for more.

[14] Proper equivalence distinguishes a well from its mirror image, but finiteness is not affected.

The uniqueness of wells, for small discriminants, has deep arithmetic consequences. As a first application, we reprove Fermat's Christmas Theorem (Theorem 8.7) without using Minkowski's Theorem (Proposition 8.8).

**Theorem 10.15 (Fermat's Christmas Theorem)** *Let $p$ be a prime number with $p \equiv 1 \bmod 4$. Then the Diophantine equation $x^2 + y^2 = p$ has a solution. In other words, $p$ can be expressed as the sum of two squares.*

PROOF: By reciprocity for $-1$, Corollary 8.6, there exist integers $u, n$ such that $u^2 + 1 = pn$. Consider the quadratic form

$$Q(x,y) = px^2 + 2uxy + ny^2.$$

The coefficients of $Q$ are integers, and its discriminant is

$$\Delta = (2u)^2 - 4pn = 4u^2 - 4(u^2 + 1) = -4.$$

Since $Q(1,0) = p > 0$ and $\Delta = -4 < 0$, the quadratic form is positive-definite.[15] By Theorem 10.12, we find that $Q$ is equivalent to $x^2 + y^2$. Since $Q$ represents $p$, so too does $x^2 + y^2$ represent $p$.[16] ∎

[15] A negative discriminant implies the form is definite. See the marginal note of Theorem 10.12.

[16] Recall that "$Q$ represents $p$" means that $Q(x,y) = p$ for some $(x,y) \neq (0,0)$.

Tracing through details in the proof above, we can make it constructive – given a prime number $p \equiv 1 \bmod 4$, we can quickly *find* integers $x, y$ such that $x^2 + y^2 = p$. The method has two steps.

First, we must find an integer $u$ such that $u^2 \equiv -1 \bmod p$. A quick way is the following – let $a$ be a quadratic nonresidue mod $p$, i.e., a number such that $\left(\frac{a}{p}\right) = -1$. Finding such a number is easy in practice – just try $a = 2, 3, \ldots$ until $a^{(p-1)/2} \equiv -1 \bmod p$. If one believes the generalized Riemann hypothesis, then such a quadratic nonresidue will occur for $a \leq \log(p)^2$ (for $p \geq 5$).[17] For such a quadratic nonresidue, we have

$$\left(a^{\frac{p-1}{4}}\right)^2 \equiv -1 \bmod p.$$

[17] See "The least quadratic non residue" by N.C. Ankeny, in *Annals of Mathematics* **55** (1952), and the recent "Conditional bounds for the least quadratic non-residue and related problems" by Y. Lamzouri, X. Li, and K. Soundararajan, in *Math. Comp.*, vol. 84 **295** (2015).

Note $(p-1)/4$ is an integer since $p \equiv 1 \bmod 4$. Letting $u$ be a representative of $a^{(p-1)/4} \bmod p$, the first step is concluded.

Next, one must find the equivalence between $px^2 + 2uxy + ny^2$ and $x^2 + y^2$. This is a matter of walking in the range topograph, from the triad $\genfrac{}{}{0pt}{}{p}{n}\!\!\!\diagup\!\!\!\diagdown\!\! p+2u+n$ to the unique well $2\!\!\!\diagup\!\!\!\diagdown\!\genfrac{}{}{0pt}{}{1}{1}\!\!\!\diagup\!\!\!\diagdown\!\! 2$ of discriminant $-4$. It can be proven that the number of arithmetic operations[18] grows no faster than $\log(p)$.

Hence the time required to find $x$ and $y$ is polynomial in $\log(p)$.

[18] Effectively, one finds the lax basis $\pm(a,b), \pm(c,d)$ at which the well occurs for $px^2 + 2uxy + ny^2$. This requires $O(\log(p))$ arithmetic steps. Inverting the matrix $\begin{pmatrix} a & b \\ c & d \end{pmatrix}$ yields the location of $p$ in the topograph of $x^2 + y^2$. It would be interesting to relate this to the standard *Hermite-Serret algorithm*.

In similar spirit, we may prove a result from the uniqueness of the well of discriminant $-3$.

**Theorem 10.16** *Let $p$ be a prime number with $p \equiv 1 \bmod 3$. Then the Diophantine equation $x^2 + xy + y^2 = p$ has a solution.*

PROOF: By quadratic reciprocity, we have[19]

$$\left(\frac{-3}{p}\right) = \left(\frac{-1}{p}\right)\left(\frac{3}{p}\right) = \left(\frac{-1}{p}\right)\left(\frac{-1}{p}\right)\left(\frac{p}{3}\right) = \left(\frac{p}{3}\right) = 1.$$

[19] By quadratic reciprocity,
$$\left(\frac{3}{p}\right) = \epsilon\left(\frac{p}{3}\right),$$
where $\epsilon = 1$ if $p \equiv 1 \bmod 4$ and $\epsilon = -1$ if $p \equiv 3 \bmod 4$. Thus $\epsilon = \left(\frac{-1}{p}\right)$.

Hence there exists an integer $v$ such that $v^2 \equiv -3 \bmod p$. Replacing $v$ by $p - v$ if necessary, we may assume that $v$ is odd. Therefore $v^2 + 3 = pn$ for some integer $n$, and moreover, $v^2 + 3 \equiv 0 \bmod 4$ since squares of odd numbers are $1 \bmod 4$. Since $p$ is odd, and $4 \mid (v^2 + 3) = pn$, we find that $4 \mid n$. Letting $m = n/4$, we have $v^2 + 3 = 4pm$.

Now consider the quadratic form

$$Q(x, y) = px^2 + vxy + my^2.$$

Its coefficients are integers and its discriminant is

$$\Delta = v^2 - 4pm = -3.$$

The form $Q$ is positive-definite, since $Q(1, 0) = p$ and its discriminant is negative.[20] By Theorem 10.12, $Q$ is equivalent to the quadratic form $x^2 + xy + y^2$. Hence $x^2 + xy + y^2$ represents $p$ as well. ∎

[20] A negative discriminant implies the form is definite. See the marginal note of Theorem 10.12.

Generalization is straightforward to other discriminants of class number one. For example, $h(-163) = 1$; every positive-definite binary quadratic form of discriminant $-163$ is equivalent to the form

$$x^2 + xy + 41y^2.$$

Moreover, quadratic reciprocity demonstrates that $-163$ is a square mod $p$ if and only if

$p \equiv 0, 1, 4, 6, 9, 10, 14, 15, 16, 21, 22, 24, 25, 26, 33, 34, 35, 36, 38, 39, 40, 41,$
$43, 46, 47, 49, 51, 53, 54, 55, 56, 57, 58, 60, 61, 62, 64, 65, 69, 71, 74, 77, 81,$
$83, 84, 85, 87, 88, 90, 91, 93, 95, 96, 97, 100, 104, 111, 113, 115, 118, 119,$
$121, 126, 131, 132, 133, 134, 135, 136, 140, 143, 144, 145, 146, 150, 151, 152,$
$155, 156, 158, 160,$ or $161 \bmod 163.$

It follows that

**Proposition 10.17** *Let $p$ be a prime number such that $p$ is congruent to a number in the above list, mod $163$. Then the Diophantine equation $x^2 + xy + 41y^2 = p$ has a solution.*

The red numbers in the list above are prime; it follows that those numbers appear in the range topograph of $x^2 + xy + 41y^2$, displayed at the opening of this chapter.

PROPER ISOMETRIES of a positive-definite binary quadratic form must arise from proper symmetries of its well. Consider a proper isometry $M$ of a quadratic form $Q$, and a well in its topograph. Wherever $M$ transports the well, one must find the same values, and thus a well again; uniqueness of the well implies that $M$ sends the well to the well. So there are not very many options.

**Theorem 10.18** *If $Q$ is a positive-definite binary quadratic form, and $Q$ has nontrivial[21] proper isometries, then either*

- *$Q$ has a 120° rotation isometry, and $Q$ is equivalent to a multiple of the form $x^2 + xy + y^2$.*

- *$Q$ has a 180° rotation isometry, and $Q$ is equivalent to a multiple of the form $x^2 + y^2$.*

[21] We consider an isometry trivial if it has no geometric effect on the domain topograph. The trivial isomorphisms of the domain topograph are
$$\begin{pmatrix} 1 & 0 \\ 0 & 1 \end{pmatrix} \text{ and } \begin{pmatrix} -1 & 0 \\ 0 & -1 \end{pmatrix}.$$

PROOF: If $Q$ has a single-well, then $M$ must be an orientation-preserving symmetry of the triad at the single-well. If $M$ does anything, it must be a 120° rotation;[22] as an isometry of $Q$, this implies that all values at the well must be equal. It follows that $Q$ is a multiple of the quadratic form taking values 1 all around the well. Hence $Q$ is equivalent to a multiple of the form $x^2 + xy + y^2$.

[22] clockwise or counterclockwise

If $Q$ has a double-well, then $M$ must be an orientation-preserving symmetry of the double-well cell. If $M$ does anything, it must be a 180° rotation; as an isometry of $Q$, this implies that the opposite values at the cell must be equal. The arithmetic progression property demonstrates that $Q$ is a multiple of a form with double-well values $2, 1, 1, 2$. Hence $Q$ is equivalent to a multiple of the form $x^2 + y^2$.

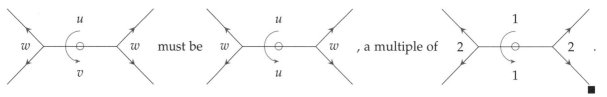

A **primitive** binary quadratic form is a form $Q(x,y) = ax^2 + bxy + cy^2$ satisfying $GCD(a, b, c) = 1$. All forms are multiples of primitive forms, and so it is natural to focus on the primitive forms. Primitivity can be seen in the range topograph, as the GCD of the values appearing. Among definite forms, we focus on *positive-definite, primitive* forms, and we classify them using *proper* equivalence.[23]

[23] In this text, remember the three Ps: primitive, proper, and positive-definite. Others count their classes differently!

DEFINITE FORMS 271

**Definition 10.19** If $\Delta < 0$, then the **class number** $h(\Delta)$ is the number of classes of primitive positive-definite forms of discriminant $\Delta$.

Corollary 10.14 states that if $\Delta < 0$ then the class number $h(\Delta)$ is finite. The method of proof demonstrates that $h(\Delta) \leq 2/3 \cdot |\Delta|$.

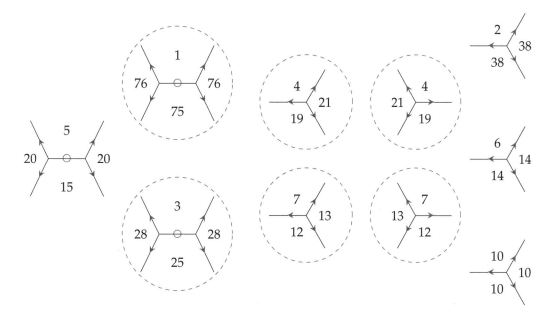

But this bound is not particularly good. Indeed, $h(-300) = 6$, far less than 200. Below is a plot of the actual class numbers for negative discriminants up to $-300$, compared to the bound above.

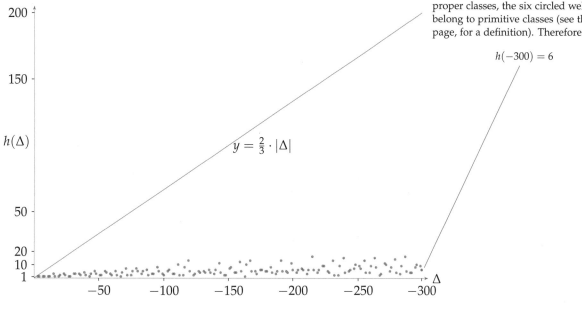

Figure 10.3: There are ten proper classes of positive-definite binary quadratic forms with discriminant $-300$. Every positive-definite binary quadratic form of discriminant $-300$ has a well, properly equivalent to one of the ten wells drawn above. Among these ten proper classes, the six circled wells belong to primitive classes (see the next page, for a definition). Therefore

$$h(-300) = 6$$

BETTER ESTIMATES for $h(\Delta)$ are proportional to $\sqrt{|\Delta|}$ rather than $|\Delta|$ itself. The first such estimate is due to Gauss, who suggests[24] that, on average, $h(\Delta)$ can be approximated by an expression $m\sqrt{|\Delta|} - b$, where $m$ and $b$ are somewhat exotic constants:

$$b = \frac{2}{\pi^2} \text{ and } m = \frac{2\pi}{7\zeta(3)}, \text{ where } \zeta(3) = \frac{1}{1^3} + \frac{1}{2^3} + \frac{1}{3^3} + \cdots.$$

[24] See the *Disquitiones*, Art. 302.

A plot of negative **fundamental**[25] discriminants $\Delta$ and their class numbers $h(\Delta)$ appears on the opposite page. Through the middle of the data, we find Gauss's average $m\sqrt{|\Delta|} - b$.

[25] Fundamental discriminants are those discriminants $\Delta$ which cannot be expressed as $f^2 \Delta'$ for a discriminant $\Delta'$ and an integer $f > 1$. By focusing on fundamental discriminants, subtleties with nonprimitive forms disappear.

But when we look at such scattered data, we are interested not just in averages but in extremes. Assuming the *Grand Riemann Hypothesis* (a grand extension of the Riemann hypothesis), one can prove that, on one hand, infinitely many values of $h(\Delta)$ lie in the green region, and on the other hand, that no values of $h(\Delta)$ lie above the green region. The region is explicitly described[26] by a complicated formula.

$$\frac{e^\gamma}{\pi}\sqrt{|\Delta|}\log\log|\Delta| \leq y \leq \frac{2e^\gamma}{\pi}\sqrt{|\Delta|}\left(\log\log|\Delta| + \lambda(|\Delta|)\right),$$

where

$$\lambda(|\Delta|) = (1/2 - \log(2)) + \frac{1}{\log\log|\Delta|}.$$

[26] This description is due to Lamzouri, Li, and Soundararajan, in *Math. Comp.* **84** (2015), refining asymptotic results of Littlewood (1927-28). There it is only mentioned for fundamental discriminants greater than $10^{10}$. But for fundamental discriminants smaller than $10^{10}$, the bound also holds thanks to the numerical computations of Buell. In these formulae, $\gamma$ denotes the Euler-Mascheroni constant.

Within the green region are the record-setters: the discriminants which "raise the bar" are circled in red and trace a path through the green region, as their class numbers are higher than all those that come before. For example, $h(-3671) = 91$, and no smaller discriminant has a class number at 91 or above.

At the bottom are the final discriminants for each class number. It is known[27] that each class number can arise from only finitely many discriminants. For example, $-163$ is the final negative discriminant with class number 1. As scattered as the data appear, they will never dip down to class number 1 ever again when $|\Delta| > 163$.[28]

[27] See "On the class number in imaginary quadratic fields", by H. Heilbronn, in *Quarterly J. of Math.* **5** (1934).

**Theorem 10.20 (Class number 1 theorem of Heegner, Stark, Baker)**
*The only negative discriminants $\Delta$ with $h(\Delta) = 1$ are the following:*

$$\Delta = -3, -4, -7, -8, -11, -12, -16, -19, -27, -28, -43, -67, -163.$$

[28] The history of this deep theorem, the solution to the Gauss Class Number 1 problem for negative discriminants, is complicated. See "The Gauss Class Number Problems," by H. M. Stark, Clay Mathematics Proceedings, Vol. 7, 2007 for more.

*(The non-fundamental discriminants are highlighted in green.)*

Similarly, when $\Delta$ is fundamental, negative and $|\Delta| > 427$, one will never find that $h(\Delta) = 2$. These are discriminants which are the last of their kind; no fundamental negative discriminant with $|\Delta| > 907$ will have class number 3.

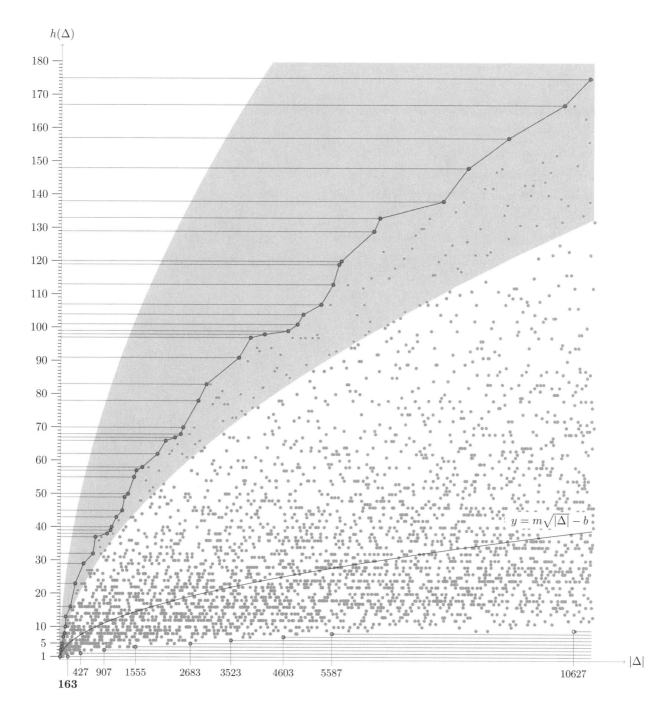

Figure 10.4: A plot of negative fundamental discriminants and their Gauss class numbers, for $3 \leq |\Delta| \leq 11000$.

We turn our attention now to much easier things – binary quadratic forms of discriminant zero. This will tie up a few loose ends and lead to the next chapter.

POSITIVE SEMIDEFINITE forms are those binary quadratic forms whose values (in the range topograph) include zero and positive integers, but never any negative integers. When zero appears in the range topograph, we follow Conway's approach to color the region blue and call it a lake.

**Definition 10.21** A **lake** is a region in the range topograph at which the value is equal to zero.

Consider a quadratic form $Q$ of discriminant $\Delta$

**Proposition 10.22** *Adjacent to a lake in a range topograph of $Q$, the values along the lakeshore form an arithmetic progression with common difference $\sqrt{\Delta}$. In particular, the discriminant $\Delta$ is a square.*

PROOF: Consider three regions adjacent to a lake.

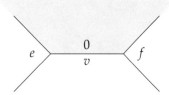

By Theorem 9.27, $e, v, f$ form an arithmetic progression, and this propagates around the lakeshore. The common difference is $f - v$. We compute the discriminant via a triad,

$$\Delta\left(\begin{array}{c}0\\ \phantom{v}\diagup\!\!\!\diagdown f\\ v\end{array}\right) = f^2 + v^2 - 2fv = (f - v)^2.$$

Hence the common difference $f - v$ in the arithmetic progression equals $\pm\sqrt{\Delta}$. In particular, $\Delta = (f - v)^2$ is a square. ∎

**Corollary 10.23** *If $Q$ is a positive semidefinite form, then $\Delta(Q) = 0$ and the range topograph of $Q$ contains a lake surrounded by a constant progression.*

PROOF: Since $Q$ is semidefinite, a lake appears in the range topograph. The arithmetic progression along the lakeshore can never include a negative number, since $Q$ is positive semidefinite. But what sort of arithmetic progression never drops below zero? Only a constant progression! Therefore we find a positive constant $a$ along the lakeshore. The previous proposition gives $\Delta = 0$. ∎

**Corollary 10.24 (Classification of forms of discriminant zero)** *If $Q$ is a positive semidefinite form, then $Q$ is equivalent to the quadratic form $ax^2$ for some positive integer $a$.*

PROOF: The cell pictured in the margin occurs in the topograph of the binary quadratic form $Q'(x, y) = ax^2$. The previous corollary implies that $Q$ is equivalent to the form $ax^2$. ∎

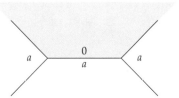

Figure 10.5: A special case occurs when $\Delta = 0$; the arithmetic progression has common difference zero, so it is a constant progression.

THE CLASS NUMBER $h(0)$ can be defined analogously to the class numbers $h(\Delta)$ for $\Delta < 0$. Let $h(0)$ denote the number of classes of primitive positive-semidefinite forms of discriminant zero. The previous result implies that $h(0) = 1$.

On the other hand, as simple as the classification is, positive semidefinite forms exhibit a new kind of isometry – a hyperbolic "rotation" around a lakeshore.

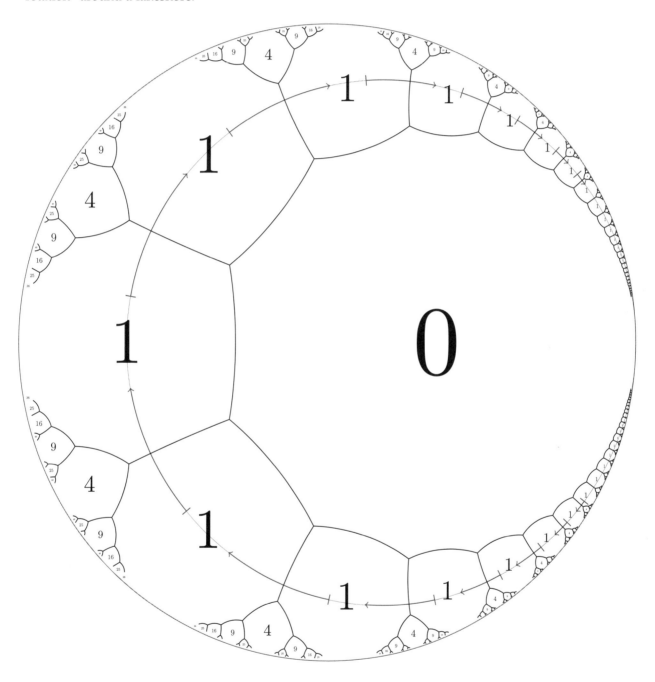

## Historical notes

The arithmetic progression rule and climbing principle, and the language of wells and double-wells, belong to Conway's approach to binary quadratic forms. This approach was first written in Conway's book "The Sensual (quadratic) Form."[29] There Conway sets out the entire language of lax vectors, bases, and superbases, arranges them into the topograph, and plots the values of binary quadratic forms.

The payoff of Conway's approach occurs with the appearance of wells and double-wells. For comparison, one finds the reduction of forms in Lagrange's *Recherches D'Arithmétique* and (explicitly named) in Gauss's *Disquisitiones*. We call a binary quadratic form $ax^2 + bxy + cy^2$ **Lagrange-reduced** if

$$|b| \leq a \text{ and } |b| \leq c.$$

Lagrange proves that every binary quadratic form is equivalent to such a reduced form.[30] The language of Lagrange is that of "transformation" rather than "equivalence", but the end result is the same as allowing proper and improper equivalence. If one fixes a discriminant[31] $\Delta$, and seeks Lagrange-reduced forms of discriminant $\Delta$, there are finitely many possibilities.

In a later Corollary (Part I.9), Lagrange obtains our estimate on the mimimum; more precisely, he proves that if $ax^2 + bxy + cy^2$ is (Lagrange) reduced, then

$$\min\{|a|, |c|\} < \sqrt{|\Delta|/3}.$$

This implies our Theorem 10.9.

In order to study the divisors of $x^2 + cy^2$ (for various $c$), Lagrange is led to study reduced forms of discrimant $-4c$ (for various $c$). Effectively, Lagrange identifies all reduced forms of discriminant $-4c$ for $1 \leq c \leq 12$. From an algebraic perspective, his method differs little from ours in the proof of Theorem 10.12.

While Lagrange may deserve credit for the algebraic results of this chapter, the framework – of forms, equivalence, and reduction – originates in the *Disquisitiones* of Gauss. Except for the fact that Gauss only considers forms $ax^2 + 2bxy + cy^2$ (with even middle coefficient), Gauss's approach is more general than all who came before and most who came after. For Gauss, equivalence is a special case of "containment" or "implication" of forms.[32] It is Gauss who introduces the langauge of equivalence (proper and improper) of forms, and of "reduced forms."

For reduction, Gauss's definition seems *ad hoc* on its own, though it simply tightens the results of Lagrange. For Gauss, a form $ax^2 +$

---

[29] See Chapter 1 of "The Sensual (quadratic) Form," by John H. Conway, assisted by Francisc Y.C. Fung, published by the Mathematical Association of America, 1997.

[30] See Théorème II and its Corollaire I of *Recherches D'Arithmétique* (1773,1775), first published in N. Mém. Berlin, collected in Lagrange's *Oeuvres* 3, 695–795.

[31] Lagrange works with the quantity $4ac - b^2$: negative the discriminant we work with.

[32] See Art. 157 of the *Disquisitiones*. In the most modern terminology, we might say that Gauss considers the entire category whose objects are binary quadratic forms $Q$ with even middle coefficient, and where a morphism $Q \to Q'$ is a two-by-two integer matrix $M$ such that $Q'(\vec{v} \cdot M) = Q(\vec{v})$ for all (row) vectors $\vec{v}$.

$2bxy + cy^2$ *of negative discriminant* is reduced when

$$|a| \leq \sqrt{|\Delta|/3} \text{ and } |a| \leq |c| \text{ and } 2|b| \leq |a|.$$

We call a form **Gauss-reduced** if it satisfies the above conditions.

The conditions for $ax^2 + bxy + cy^2$ to be Lagrange- or Gauss-reduced are roughly the same as requiring a well to occur at home basis.[33] But the well exhibits structure that is hidden in the algebra – the possible symmetries at a well or double-well and complications with ambiguous forms. Conway's topograph makes the reduction of forms transparent, simplifying the proofs given in the *Disquisitiones*.

As Gauss considers forms as creatures of interest, he uses biological nomenclature, placing forms into *genera*, and subdividing each genus into *classes*.[34] Gauss's classes are determined by proper equivalence among binary quadratic forms. The notion of "genus" is a bit beyond the scope of this text, but we refer to the survey of Lemmermeyer[35] and Cox's book "Primes of the form $x^2 + ny^2$."[36]

The number of classes of a given discriminant $\Delta$ is the class number $h(\Delta)$, and Gauss seems the first to study it. Art. 301-303 of the *Disquisitiones* draws conclusions from computational evidence and heuristics. A footnote of Gauss reads,

> While this was being printed we worked out the table [of classifications of forms of discriminants] up to $-3000$ and also for the whole tenth thousand, for many separate hundreds, and for many carefully selected individual determinants.[37]

Of his own extensive observations, Gauss writes that "*rigorous* proofs... seem to be very difficult."[38] And so they are, and have occupied the attention of mathematicians ever since.

For Gauss, a key to understanding $h(\Delta)$ was the understanding of *ternary* quadratic forms – quadratic forms in three variables. Indeed, Gauss understood that $h(\Delta)$ is related to the number of representations of $\Delta$ or $\Delta/4$ as the sum of three squares.[39] In this way, the statistics of $h(\Delta)$ relate to the statistics of lattice points on spheres.

The study of $h(\Delta)$ changed course after the *class number formulae* of Dirichlet (1839).[40] This formula implies, for negative fundamental $\Delta$,

$$h(\Delta) = \frac{w\sqrt{|\Delta|}}{\pi} \cdot L(1, \chi_\Delta),$$

where $w$ is the number of proper isometries of a quadratic form of discriminant $\Delta$ (so $w = 1$ for $|\Delta| > 4$), and $L(1, \chi)$ is the *Dirichlet L-function* evaluated at 1. After Dirichlet, the analysis of L-functions became central to number theory, and progress therein has been the source of our best estimates on $h(\Delta)$.

[33] See the exercises of this chapter, where the connection is made more precise.

[34] He also places forms into *orders*, according to the greatest common divisors of their coefficients.

[35] "The development of the principal genus theorem" by Franz Lemmermeyer, Chapter VIII.3 of "The shaping of arithmetic after C.F. Gauss's *Disquisitiones Arithmeticae*," Springer-Verlag (2007).

[36] See Chapter 1 of "Primes of the form $x^2 + ny^2$: Fermat, class field theory, and complex multiplication," by David A. Cox (1989) for a wonderful historical treatment.

[37] Art. 303 of the *Disquisitiones*, translation by Arthur A. Clarke.

[38] loc. cit.

[39] See Art. 292 of the *Disquisitiones*. Gauss cites earlier work of Legendre, "Recherches d'Analyse indéterminée", in *Histoire de l'Acadamie Royale des Sciences de Paris* (1785).

[40] Compare to §8 of Dirichlet's "Recherches sur diverses applications de l'analyse infinitésimal à la théorie des nombres," found in *J. reine angew. Math.* **19** (1839) or in his *Mathematische Werke*, vol. 1. Dirichlet considers forms with even middle term, like Gauss.

## Exercises

1. Solve the Diophantine equation $x^2 + 2xy + 3y^2 = n$ when $n = 1$, $n = 2$, $n = 3$, and $n = 4$.

2. Solve the Diophantine equation $2x^2 + 3xy + 13y^2 = n$ when $n = 27$, $n = 48$, and $n = 60$.

3. What is the smallest positive integer $n$ that can be expressed in the form $306x^2 + 1000xy + 817y^2$?

4. Prove that every positive-definite binary quadratic form of discriminant $-12$ is equivalent to $x^2 + 3y^2$ or to $2x^2 + 2xy + 2y^2$.

5. Find two positive-definite binary quadratic forms of discriminant $-15$ which are *not* properly equivalent.

6. The following three quadratic forms have the same discriminant.[41]

   $$Q_1(x,y) = 41x^2 + 70xy + 30y^2, \quad Q_2(x,y) = 7x^2 + 36xy + 47y^2, \quad Q_3(x,y) = 23x^2 + 76xy + 63y^2.$$

   Which are properly equivalent? Place them into classes accordingly.

   [41] This exercise is adapted from Gauss's *Disquisitiones*, Art. 173.

7. Prove that $h(-163) = 1$ and $h(-427) = 2$.

8. Fermat stated the following three results. Use the topograph and quadratic residues as in Theorem 10.15 to prove them.

   (a) If $p \equiv 1 \bmod 6$ then $p$ is represented by $x^2 + 3y^2$.

   (b) If $p \equiv 1$ or $3 \bmod 8$ then $p$ is represented by $x^2 + 2y^2$.

   (c) If $p \equiv \pm 1 \bmod 8$ then $p$ is represented by $x^2 - 2y^2$.

9. Write a computer program, which takes as input a negative integer $\Delta$ and which outputs a list of all wells of discriminant $\Delta$.

10. Let $\Delta$ be a negative integer. The number of wells of discriminant $\Delta$ is bounded by $|\Delta|/3$. Improve this bound as much as possible.

11. Let $\Delta$ be a negative integer. Prove that if $\Delta$ is odd, then there are no double-wells of discriminant $\Delta$. Prove that if $\Delta = -4D$ is a multiple of 4, then the number of double-wells[42] of discriminant $\Delta$ is the number of positive divisors of $D$ less than or equal to $\sqrt{D}$.

    [42] When counting double-wells, count ordered triples $(u, v, w)$ such that $u \leq v \leq w$ and $w = u + v$ and $u^2 + v^2 + w^2 - 2uv - 2vw - 2wu = \Delta$.

12. The following exercise relates the notions of well and double-well to classical reduction theory. A positive-definite binary quadratic form $Q(x,y) = ax^2 + bxy + cy^2$ is called **reduced** if $|b| \leq a \leq c$ **and** if either $|b| = a$ or $a = c$, then $b \geq 0$.

(a) Demonstrate that a reduced positive-definite binary quadratic form has a well at home-basis. How does the well depend on whether $b > 0$, $b < 0$, or $b = 0$?

(b) Consider a single-well in which the values $u, v, w$ are all distinct, with $u < v < w$. Prove that this single-well occurs in the range topograph of a reduced form at home basis. How does the reduced form depend on the orientation[43] of $u, v, w$?

[43] I.e., whether $u, v, w$ occur clockwise or counterclockwise around the triad.

(c) Prove that every positive-definite binary quadratic form is properly equivalent to exactly one reduced form

13. Consider a zig-zag path through a range topograph, in which all displayed values are positive and climbing as below.

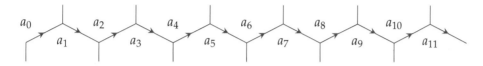

Prove that the values $a_n$ grow exponentially in $n$. In other words, find positive real numbers $k, b$ for which $b > 1$ and

$$a_n \geq k \cdot b^n \text{ for all } n \geq 0.$$

14. Does there exist a Gaussian integer whose norm is 57? An Eisenstein integer?

15. Describe algebraically the conditions on a matrix $M = \begin{pmatrix} a & b \\ c & d \end{pmatrix}$, for $M$ to be an isometry of $Q(x, y) = x^2 + y^2$. Use these conditions to demonstrate that $Q$ has exactly four proper isometries.

16. Let $Q(x, y) = x^2$. Describe the proper isometries of $Q$.

17. Let $Q(x, y) = ax^2 + bxy + cy^2$, for real numbers $a, b, c$. How does the shape of the curve $Q(x, y) = 1$ (drawn in the plane) depend on the discriminant $\Delta(Q)$?

18. Let $Q(x, y) = ax^2 + bxy + cy^2$ be a binary quadratic form with $\Delta(Q) < 0$ and $a > 0$. Prove by analytic means that $Q(x, y) > 0$ for all pairs of real numbers $(x, y)$ except $(0, 0)$.

19. Let $Q(x, y) = ax^2 + bxy + cy^2$ be a binary quadratic form with $\Delta(Q) = 0$ and $a > 0$. Check that

$$Q(x, y) = a \left( x + \frac{b}{2a} y \right)^2.$$

Deduce that every value on the range topograph is zero or positive. Use this to prove that a lake occurs on the range topograph. Hint: look around the lowest value in the range topograph.

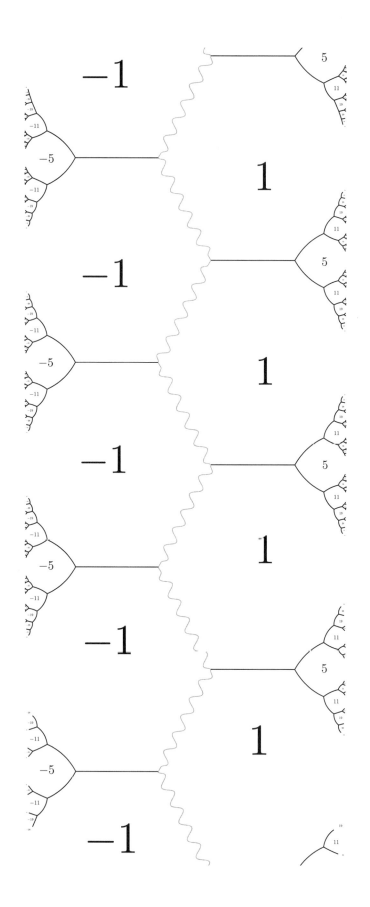

# 11
# Indefinite Forms

For many quadratic forms, we find both positive and negative values in the topograph. These forms we call **indefinite**. The features of indefinite forms – sign changes and zero – are drawn on the range topograph with **water**. Recall that regions in the range topograph with value zero are called **lakes** and colored blue. Extending this metaphor, Conway defines the river.

**Definition 11.1** The **river** is the set of edges in the range topograph which separate positive values from negative values.

Below is a range topograph with two lakes and a river.

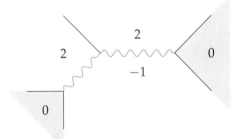

On the opposite page is the topograph of $Q(x,y) = x^2 + xy - y^2$. It exhibits an endless periodic river – the pattern of 1s and -1s repeats as one travels upwards or downwards. Indeed, as the pattern repeats once, it must repeat again and again in both directions, since the range topograph is determined by following the arithmetic progression rule throughout. A Diophantine consequence is that the equation $x^2 + xy - y^2 = 1$ has infinitely many solutions.

In this chapter, we will see that for indefinite forms, range topographs without lakes have endless periodic rivers. The Diophantine consequences include the solution of **Pell's equation** $x^2 - Ny^2 = 1$ for all positive nonsquare $N$.

Proposition 10.22 states that if a lake occurs in a range topograph, then its discriminant is a square. The converse can be proven algebraically.

**Proposition 11.2** *If $Q$ is a binary quadratic form whose discriminant $\Delta$ is a square, then the range topograph of $Q$ has at least one lake.*

PROOF: Let $Q(x,y) = ax^2 + bxy + cy^2$ be a binary quadratic form such that $\Delta = b^2 - 4ac$ is a square. If $a = 0$, then
$$Q(1,0) = a(0)^2 + b(1)(0) + c(0)^2 = 0.$$
Hence if $a = 0$, then the range topograph has a lake at $(1,0)$.

If $a \neq 0$, let $x = \sqrt{\Delta} - b$ and let $y = 2a$.[1] We compute directly,
$$Q(x,y) = a(\sqrt{\Delta} - b)^2 + b(\sqrt{\Delta} - b)(2a) + c(2a)^2$$
$$= a(\Delta - 2b\sqrt{\Delta} + b^2) + a(2b\sqrt{\Delta} - 2b^2) + a(4ac)$$
$$= a\left(\Delta - b^2 + 4ac\right) = 0.$$

Let $g = \mathrm{GCD}(x,y)$. Since $a \neq 0$, we find that $y \neq 0$ and $g \neq 0$. Scaling by $g$, and applying Proposition 9.25, we find two facts:
$$Q\left(\frac{x}{g}, \frac{y}{g}\right) = \frac{1}{g^2} Q(x,y) = 0, \text{ and } \pm\left(\frac{x}{g}, \frac{y}{g}\right) \text{ is a primitive lax vector.}$$
Hence the range topograph of $Q$ has a lake at $\pm(x/g, y/g)$. ∎

**Proposition 11.3** *If the range topograph of a binary quadratic form $Q$ has a lake, then $Q$ is equivalent to $ax^2 + \sqrt{\Delta}xy$, for some integer $a$. If $\Delta \neq 0$, then one may find such an integer $a$ satisfying $0 \leq a < \sqrt{\Delta}$.*

PROOF: Alongside a lakeshore, one finds an arithmetic progression with difference $\sqrt{\Delta}$. If $\Delta = 0$, all values along the lakeshore are the same, and call this value $a$. The result follows from Corollary 10.24.

If $\Delta \neq 0$, then in the progression along the lakeshore, the smallest non-negative value is an integer $a$ satisfying $0 \leq a < \sqrt{\Delta}$. Next to $a$ along the lakeshore, we find the value $a + \sqrt{\Delta}$. Thus the topograph of $Q$ contains the triad $\overset{a}{\underset{0}{\prec}}(a + \sqrt{\Delta})$ (or its mirror image).

This triad also appears in the range topograph of $ax^2 + \sqrt{\Delta}xy$ (at home basis). Thus $Q$ is equivalent to $ax^2 + \sqrt{\Delta}xy$. ∎

When $\Delta$ is a positive square, the **class number** $h(\Delta)$ is the number of classes of primitive binary quadratic forms[2] of discriminant $\Delta$.

**Corollary 11.4** *If $\Delta$ is a positive square, then $h(\Delta) \leq 2\sqrt{\Delta}$.*

PROOF: Every form of positive square discriminant $\Delta$ contains the triad $\overset{a}{\underset{0}{\prec}}(a + \sqrt{\Delta})$ or $(a + \sqrt{\Delta})\overset{a}{\underset{0}{\succ}}$, for some integer $a$ satisfying $0 \leq a < \sqrt{\Delta}$. The result follows. ∎

---

[1] Where do these values of $x$ and $y$ come from? The quadratic formula gives them. Indeed, to find a solution to $ax^2 + bxy + cy^2 = 0$ (with $y \neq 0$), it suffices, by dividing through by $y^2$, to find a solution to the equation
$$a(x/y)^2 + b(x/y) + c = 0.$$
A solution $x/y$ can be found with the quadratic formula:
$$\frac{x}{y} = \frac{-b \pm \sqrt{\Delta}}{2a}.$$

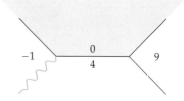

Figure 11.1: The arithmetic progression $-1, 4, 9$ has common difference $5$. The discriminant equals $5^2 = 25$.

[2] No mention of positivity is appropriate, since such forms have positive and negative values.

INDEFINITE FORMS 283

**Theorem 11.5 (The discriminant counts the lakes)** *Let Q be a nonzero[3] binary quadratic form with discriminant $\Delta$. The number of lakes in its range topograph equals*

$$\begin{cases} 0 & \text{if } \Delta \text{ is not a square;} \\ 1 & \text{if } \Delta = 0; \\ 2 & \text{if } \Delta \text{ is a nonzero square.} \end{cases}$$

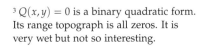

[3] $Q(x,y) = 0$ is a binary quadratic form. Its range topograph is all zeros. It is very wet but not so interesting.

PROOF: By Proposition 10.22, if $\Delta$ is not a square, then the range topograph of $Q$ cannot have a lake. Conversely, by Proposition 11.2, if the range topograph of $Q$ has a lake, then $\Delta$ must be a square. It remains to prove that there is exactly one lake if $\Delta = 0$, and there are exactly two lakes if $\Delta$ is a positive square. In either case, $Q$ is equivalent to $ax^2 + \sqrt{\Delta}xy$, for some integer $a$.

If $\Delta = 0$, then $Q$ is equivalent to $ax^2$. But the only lax primitive vector $\pm(x,y)$ for which $ax^2 = 0$ is $\pm(0,1)$. Hence the only lake in the range topograph of $ax^2$ is at $\pm(0,1)$. By equivalence, there is only one lake in the range topograph of $Q$.

If $\Delta$ is a positive square, then $Q$ is equivalent to $ax^2 + \sqrt{\Delta}xy$. The lakes of $ax^2 + \sqrt{\Delta}xy$ correspond to lax primitive vectors $\pm(x,y)$ for which $ax^2 + \sqrt{\Delta}xy = 0$. Factoring, this is equivalent to

$$x(ax + \sqrt{\Delta}y) = 0.$$

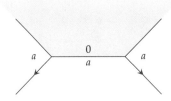

Figure 11.2: Alternatively, when $\Delta = 0$, the climbing principle implies that no other lakes will occur. The values climb as one walks away from the lake, and so no other lake can occur.

This equality holds if $x = 0$, or if $x/y = -\sqrt{\Delta}/a$. The only lax primitive vector with $x = 0$ is $\pm(0,1)$. The only lax primitive vector $\pm(x,y)$ which satisfies $x/y = -\sqrt{\Delta}/a$ is the result of reducing[4] the fraction:

$$x = \frac{-\sqrt{\Delta}}{\text{GCD}(\sqrt{\Delta},a)}, \quad y = \frac{a}{\text{GCD}(\sqrt{\Delta},a)}.$$

[4] If $Q$ is primitive, then $\text{GCD}(\sqrt{\Delta},a) = 1$ by looking at the triad from the previous page. In this case the lakes are located at $\pm(0,1)$ and $\pm(-\sqrt{\Delta},a)$.

Hence exactly two lakes occur in the range topograph of $Q$: one at $(\pm 1, 0)$ and the other at $\pm(x,y)$ with $x,y$ as above. ∎

On the next pages, we study the topography of lakes and rivers.

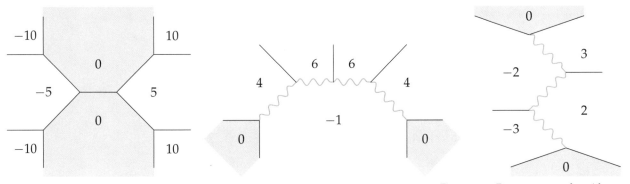

Figure 11.3: Range topographs with $\Delta = 25$. At left, a double-lake.

RIVERS separate positive values in the topograph from negative values. Their basic features follow from a few thought exercises.

**Proposition 11.6 (The topology of rivers)** *Rivers cannot suddenly stop, nor can they fork. Their flow can only terminate in a lake.*

PROOF: Consider a triad containing a segment of river; thus it contains a positive value, a negative value, and another value which we call $f$. Either $f < 0$ or $f = 0$ or $f > 0$. The results are below.

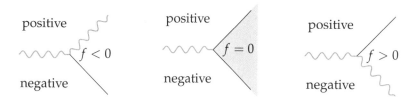

∎

**Proposition 11.7** *If a positive value and a negative value occur in a range topograph, then the range topograph must contain a lake or river (or both).*

PROOF: By Theorem 9.14, one may walk from a positive value to a negative value in the range topograph. Along the way, consider the values on one side of the path.

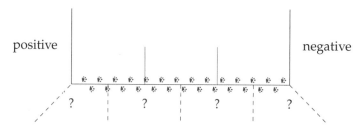

Along the path, one must find a lake, or else one must find a direct transition from positive to negative – a river. One of the dashed lines must be a river, or one of the ?-marked regions must be a lake. ∎

**Corollary 11.8** *Let $Q$ be a quadratic form of discriminant $\Delta$. If $\Delta > 0$ and $\Delta$ is not a square, then the range topograph of $Q$ contains an endless river.*

PROOF: Since $\Delta > 0$, the form $Q$ is not definite nor semidefinite, and since $\Delta$ is nonsquare the range topograph does not contain a lake. Therefore both positive and negative values occur in the range topograph of $Q$, and zero does not occur. A river must occur by Proposition 11.7. The river cannot terminate since there are no lakes, and so the river is endless. ∎

**Proposition 11.9 (Uniqueness of the river)** *A topograph can have at most one river.*

PROOF: If there were two rivers, then walk from one to the other.

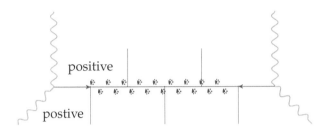

Figure 11.4: If the path from one river to another is on the positive side of the river, then the values become larger and larger as one proceeds eastward. The topograph pictured here cannot occur.

As one steps away from the river, the values on either side are both positive or both negative. By the climbing principle, the values must grow (more positive or more negative) as one continues along this path. Thus one cannot follow this path to another river. ∎

**Proposition 11.10** *If $\Delta$ is a positive square, then there is a double-lake (two lakes sharing an edge) and no river or there are two lakes joined by a river.*

PROOF: When $\Delta$ is a positive square, there are two lakes in the topograph. In the case of a double-lake, the climbing principle shows that there cannot be a river.

If there are two nonadjacent lakes, then the nonconstant arithmetic progression around each lake does not include zero. Where each progression changes sign, a river segment protrudes from the lakeshore.

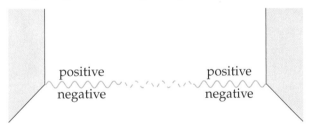

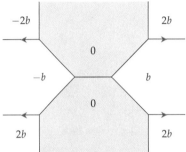

Figure 11.5: A double-lake must have numbers $\pm b$ on either side, by the arithmetic progression rule. Such a form is equivalent to $\sqrt{\Delta}xy$. The arithmetic progressions along the lakeshores guarantee that positive numbers can be found on one shore of the double-lake and negative numbers on the other shore. The climbing principle guarantees that the numbers grow more positive or more negative as one travels away from the double-lake.

Proposition 11.9 joins these river segments into a single river. ∎

**Proposition 11.11** *A topograph with a river has positive discriminant.*[5]

PROOF: Consider a cell along the river, with common difference $h$.

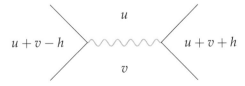

The discriminant is $\Delta = h^2 - 4uv$. Since $u$ and $v$ have opposite signs, $-4uv > 0$. Since $h^2 \geq 0$, we find that $\Delta > 0$. ∎

[5] A table at the end of this chapter summarizes the connections between discriminant and the features expected on the topograph. We have now proven all the results needed to verify this table.

286

A closer analysis of river segments demonstrates that there are only finitely many of a given discriminant.

**Lemma 11.12** *Fix a positive integer $\Delta$. Then there are only finitely many distinct river segments with discriminant $\Delta$.*

PROOF: Consider a river segment of discriminant $\Delta$.

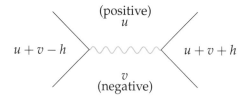

Thus $h^2 - 4uv = \Delta$. As $u > 0$ and $v < 0$, we have $-4uv > 0$. Hence

$$0 \leq h^2 < \Delta \text{ and } 0 < |u| \cdot |v| \leq \frac{\Delta}{4}.$$

It follows that there are no more than $\sqrt{\Delta}$ possible values of $h$. Similarly, the number of possible values of $u$ is bounded by $\Delta/4$. The value of $h$ and the value of $u$ determine the value of $v$, when $\Delta$ is fixed. So when $\Delta > 0$, there can be no more than $\frac{1}{4}\Delta^{3/2}$ river segments of discriminant $\Delta$. ∎

**Theorem 11.13 (Periodicity of the river)** *If $\Delta$ is positive and nonsquare, then every range topograph of discriminant $\Delta$ has an endless **periodic** river.*

PROOF: When $\Delta$ is positive and nonsquare, every range topograph of discriminant $\Delta$ has an endless river. Following the river, from segment to segment, one must find a repetition by Lemma 11.12. But, once one finds a repetition, the pattern repeats for ever after. ∎

The Diophantine corollary is powerful.

**Corollary 11.14** *Let $Q(x,y)$ be a binary quadratic form of positive, nonsquare discriminant. Let $N$ be a nonzero integer. If the Diophantine equation $Q(x,y) = N$ has one solution, then it has infinitely many solutions.*

PROOF: Given $Q(x,y) = N$, let $g = \mathrm{GCD}(x,y)$, and let $u = x/g$ and $v = y/g$. Let $n = Q(u,v)$. Then $\pm(u,v)$ is a primitive lax vector and

$$N = Q(x,y) = Q(gu, gv) = g^2 Q(u,v) = g^2 n.$$

Since the river is periodic, the entire topograph is periodic; the value $n$ appears infinitely many times in the topograph. Each time $n$ appears at a lax vector $\pm(u', v')$, we find a new solution $x' = gu'$, $y' = gv'$ satisfying $Q(x', y') = N$. ∎

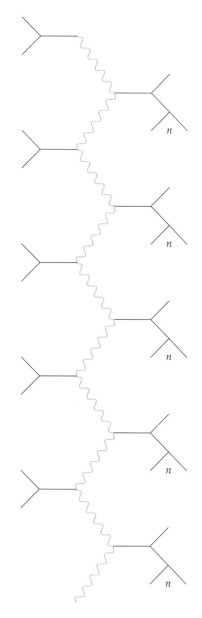

Figure 11.6: If a number $n$ occurs once on the topograph, then it will occur again and again, at the same location relative to the river.

FOR CONTRAST, we solve two quadratic Diophantine equations.

**Problem 11.15** Solve the Diophantine equation $x^2 + 3y^2 = 7$.

SOLUTION: The quadratic form $Q(x,y) = x^2 + 3y^2$ is positive-definite. Around home basis, we find its double-well, and four solutions at two lax vectors:

$$(x, y) = (2, 1) \text{ or } (-2, 1) \text{ or } (2, -1) \text{ or } (-2, -1).$$

The climbing principle implies that no other solutions exist. ✓

**Problem 11.16** Solve the Diophantine equation $x^2 - 3y^2 = 7$.

SOLUTION: The quadratic form $Q(x,y) = x^2 - 3y^2$ is indefinite, of nonsquare positive discriminant 12. Hence its topograph has no lakes and an infinite periodic river. We plot it below.

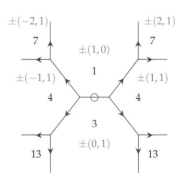

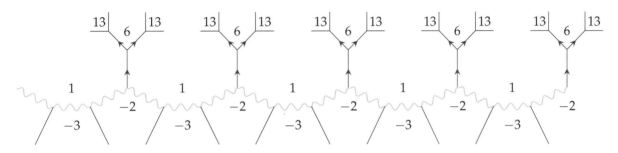

If there existed a primitive lax vector $\pm(x, y)$ for which $Q(x, y) = 7$, it would be on the positive side of the river. But the climbing principle and the periodicity of the river demonstrate that 7 never appears. Hence $x^2 - 3y^2 = 7$ has no solutions. ✓

On the other hand, observe that $x^2 - 3y^2 = 1$ has infinitely many solutions. This is a special case of Pell's Equation.

**Theorem 11.17 (Infinitude of solutions to Pell's equation)** *If $N$ is a positive nonsquare, then the Diophantine equation $x^2 - Ny^2 = 1$ has infinitely many solutions.*

PROOF: The discriminant of the quadratic form $Q(x, y) = x^2 - Ny^2$ is $4N$. As long as $N$ is a positive nonsquare, $4N$ is a positive nonsquare. Hence the topograph of $Q$ contains an endless periodic river. Since $x^2 - Ny^2 = 1$ has one solution (namely $x = 1, y = 0$), the Diophantine equation has infinitely many solutions. ■

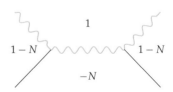

The topograph not only illustrates the existence of infinitely many solutions – it provides a method for finding them in sequence by following the river.

PROPER ISOMETRIES of an indefinite binary quadratic form must arise from proper symmetries of its river. Let $Q$ be a quadratic form with positive nonsquare discriminant, and $M$ a proper isometry of $Q$. Then $M$ sends river to river, positives to positives; therefore $M$ must translate the topograph along the river. If we stretch the river straight and assign each edge length 1, then $M$ must be a river-translation by an integer number of units.

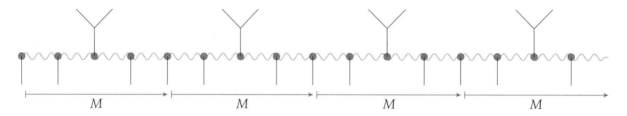

Figure 11.7: Pictured above, $M$ is a river-translation by 4 units.

If $t$ is the *smallest* positive length of a river-translation, then *every* isometry is a translation by a multiple of $t$.[6]

More interesting are the **improper isometries** of $Q$. Geometrically,[7] these are the symmetries of the range topograph which *reverse* orientation – the mirror symmetries. **Ambiguous forms** are quadratic forms which possess an improper isometry. Ambiguous indefinite forms therefore have *both* a translational symmetry $M$ of their river as well as a mirror symmetry $T$.

Like proper isometries, the improper ones must also send river to river, positives to positives. To reverse orientation, an improper isometry must transform a flow in one direction along the river to a flow in the opposite direction. It follows that if an improper isometry $T$ sends a point $p$ on the river to a point $q$, then it must act as a mirror reflection about their midpoint $m$.

[6] If one can translate by $a$ units left or right, and one can translate $b$ units left or right, then one can translate $GCD(a,b)$ units left or right. In this fashion, one can find a smallest distance of translation, which is a divisor of all others.

[7] Algebraically, these are integer matrices $T = \begin{pmatrix} a & b \\ c & d \end{pmatrix}$ such that $Q(x,y) = Q(ax+cy, bx+dy)$ and whose determinant is $-1$.

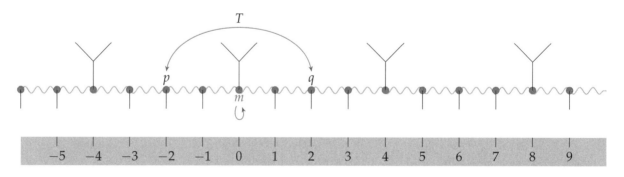

Place a ruler along the river now, with 0 at the midpoint $m$. The improper isometry $T$ now takes a point with coordinate $x$ on the ruler to the point with coordinate $-x$. We write $T(x) = -x$.

Figure 11.8: $T$ reflects the topograph, switching $p$ and $q$, and fixing their midpoint. Their midpoint may lie at a vertex, as shown, or in the middle of an edge.

IN COORDINATES, an ambiguous indefinite form must have a (minimal) river-translation $M$ and a mirror reflection $T$, given by

$$M(x) = x + t \text{ and } T(x) = -x.$$

Composing them (doing $T$ then $M$) yields a new isometry $MT(x) = -x + t$. Observe that $MT$ is a new reflection, with mirror at $t/2$. Continuing in this way, we find an infinite series of reflections across mirrors at $-3t/2, -t/2, 0, t/2, 3t/2$, etc.. In fact, there can be no other isometries.

**Theorem 11.18** *If $Q$ is an ambiguous indefinite form, then for some integer $t$, the isometries of $Q$ consist of river-translations by each multiple of $t$, and reflections across a series of mirrors located $t/2$ units apart.*

PROOF: It remains to see that there can be no other isometries of $Q$ besides the ones mentioned above. There can be no other river-translations, if we choose $t$ minimally. If there were another reflection, call it $A$, across a mirror located at $a$, then formulaically $A(x) = 2a - x$. Composing $A$ and $T$ yields a new isometry $AT(x) = 2a + x$. But this is a river-translation by $2a$ units. Hence $2a$ is a multiple of $t$, and so $a$ is a multiple of $t/2$. Hence $A$ is already among the given series of mirrors. ∎

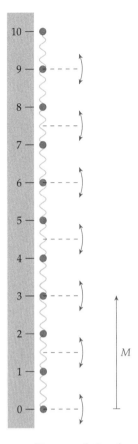

Figure 11.9: River-translations by 3 units yield mirrors at intervals of $1^{1}/2$ units. Half of the mirrors intersect the river at vertices, while the other half intersect at the midpoints of edges.

In an ambiguous form, a mirror can hit the river in three ways.

Type H            Type Y            Type λ

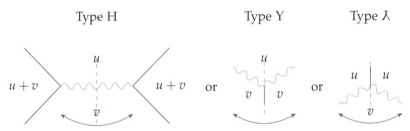

Such a symmetric cell or triad of the river will be called an **ambiguous river segment**. The river of an ambiguous form will have exactly two inequivalent ambiguous river segments; one will be located at 0 and the other at $t/2$ when we line up our ruler appropriately.

If one computes the discriminant at an ambiguous river segment, one finds the following formulae in types H, Y, λ, respectively.

$$\Delta = -4uv \text{ or } \Delta = u(u - 4v) \text{ or } \Delta = v(v - 4u).$$

In type H, for example, we find that ambiguous river segments correspond precisely to positive divisors of $\Delta/4$. In practice, one can use the discriminant formulae above to enumerate the ambiguous forms of discriminant $\Delta$ by the divisors of $\Delta$ (or of $\Delta/4$). Details and applications are in the exercises.

ALL RIVERS BEND and these bends can be used to classify indefinite forms. The starting point is a closer analysis of the values at an infinity-gon in the range topograph.

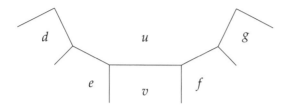

**Proposition 11.19** *Around a value $u$, the values on a range topograph form a quadratic progression with* **acceleration**[8] $2u$. *In other words, the sequence of differences form an arithmetic progression with constant difference $2u$.*

PROOF: The arithmetic progression property implies[9]

$$(v - e) = (e - d) + 2u,$$
$$(f - v) = (v - e) + 2u,$$
$$(g - f) = (f - v) + 2u.$$

Hence we find that the sequence of differences for $d, e, v, f, g$ is an arithmetic progression with common difference $2u$.

[8] Quadratic progressions with acceleration $2u$ are sequences indexed by $n$ of the general form

$$s_n = un^2 + hn + v.$$

Here the index $n$ measures the location along the infinifty-gon, and $s_n$ is the value at that location.

[9] Since $e, u+v, f$ form an arithmetic progression,

$$f - (u + v) = (u + v) - e.$$

It follows that

$$(f - v) = (v - e) + 2u.$$

The other cases are the same.

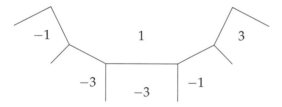

■

The values of a quadratic progression fit a parabolic curve, if one tick left or right corresponds to one step in the progression. For example, the quadratic progression below is graphed in the margin.

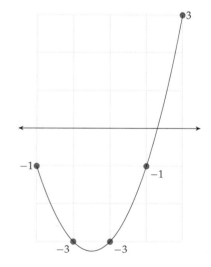

A degenerate case occurs when $u = 0$. In this case, the values around $u$ form a "quadratic" progression with acceleration zero; this is just a linear (or arithmetic) progression. We have seen this case when we observed that the values around a lake form an arithmetic progression. We will be interested in the case $u \neq 0$ in what follows.

**Corollary 11.20** *A river can occupy only finitely many edges from an infinity-gon.*

PROOF: Follow a river around the edges of an infinity-gon.

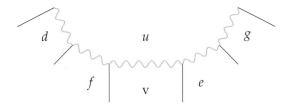

If the value $u$ is positive, then the values across the river are negative. But in this case, the values $d, f, v, e, g$ form a quadratic progression, accelerating upwards. Hence as one travels far enough, to the left and to the right, the values across from $u$ must transition from negative to positive. At these transitions, the river bends.

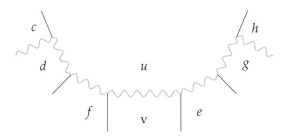

Figure 11.10: If $c > 0$ and $h > 0$, the river bends as pictured here.

When $u < 0$, the same reasoning applies with the quadratic progression accelerating downwards. ∎

RIVERBENDS are the landmarks which allow us to understand indefinite binary quadratic forms without lakes. A quadratic form $Q$ with positive nonsquare discriminant $\Delta$ must contain a single neverending river – one which bends back and forth along its path. At each bend, we find two positive and two negative values. There are two possible orientations for a riverbend, drawn below.

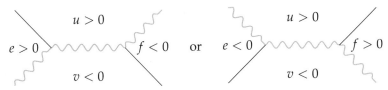

Riverbends are the landmarks in the topographs of indefinite forms, just as wells are the landmarks for definite forms. But a crucial difference makes indefinite forms more difficult and more interesting to study – an indefinite form can have multiple riverbends along its river while definite forms have a unique well.

**Theorem 11.21 (Minimum value bound for indefinite forms)** *Let $Q$ be a quadratic form with positive nonsquare discriminant $\Delta$. Then the smallest (in absolute value) nonzero value of $Q$ is no greater than $\sqrt{\Delta/5}$ (in absolute value).*[10]

PROOF: Such a quadratic form contains a riverbend in its topograph. At the riverbend, we compute $\Delta = (u-v)^2 - ef$. Expanding, we find that
$$\Delta = u^2 - uv - vu + v^2 - ef.$$

In either orientation, $u$ and $v$ have opposite signs, and $e$ and $f$ have opposite signs. It follows that the above equality describes the positive integer $\Delta$ as a sum of five positive integers. Hence one of the five positive integers must be no greater than $\Delta/5$. Taking absolute values, we find that

$$|u| \cdot |u| \leq \Delta/5 \text{ or } |u| \cdot |v| \leq \Delta/5 \text{ or } |v| \cdot |v| \leq \Delta/5 \text{ or } |e| \cdot |f| \leq \Delta/5.$$

In each case, the product of two positive integers is bounded by $\Delta/5$; hence one of the two integers must be no greater than $\sqrt{\Delta/5}$.

$$0 < |u| \leq \sqrt{\frac{\Delta}{5}} \text{ or } 0 < |v| \leq \sqrt{\frac{\Delta}{5}} \text{ or } |e| \leq \sqrt{\frac{\Delta}{5}} \text{ or } |f| \leq \sqrt{\frac{\Delta}{5}}. \blacksquare$$

Four key facts now allow us to enumerate riverbends (of either orientation) with a given positive nonsquare discriminant.

1. Since $u \geq 1$, $v \leq -1$, $ef \leq -1$ and $\Delta = (u-v)^2 - ef$, we find that
$$2 \leq (u-v) \leq \sqrt{\Delta - 1}.$$

2. For each possible value of $(u-v)$, there are $(u-v-1)$ possible values of $u$ and $v$ satisfying $u \geq 1$, $v \leq -1$.[11]

3. For each value of $u - v$, $ef = (u-v)^2 - \Delta$.

4. The sequence $e, (u+v), f$ is an arithmetic progression.

**Corollary 11.22 (Finiteness of class number for indefinite forms)** *Given a positive nonsquare $\Delta$, the number of riverbends of discriminant $\Delta$, and hence the class number $h(\Delta)$, is less than $2\Delta - 2$.*

PROOF: There are no more than $\sqrt{\Delta - 1}$ choices for $(u-v)$ and for each, fewer than $\sqrt{\Delta - 1}$ possible values of $u$ and $v$. Hence there are fewer than $\Delta - 1$ possible values for $u$ and $v$. The values of $u$ and $v$ determine at most two possible values of $e$ and $f$, one of each orientation, since the pair of equations:
$$ef = (u-v)^2 - \Delta \text{ and } \frac{e+f}{2} = u+v$$

describe a line and a hyperbola which intersect in at most two integer points. So there are fewer than $2(\Delta - 1)$ riverbends. $\blacksquare$

---

[10] I learned this proof from Gordan Savin. With Mladen Bestvina, Savin found an analogous proof for bounding minimal values of indefinite binary Hermitian forms.

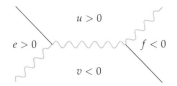

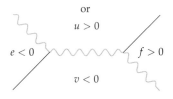

Figure 11.11: The discriminant of $Q$ can be computed at a riverbend of either orientation:
$$\Delta(Q) = (u-v)^2 - ef.$$

[11] For example, if $u - v = 3$ then $u = 2, v = -1$ or $u = 1, v = -2$.

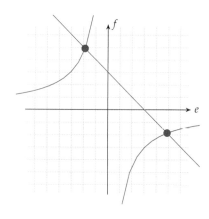

The system of equations $ef = (u-v)^2 - \Delta$ and $\frac{1}{2}(e+f) = u+v$ can be solved, expressing $e$ and $f$ in terms of $u$ and $v$.

$$e = u + v \pm \sqrt{\Delta + 4uv}.$$

$$f = u + v \mp \sqrt{\Delta + 4uv}.$$

Therefore, for values of $u, v$ give a riverbend, it is necessary that $\Delta + 4uv$ is a square.

**Problem 11.23** Find all riverbends of discriminant 37.

SOLUTION: Consider a riverbend of discriminant 37 of either orientation. Since $\sqrt{37-1} = 6$, we find that $2 \leq (u-v) \leq 6$. We tabulate the possibilities in the margin.

Three possible values of $u, v$ occur in which $\Delta + 4uv$ is a square. Each comes with two choices for orientation, from which we find six possible riverbends.

| $u-v$ | $u$ | $v$ | $\Delta+4uv$ | |
|---|---|---|---|---|
| 2 | 1 | −1 | 33 | ✠ |
| 3 | 2 | −1 | 29 | ✠ |
| 3 | 1 | −2 | 29 | ✠ |
| 4 | 3 | −1 | 25 | ✓ |
| 4 | 2 | −2 | 21 | ✠ |
| 4 | 1 | −3 | 25 | ✓ |
| 5 | 4 | −1 | 21 | ✠ |
| 5 | 3 | −2 | 13 | ✠ |
| 5 | 2 | −3 | 13 | ✠ |
| 5 | 1 | −4 | 21 | ✠ |
| 6 | 5 | −1 | 17 | ✠ |
| 6 | 4 | −2 | 5 | ✠ |
| 6 | 3 | −3 | 1 | ✓ |
| 6 | 2 | −4 | 5 | ✠ |
| 6 | 1 | −5 | 17 | ✠ |

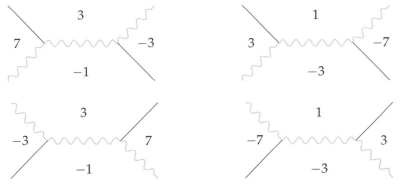

If we begin with one of the riverbends above, and follow the river through an entire period, one obtains the range topograph below.

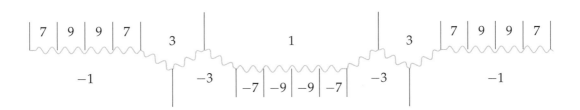

All six riverbends occur on the same river! It follows that there is essentially one binary quadratic form of discriminant 37. Every binary quadratic form of discriminant 37 is properly equivalent to $3x^2 + xy - 3y^2$. In other words, the class number of 37 is one.

THE CLASS NUMBER is subtler for indefinite forms than for definite forms. If $\Delta$ is a positive nonsquare, the **class number** $h(\Delta)$ is the number of classes of primitive binary quadratic forms of discriminant $\Delta$. As before, we place forms into classes using *proper* equivalence.[12]

For definite forms, wells correspond to classes of forms. Every positive-definite form has a unique well, so assembling a list of wells suffices to assemble a list of classes. For indefinite forms, many riverbends can occur in the same form. So the determination of a class number requires one to determine all possible riverbends of a given discriminant, then to place them onto rivers.

Corollary 11.22 implies that when $\Delta$ is a positive nonsquare,
$$h(\Delta) < 2\Delta - 2.$$
But this bound can be improved dramatically in two ways. First is that the *actual* number of riverbends of discriminant $\Delta$ tends to grow like $\sqrt{\Delta}$ rather than $\Delta$. On the opposite page, $r(\Delta)$ denotes the **riverbend number** – the number of riverbends of discriminant $\Delta$.

The plot at the top illustrates the values of $r(\Delta)$ for fundamental discriminants up to 10000. The data, while scattered, bears resemblance to the analogous data counting wells (Figure 10.4). In the meat of the data is a best-fit[13] curve of the form $y = m\sqrt{\Delta} + b$.

But the large number of riverbends, for each discriminant, tend to collapse onto a small number of rivers, and hence $h(\Delta)$ tends to be small. Said another way, the *number of riverbends* of a given discriminant tends to grow proportionally to $\sqrt{\Delta}$, but also the *number of bends per river* tends to grow proportionally to $\sqrt{\Delta}$; so the number of rivers tends to be small (with growth proportional to $\log(\Delta)^2$, on average).

The red dot at the very top right corresponds to the fact that there are 346 riverbends of discriminant 9241. The red stripe far below it indicates that there is only *one* class. There is exactly one river of discriminant 9241 and it contains all 346 bends.

The following conjecture has its origins in Art. 304 of Gauss's *Disquisitiones*.

**Conjecture 11.24 (Gauss class number 1 conjecture)** *There are infinitely many positive nonsquare discriminants $\Delta$ for which $h(\Delta) = 1$.*

The Cohen-Lenstra heuristics[14] give much finer estimates for how often we should expect different values of $h(\Delta)$. They are easiest to state for prime discriminants and odd[15] class numbers; among prime discriminants, their heuristics predict that $h(\Delta) = 1$ about 75.446% of the time; that $h(\Delta) = 3$ about 12.574% of the time; that $h(\Delta) = 5$ about 3.772% of the time. In particular, specific values of $h(\Delta)$ should occur for infinitely many positive values of $\Delta$, in constrast to the case of negative discriminants (Theorem 10.20).

[12] For those who work with quadratic number fields: the class number $h(\Delta)$ we work with coincides with the *narrow* class number of the corresponding real quadratic field $\mathbb{Q}(\sqrt{\Delta})$. Our "riverbend number" should be related to the product $h(\Delta) \cdot \log(\epsilon^+)$ where $\epsilon^+$ is a fundamental totally positive unit, but this may be difficult to make precise.

[13] the result of least-squares linear regression of $r(\Delta)$ against $\sqrt{\Delta}$.

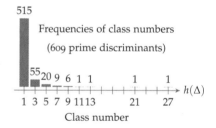

Figure 11.12: Frequencies of class numbers, $h(\Delta)$, when $\Delta$ is a positive prime between 1 and 10000.

[14] See "Heuristics on class groups of number fields" by H. Cohen and H.W. Lenstra, Jr., in *Number Theory Noordwijkerhout*, published as Lecture Notes in Math., vol. 1068 (1984).

[15] Powers of two appear in class numbers when $\Delta$ is nonprime, due to ambiguous forms; the difference at $h(\Delta) = 2, 4, 8, 16$ is visible to the right. See "Class numbers of real quadratic number fields" by Ezra Brown, in *Trans. AMS*, vol. 190 (1974) for more.

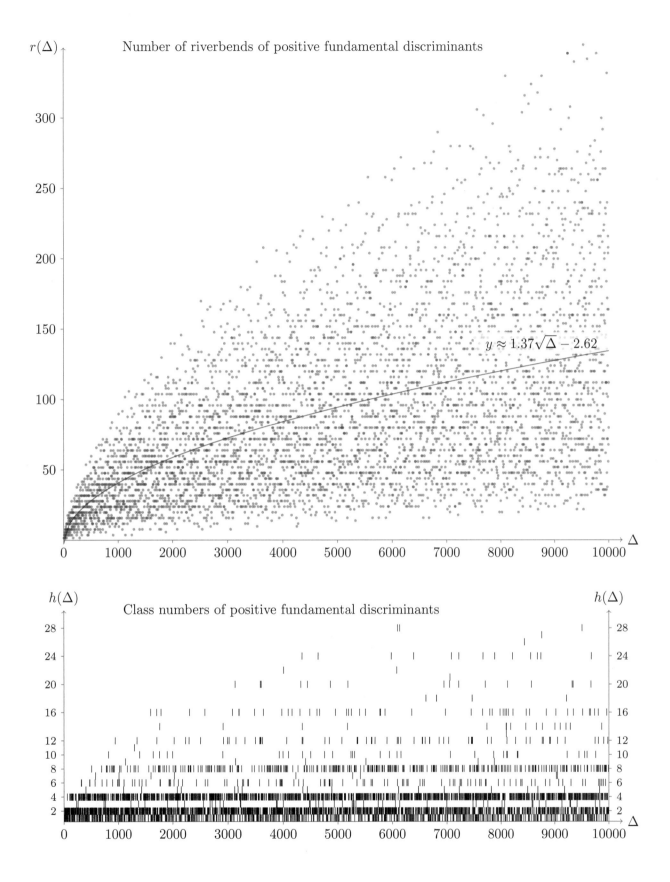

THE MARKOV INVARIANT is a "dimensionless" invariant of binary quadratic forms which captures the size of the minimal value *relative* to the discriminant. If $Q$ is a binary quadratic form, define $\min|Q|$ to be the smallest value of $|Q(x,y)|$ as $(x,y)$ ranges over *nonzero* vectors.[16] Theorem 11.21 states that when $\Delta$ is a positive nonsquare, and $Q$ is a quadratic form of discriminant $\Delta$,

$$\min|Q| \leq \sqrt{\frac{\Delta(Q)}{5}}.$$

The **Markov invariant** of $Q$ is defined to be the quotient

$$\mu(Q) = \frac{\sqrt{\Delta(Q)}}{\min|Q|}.$$

Note that if $Q$ is scaled by a positive constant $\lambda$, then $\min|Q|$ is scaled by $\lambda$ and $\Delta(Q)$ is scaled by $\lambda^2$. It follows that the Markov invariant is unchanged by scaling a quadratic form.

Theorem 11.21 gives a bound for Markov invariants.

$$\mu(Q) \geq \sqrt{5} \text{ for all indefinite } Q.$$

The **(integer) Markov spectrum** is the set of real numbers $\mu(Q)$ that arise from indefinite binary quadratic forms.[17] Below we place a dot for each quadratic form of positive fundamental discriminant up to 20000. The location of the dot is its Markov invariant.

[16] Looking at primitive lax vectors suffices.

[17] The Markov spectrum typically refers to the numbers $\mu(Q)$ that result from indefinite forms $Q(x,y) = ax^2 + bxy + cy^2$ with $a, b, c$ *real* numbers. By scale-invariance, the integer Markov spectrum coincides with the rational Markov spectrum (using rational coefficients) which is dense in the real Markov spectrum.

One could study the analogous Markov spectrum for definite forms. There $\mu(Q) \geq \sqrt{3}$; but it can be shown that the Markov spectrum in this case is dense in the interval $(\sqrt{3}, \infty)$ and so the most interesting phenomena do not occur.

Figure 11.13: The Markov spectrum, for indefinite integer binary quadratic forms $Q$ with $\Delta(Q) < 20000$.

**Theorem 11.25 (Location of the Hall Ray (1975))** *Let $\mu_0 \approx 4.528$ be the constant in the margin. If $\mu_1, \mu_2$ are real numbers satisfying $\mu_0 < \mu_1 < \mu_2$, then there exists a binary quadratic form $Q$ such that $\mu_1 < \mu(Q) < \mu_2$.*

In other words, the Markov spectrum is *dense* in "Hall's ray" to the right of $\mu_0$. There are no interval gaps in that region.

Markov completely understood the left side of the spectrum.

**Theorem 11.26 (Classification of Markov forms (1879))** *If $\mu(Q) < 3$ then $\mu(Q)$ belongs to the "Markov sequence"*[18]

$$\mu(Q) = \sqrt{9 - \frac{4}{m^2}} \text{ for a Markov number } m = 1, 2, 5, 13, 29, \ldots.$$

*For each Markov number $m$, there is a unique class of indefinite forms with Markov invariant $\mu = \sqrt{9 - 4/m^2}$.*[19]

The constant was whittled down from 6 to 5.1 (Marshall Hall, "The Markoff spectrum", *Acta Arith.* (1971)) and eventually by G.A. Freiman ("Diophantine Approximations and the Geometry of Numbers" (1975)) to

$$\mu_0 = 4 + \frac{253589820 + 283748\sqrt{462}}{491993569}.$$

[18] Described on the next page.

[19] For example, every indefinite form $Q$ with $\mu(Q) = \sqrt{5}$ is equivalent to the form $x^2 + xy - y^2$. Every indefinite form $Q$ with $\mu(Q) = \sqrt{8}$ is equivalent to the form $x^2 - 2y^2$.

**Markov triples** are unordered triples of positive integers $\{m, n, r\}$ which satisfy the Diophantine equation

$$m^2 + n^2 + r^2 = 3mnr.$$

The largest element in the triple is typically called $m$. The **Markov numbers** are those $m$'s that arise as largest elements in Markov triples. For example, $\{1, 1, 1\}$ and $\{2, 1, 1\}$ are Markov triples, yielding the Markov numbers 1 and 2, respectively. The next Markov number one finds is 5, arising from the triple $\{5, 2, 1\}$.[20]

Solutions to Diophantine equations rarely grow on trees, but the Markov triples are a lucky exception. If $\{m, n, r\}$ is a Markov triple with distinct $m, n, r$, then $\{m', n, r\}$ and $\{m, n', r\}$ and $\{m, n, r'\}$ are Markov triples when $m', n', r'$ are defined by

$$m' = 3nr - m \text{ and } n' = 3mr - n \text{ and } r' = 3mn - r.$$

Accordingly we attach most Markov triples to three others.

[20] The big open conjecture on Markov numbers is:

**Conjecture 11.27** *Every Markov number arises from a* unique *Markov triple.*

E.g., there is only one Markov triple whose greatest value is 13. More can be found in "Markov's Theorem and 100 years of the uniqueness conjecture" by Martin Aigner (2013).

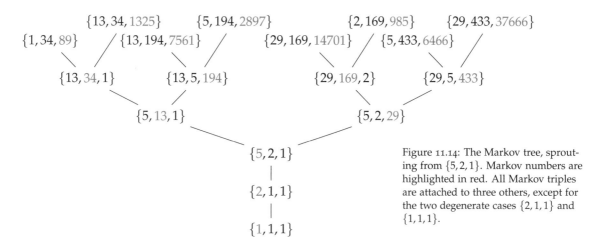

Figure 11.14: The Markov tree, sprouting from $\{5, 2, 1\}$. Markov numbers are highlighted in red. All Markov triples are attached to three others, except for the two degenerate cases $\{2, 1, 1\}$ and $\{1, 1, 1\}$.

**Theorem 11.28 (Markov triples occur uniquely on the tree)** *All Markov triples appear exactly once on the tree above.*[21]

In this way, all points $\mu$ in the Markov spectrum satisfying $\mu < 3$ are understood – at least as well as the Markov triples above.

The portion of the Markov spectrum between 3 and $\mu_0 \approx 4.528$ is much less understood. Starting with Perron, it is known that there are gaps in which the Markov spectrum has no values. Gaps, accumulation points, and (possible) intervals of density may occur in the Markov spectrum; in is way, the arithmetic of indefinite quadratic forms becomes entwined with questions of measure and fractal (Hausdorff) dimension, far beyond the scope of this text.

[21] For a proof of this and other results, see J.W.S. Cassels, "An introduction to Diophantine approximation," Chapter II, Cambridge University Press (1957). See the exercises for more.

AN APPLICATION of indefinite quadratic forms will conclude our treatment. SQUFOF is an algorithm, described by Daniel Shanks,[22] which uses indefinite forms to factor large numbers. For numbers with 10-18 digits, especially those without small factors, SQUFOF is typically the fastest method of factorization and requires little memory. For simplicity, suppose $N$ is a number which we know can be expressed as the product of two distinct prime numbers, $N = pq$, but we know neither $p$ nor $q$. The first observation that underlies SQUFOF is that two forms

$$Q_1(x,y) = x^2 - Ny^2 \text{ and } Q_2(x,y) = px^2 - qy^2$$

have the same positive nonsquare discriminant: $\Delta = 4N$. SQUFOF works in three steps, beginning in the range topograph of $Q_1$.

[22] Our exposition is primarily based on the in-depth history and complexity analysis by Jason Gower and Samuel Wagstaff Jr., "Square form factorization," in *Math. of Computation*, vol. 77, **261** (2008).

1. Travel along the river until you find a "square form" – one in which a square number $a^2$ is found on the river.

2. Jump to another river segment of the same discriminant by "spreading the square" – changing the triple of numbers $(a^2, b, -c)$ in the figure to the triple of numbers $(ac, b, -a)$.

3. Follow the river again to find an ambiguous river segment.

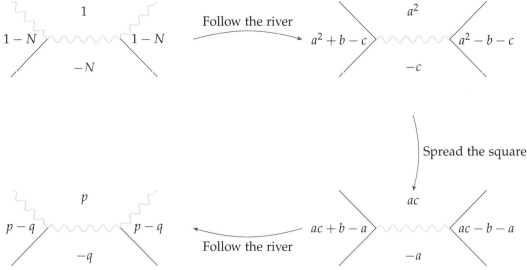

Remarkably – for reasons first suggested by Shanks – these steps often[23] terminate quickly at the river segment where the desired factors $p, q$ are located. Moreover, this process can be accelerated so that the number of arithmetic operations needed is proportional to $N^{1/4} = \sqrt{\sqrt{N}}$. This is far fewer than brute force factorization – where we look for factors up to $\sqrt{N}$.

[23] More than 99% of the time apparently. SQUFOF does not work 100% of the time, but those who care to factor large numbers tend to be pragmatic.

BINARY QUADRATIC FORMS arise naturally from the Diophantine perspective. After one knows how to solve Diophantine equations of the form $ax + by = N$, the next in line are equations of the form $ax^2 + bxy + cy^2 = N$. Conway's topograph provides us with a method of solution for *all* such equations – a method that can often be carried out by hand, and that can be adapted easily enough for a computer. The method depends on the type of form one begins with – indefinite or definite or semidefinite. This type can be determined quickly from the discriminant in the table below.

| Discriminant | Values represented | Features |
| --- | --- | --- |
| Negative | All positve or all negative | A unique well |
| Zero | All nonnegative or all nonpositive | A unique lake |
| Positive nonsquare | Positive and negative, never zero | An endless river, no lakes |
| Positive square | Positve, negative, and zero | Two lakes joined by a river, or a double-lake |

Table 11.1: Interpreting the discriminants of nonzero binary quadratic forms. This table is the consequence of results from this and the previous two chapters.

We can recover Euler's solution of Pell's equation, $31x^2 + 1 = y^2$ with the topograph. For the equation is the same as $31x^2 - y^2 = -1$ and we are led to find $-1$ on the topograph of $Q(x,y) = 31x^2 - y^2$. The discriminant is 124, a positive nonsquare, and so we follow the periodic river. The solution $x = 273, y = 1520$ is found in due time.

DIOPHANTINE EQUATIONS motivate some number theorists, and we have found intricate structures in studying just those equations of degree 1 and 2, in only two variables. One can move to more complicated equations by introducing more variables or by increasing the degree. The Markov equation $m^2 + n^2 + r^2 = 3mnr$ is a degree three Diophantine equation in three variables, and is the source of unsolved questions. By degree 4, one runs into a wall – it is proven that there is no algorithm to solve general Diophantine equations of degree 4 and higher.[24]

But Diophantine equations are just the beginning, and most number theorists stray far from basic questions. Class numbers, L-functions, Galois representations, motives, ergodic actions, and perfectoid spaces form an ever-growing bestiary for the number theorist. What number theorists have in common, then, is a delight every time we return with some more insight into ancient questions.

[24] This is the story of Hilbert's 10th problem, and the undecidability of $\mathbb{Z}$. For a survey of such "undecidability," we recommend Bjorn Poonen's "Undecidability in number theory," in *Notices Amer. Math. Soc.*, **55**, no. 3 (2008).

## Historical notes

When $Q(x,y)$ is an indefinite form, with positive nonsquare discriminant, Conway's topograph allows one to find a solution to $Q(x,y) = N$ (when a solution exists), and to follow the periodic river to an infinitude of solutions. A different but largely successful approach was undertaken in India, first by Brahmagupta and later by Bhaskara II.

Recall[25] the identity, due to Fibonacci (1225) in the west,

$$(a^2 + b^2) \cdot (c^2 + d^2) = (ac - bd)^2 + (ad + bc)^2.$$

[25] See the previous historical notes, especially Equation (9.1).

A more general identity was found by Brahmagupta, in his *Brāhma-sphuṭa-siddhānta* (628CE).[26] Verses of the 18th chapter are related to equations of the form

$$Ny^2 + K = x^2. \qquad (11.1)$$

[26] An excellent survey can be found in Kim Plofker's book, "Mathematics in India," published by Princeton University Press (2009). See Section 5.1.3 for more on what follows.

In Verses 18.64–65, Brahmagupta writes,

> Of the square of an optional number multiplied by the *gunaka* and increased or decreased by another optional number, (extract) the square-root. (Proceed) twice. The product of the first roots multiplied by the *gunaka* together with the product of the second roots will give a (fresh) second root; the sum of their crossproducts will be a (fresh) first root. The (corresponding) *interpolator* will be equal to the product of the (previous) *interpolatpors*.[27]

[27] Translation from Datta and Singh, "History of Hindu mathematics: a source book", Part II, p.146 (1962).

The *gunaka* corresponds to $N$, and the *interpolator* to $K$ in our equation above. Proceeding twice suggests we consider two solutions

$$Na^2 + K_1 = b^2 \text{ and } Nc^2 + K_2 = d^2.$$

The *first roots* are $a, c$, and the *second roots* are $b, d$. The *fresh second root* is then $Nac + bd$ and the *fresh first root* is $ad + cb$. Brahmagupta's identity now reads[28]

$$N(ad + cb)^2 + K_1 K_2 = (Nac + bd)^2. \qquad (11.2)$$

[28] When $N = -1$ and $K = K_1 = K_2 = 1$, this gives the identity of Fibonacci above

From this formula, one may often produce infinitely many solutions to the Diophantine equation $x^2 - Ny^2 = 1$, once one nontrivial solution has been found.

A complete method of solving equations of the form $Ny^2 + 1 = x^2$ can be found in the *Bījagaṇita* of Bhāskara II (1114–1185CE). The key point is finding one nontrivial $((x,y) \neq (0, \pm 1))$ solution, from which others can be found by Brahmagupta's identity (11.2). The *cyclic method* of Bhakara II is discussed briefly in Chapter 6 of Plofker's text cited above.[29] It may be interesting to revisit the cyclic method and relate it to Conway's topograph.

[29] For details on the cyclic method, see A. A. Krishnaswami Ayyangar's "New light on Bhaskara's Chakravala or cyclic method of solving indeterminate equations of the second degree in two variables," in *J. Indian Math. Soc.* **18** (1929–30). Also, see Anne Bauval's recent (2014) "An elementary proof of the halting property for Chakravala algorithm," available on the ArXiv, https://arxiv.org/pdf/1406.6809v1.pdf.

The Diophantine equation $x^2 - Ny^2 = 1$, equivalent to (11.1) with $K = 1$, is typically called Pell's equation after Euler's (questionable) attribution (1767).[30] In this article, Euler presents a method for finding a solution to Pell's equation, and tabulates nontrivial solutions $(x, y)$ for all nonsquare $N$ up to 99, and a few other examples ($N = 103, 109, 113, 157, 167$) for good measure. For $N = 61$, his solution is

$$x = 1766319049, \quad y = 226153980.$$

Six hundred years before Euler, Bhaskara II writes

> What is that number whose square multiplied by 67 or 61 and then added by unity becomes capable of yielding a square-root? Tell me, O friend, if you have a thorough knowledge of the method of the Square-nature.[31]

Bhaskara II goes on to find precisely the solutions Euler found.[32]

Controversies arise in studying the relationships between the mathematics of ancient Greece, of ancient and medieval India, of Islam at the House of Wisdom, and later in Europe. Questions of influence often become arguments over priority tinged with nationalism and ethnocentrism.[33] Points of contact are undeniable but textual citations are rarer, and dating Sanskrit texts is difficult. Kim Plofker carries out a careful treatment, honest about uncertainties, in her book.[34]

After Euler, Lagrange and Gauss settled (by reduction theory) the general problem of solving Diophantine problems of the form $ax^2 + bxy + cy^2 = N$; the history for indefinite forms does not differ substantially from the history for definite forms in the early 19th century. But, as noticed by Lagrange and Gauss, indefinite forms present different phenomena, and so the history diverges later.

A Russian tradition of number theory in St. Petersburg flourished around Chebyshev in the 1870s.[35] Following Chebyshev were Korkin, Zolotarev, and A.A. Markov. Zolotarev (1847 – 1878) has been mentioned earlier in the context of quadratic reciprocity; here we mention Zolotarev's work with Korkin (1837–1908) on the reduction theory of quadratic forms in many variables. Their joint paper of 1873[36] estimates the minimum value of a positive-definite quadratic form in $n$ variables, and they go on to study *extremal* forms.

The study of quadratic forms by examination of their minima continued in Markov's master's dissertation, "On binary quadratic forms of positive determinant."[37] He follows an amazing thread, from reduction theory to the minima of indefinite forms, to the study of extremal forms with Markov invariant $\mu(Q) < 3$ (i.e., with large minimum, relative to the discriminant). This leads him to the study of the Markov equation $m^2 + n^2 + r^2 = 3mnr$, which we have covered all too quickly in this chapter, but is still a subject of research.[38]

[30] See Euler's "De usu novi algorithmi in problemate Pelliano solvendo," published in *Novi Commentarii academiae scientiarum Petropolitanae* **11** (1767). Available as E323 at the Euler Archive, http://eulerarchive.maa.org/pages/E323.html

[31] Translation from Datta and Singh, "History of Hindu mathematics: a source book", Part II, p.166 (1962).

[32] Their solutions are minimal among the nontrivial solutions.

[33] Recent sources mount a multicultural defense, but many still fall into the trap of reducing history to arguments over priority.

[34] See, in particular, Chapter 8 of "Mathematics in India".

[35] An excellent historical resource, followed here, is "The St. Petersburg School of Number Theory" by B.N. Delone, now translated by Robert Burns and published in English by the American Mathematical Society (2005).

[36] See "Sur les formes quadratiques," in *Math. Ann.* **6** (1873).

[37] His dissertation was completed in 1880, supervised by Korkin and Zolotarev. It was published as "On binary quadratic forms with positive determinant" in *Uspekhi Mat. Nauk*, vol. 3, **5**(27) (1948).

[38] For example, see "Markoff triples and strong approximation," by Jean Borgain, Alex Gamburd, and Peter Sarnak (2015), available on the ArXiv at http://arxiv.org/abs/1505.06411.

## Exercises

1.  How many lakes do you expect in the range topograph of the following binary quadratic forms?

    $$Q_1(x,y) = x^2 + y^2, \quad Q_2(x,y) = x^2 - y^2, \quad Q_3(x,y) = 3xy, \quad Q_4(x,y) = x^2 + 5xy.$$

2.  Consider the Diophantine equation $x^2 - 11y^2 = 1$. Find a solution in which $x > 1$.

3.  For which integers $n$ between 1 and 10 does the Diophantine equation $x^2 - 13y^2 = n$ have a solution?

4.  Prove that every binary quadratic form of discriminant 12 is equivalent to $x^2 + 2xy - 2y^2$ or to $2x^2 + 2xy - y^2$.

5.  Find a solution to the Diophantine equation $2x^2 + 5xy + y^2 = 13$, or prove that there is no solution.

6.  Write down the simplest nontrivial proper isometry of $Q(x,y) = x^2 - 11y^2$.

7.  Let $\Delta$ be a positive square number. Recall that $h(\Delta)$ is the number of classes of primitive binary quadratic forms of discriminant $\Delta$. Demonstrate that every such form has a double-lake or else has a unique *rivermouth*: a triad around which the values are negative-positive-zero in clockwise orientation. Use this to derive an exact formula for $h(\Delta)$.

8.  Given a positive nonsquare $\Delta$, we have proven that the number of riverbends of discriminant $\Delta$ is bounded by $2\Delta - 2$. Improve this bound as much as possible.

9.  Prove that if $Q$ is a primitive indefinite binary quadratic form, and $\mu(Q) = \sqrt{5}$, then $Q$ is equivalent to $x^2 + xy - y^2$.

10. Prove that the number of distinct bends along an endless river is even.

11. Consider a topograph with two adjacent lakes, i.e., two lakes with an edge in common. Demonstrate that any two such topographs with the same discriminant must be equivalent to each other.

12. Suppose that $N = pq$ is the product of two odd primes (as in SQUFOF). Sketch the possible ambiguous cells along the river with discriminant $4pq$.

13. Let $\Delta$ be a fundamental[39] positive discriminant. Enumerate the classes of ambiguous indefinite forms of discriminant $\Delta$. The answer should depend on the prime decomposition of $\Delta$.

[39] Recall that fundamental discriminants are those discriminants $\Delta$ which cannot be expressed as $f^2 \Delta'$ for a discriminant $\Delta'$ and an integer $f > 1$.

14. The following sequence of exercises relates riverbends to classical reduction theory. Let $Q(x,y) = ax^2 + bxy + cy^2$ be a binary quadratic form of positive nonsquare discriminant $\Delta$. Such a form is called **reduced** if $0 < b < \sqrt{\Delta}$ and $\sqrt{\Delta} - b < 2|a| < \sqrt{\Delta} + b$.

   (a) Prove that if $Q$ is reduced, then $a$ and $c$ have opposite signs. Hint: prove that $-4ac$ is positive.

   (b) Prove that if $Q$ is reduced, then $Q'(x,y) = cx^2 + bxy + ay^2$ is also reduced.

   (c) Prove that if $Q$ is reduced, then $|a| + |c| < \sqrt{\Delta}$. For this, prove the following general fact: if $u < x < y < v$ are positive real numbers, and $uv = xy$, then $(x+y) < (u+v)$. Hint: it is obvious that $(y-x)^2 < (v-u)^2$. Now plug in $u = \sqrt{\Delta} - b$, $v = \sqrt{\Delta} + b$, $x = 2|a|$ and $y = 2|c|$ (or vice-versa if $|a| > |c|$).

   (d) Prove $Q$ is reduced if and only if there is a riverbend at home basis in the range topograph of $Q$. Hint: for one direction, use the previous part to show that the product $(a+c+b)(a+c-b)$ is negative.

   (e) Suppose that $Q$ is reduced. Let $a' = c$, and let $b'$ be an integer between $\sqrt{\Delta} - 2|c|$ and $\sqrt{\Delta}$ for which $b' + b$ is a multiple of $2c$. Define
   $$c' = \frac{(b')^2 - \Delta}{4c}.$$
   Prove that these conditions determine unique integers $a', b', c'$, and that $Q'(x,y) = a'x^2 + b'xy + c'y^2$ is a reduced quadratic form.[40]

   (f) Following the exercises of Chapter 9, check that $Q'$ is a neighbor of $Q$, hence properly equivalent to $Q$. What is the relationship between $Q$ and $Q'$ on the topograph?

[40] The process which takes $Q$ to $Q'$ is called the **reduction operator**.

15. The following exercises place all Markov triples on a tree.

   (a) Check that if $\{m,n,r\}$ is a Markov triple, and $m' = 3nr - m$, then $\{m',n,r\}$ is also a Markov triple.

   (b) If furthermore, $m > n > r \geq 1$, prove that $m' < m$. For this,[41] consider the quadratic function $f(x) = x^2 + n^2 + r^2 - 3xnr$; demonstrate that $f(n) < 0$, and so $n$ lies between the two roots $m$ and $m'$ of $f$.

   (c) Use this to demonstrate that all Markov triples occur on the tree in Figure 11.14.

   (d) (Challenge) Prove that every Markov triple occurs exactly once on the tree in Figure 11.14.

[41] This proof is based on Part II of "The Markoff Chain," by J.W.S. Cassels, vol. 50, 3 (1949).

# Index of Theorems

| | | |
|---|---|---|
| Proposition 0.6 | Summation up to $N$ | 5 |
| Proposition 0.12 | Counting pairs | 8 |
| Proposition 0.13 | Summation up to $N-1$ | 9 |
| Proposition 0.14 | Counting multiples | 10 |
| Proposition 0.15 | Counting squares | 10 |
| Proposition 0.16 | Counting digits | 11 |
| Proposition 0.17 | Counting bits | 11 |
| Proposition 0.18 | Traditional division with remainder | 13 |
| Proposition 0.19 | Division with minimal remainder | 13 |
| Proposition 0.26 | Reflexive property of $\mid$ | 16 |
| Proposition 0.27 | Antisymmetric property of $\mid$ | 16 |
| Proposition 0.28 | Transitive property of $\mid$ | 16 |
| Corollary 0.31 | The two out of three principle for divisibility | 16 |
| Proposition 1.6 | Euclidean algorithm runtime | 29 |
| Theorem 1.7 | Euclid's algorithm gives the GCD | 30 |
| Theorem 1.14 | Solubility of linear Diophantine equations | 33 |
| Theorem 1.21 | General solution of linear Diophantine equations | 36 |
| Theorem 1.23 | GCD-LCM product formula | 39 |
| Corollary 1.25 | General solution of linear Diophantine equations | 40 |
| Proposition 1.27 | Scaling property of GCD and LCM | 41 |
| Theorem 2.3 | Infinitude of primes | 50 |
| Theorem 2.4 | Infinitude of primes in arithmetic progressions | 51 |
| Theorem 2.5 | Green-Tao Theorem | 51 |
| Conjecture 2.8 | Riemann hypothesis, prime error-term formulation | 53 |
| Conjecture 2.9 | Twin prime conjecture | 54 |
| Theorem 2.11 | Zhang-Maynard bounded gaps theorem | 54 |
| Lemma 2.12 | Euclid's Lemma | 56 |
| Corollary 2.13 | Euclid's Lemma for primes | 56 |
| Theorem 2.15 | Uniqueness of prime decomposition | 57 |
| Conjecture 2.17 | Goldbach's conjecture | 58 |
| Theorem 2.18 | Chen's theorem | 58 |
| Theorem 2.19 | Ternary Goldbach | 58 |
| Theorem 2.27 | Probability of coprimality | 62 |
| Theorem 2.31 | Divisor-power-sums are multiplicative | 65 |

| | | |
|---|---|---|
| Proposition 2.33 | Geometric series formula | 66 |
| Theorem 2.37 | Mersenne primes yield perfect numbers | 68 |
| Theorem 2.38 | Even perfect numbers arise from Mersenne primes | 69 |
| Theorem 3.1 | Reduction of fractions | 76 |
| Theorem 3.3 | Algebraic characterization of constructible numbers | 80 |
| Theorem 3.4 | Infinitude of Pythagorean triples | 81 |
| Proposition 3.5 | Irrationality of surds | 82 |
| Theorem 3.8 | Rational Root Theorem | 83 |
| Proposition 3.11 | Mediant fractions lie between | 86 |
| Theorem 3.15 | Kissing fractions have tangent Ford circles | 88 |
| Theorem 3.21 | Integers generate all reduced fractions via mediants | 91 |
| Theorem 3.23 | Dirichlet approximation theorem | 93 |
| Theorem 3.24 | Fermat's Last Theorem | 95 |
| Theorem 3.25 | Thue-Siegel-Roth Theorem | 95 |
| Proposition 4.3 | Polar multiplication of complex numbers | 101 |
| Proposition 4.5 | Conjugation is a field automorphism | 103 |
| Conjecture 4.9 | Gauss's Circle Problem | 106 |
| Theorem 4.11 | Gaussian/Eisenstein division with remainder | 108 |
| Theorem 4.13 | Prime decomposition for Gaussian/Eisenstein integers | 110 |
| Theorem 4.18 | Lifting primes | 114 |
| Theorem 4.20 | Lowering primes | 115 |
| Theorem 4.24 | Fermat's Christmas Theorem | 116 |
| Proposition 5.3 | Well-definedness of arithmetic mod $m$ | 130 |
| Proposition 5.9 | Divisibility by 3 and 9 | 133 |
| Theorem 5.15 | Lagrange's Four-Square Theorem | 135 |
| Proposition 5.18 | Solubility of linear congruences | 137 |
| Theorem 5.20 | Existence of modular inverses | 138 |
| Corollary 5.24 | Cancellation Property | 139 |
| Theorem 5.28 | Degree-formula for products of polynomials | 142 |
| Proposition 5.32 | Ultrametric triangle inequality | 143 |
| Theorem 5.35 | Division with remainder for polynomials mod $p$ | 144 |
| Theorem 5.36 | Unique factorization of polynomials mod $p$ | 145 |
| Theorem 5.39 | Roots are bounded by the degree | 145 |
| Theorem 5.40 | Infinitude of irreducible polynomials mod $p$ | 146 |
| Theorem 5.41 | Prime number theorem for polynomials mod $p$ | 146 |
| Proposition 6.2 | Cycle length for addition mod $m$ | 154 |
| Lemma 6.9 | Dynamics of multiplication mod $m$ | 157 |
| Theorem 6.10 | The Fermat-Euler Theorem | 158 |
| Proposition 6.21 | Roots Of One property of primes | 162 |
| Lemma 6.24 | Totient sum formula | 164 |
| Theorem 6.26 | Existence of primitive roots mod $p$ | 165 |
| Theorem 7.2 | Chinese Remainder Theorem | 175 |
| Theorem 7.9 | The totient is a multiplicative function | 178 |
| Lemma 7.16 | Lifting multiplicative inverses | 183 |

# INDEX OF THEOREMS

| | | |
|---|---|---|
| Theorem 7.20 | Lifting square roots | 185 |
| Theorem 8.2 | Wilson's Theorem | 194 |
| Theorem 8.5 | Euler's Criterion for squareness mod $p$ | 198 |
| Corollary 8.6 | Reciprocity for $-1$ | 199 |
| Theorem 8.7 | Fermat's Christmas Theorem | 200 |
| Proposition 8.8 | Minkowski's theorem in the plane | 200 |
| Theorem 8.10 | Reciprocity for 2 | 202 |
| Proposition 8.12 | Permutations are generated by transpositions | 205 |
| Theorem 8.14 | The transposition interpretation of sgn | 207 |
| Corollary 8.15 | Composition formula for sgn | 207 |
| Theorem 8.19 | The inversion interpretation of sgn | 209 |
| Lemma 8.21 | Zolotarev's Lemma | 210 |
| Theorem 8.22 | Quadratic Reciprocity | 211 |
| Theorem 8.35 | Lagrange's Four-Square Theorem | 221 |
| Theorem 9.5 | Area interpretation of the determinant | 228 |
| Theorem 9.6 | Pick's theorem for parallelograms | 229 |
| Theorem 9.8 | Determinant criterion for bases | 230 |
| Theorem 9.14 | Connectedness of the topograph | 236 |
| Theorem 9.19 | Orientation of the topograph by determinants | 239 |
| Theorem 9.24 | Uniformity of the topograph | 241 |
| Theorem 9.27 | Arithmetic progression rule | 246 |
| Theorem 9.33 | Triad criterion for equivalence | 249 |
| Theorem 9.40 | Discriminants are equal at all cells and triads | 251 |
| Corollary 9.41 | Equivalent forms have equal discriminants | 252 |
| Theorem 10.1 | Conway's climbing principle | 259 |
| Theorem 10.6 | Existence and uniqueness of wells | 262 |
| Theorem 10.7 | The topograph contains no loops | 263 |
| Theorem 10.9 | Minimum value bound for definite forms | 265 |
| Corollary 10.14 | Finiteness of class number for definite forms | 267 |
| Theorem 10.15 | Fermat's Christmas Theorem | 268 |
| Theorem 10.20 | Class number 1 theorem of Heegner, Stark, Baker | 272 |
| Corollary 10.24 | Classification of forms of discriminant zero | 274 |
| Theorem 11.5 | The discriminant counts the lakes | 283 |
| Proposition 11.6 | The topology of rivers | 284 |
| Proposition 11.9 | Uniqueness of the river | 285 |
| Theorem 11.13 | Periodicity of the river | 286 |
| Theorem 11.17 | Infinitude of solutions to Pell's equation | 287 |
| Theorem 11.21 | Minimum value bound for indefinite forms | 292 |
| Corollary 11.22 | Finiteness of class number for indefinite forms | 292 |
| Conjecture 11.24 | Gauss class number 1 conjecture | 294 |
| Theorem 11.25 | Location of the Hall Ray (1975) | 296 |
| Theorem 11.26 | Classification of Markov forms (1879) | 296 |
| Theorem 11.28 | Markov triples occur uniquely on the tree | 297 |

# Index of Terms

$p$-adic distance, 190
$p$-adic norm, 190

absolute value
   of a complex number, 102
   of a polynomial, 142
abundant, 71
acceleration
   of a quadratic progression, 290
algebraic, 83
aliquot parts, 67
ambiguous
   binary quadratic form, 257, 288
   river segment, 289
amicable, 73
associate
   quadratic form, 256
authentication, 166
automorphism
   composite, 241
   inverse, 241
   of the domain topograph, 240

basis, 230
   home, 232
   lax, 232
   lax, adjacent, 233
bijection, 1
binary quadratic form, 244
bits, 11

Carmichael number, 162
ceiling, 10

cell, 237
Chebyshev's bias, 116
choose, 8
cipher, 166
ciphertext
   RSA, 187
class
   of binary quadratic forms, 252
class number, 252
   definite form, 271
   for square discriminants, 282
   indefinite form, 294
commensurable, 82
common divisor, 31
common multiple, 35
complementary angle, 79
complex conjugate, 103
composite, 48
congruent, 130
   integers, 127
constructible number, 79
coprime, 62
cycle diagram
   of a permutation, 204
cycle length
   mod p, 164
cycle number
   mod p, 164

deficient, 71
definite
   negative, 262
   positive, 262

degree
   of a polynomial, 140
deprecated, 32
descent
   of congruences, 182
determinant, 227
Diffie-Hellman protocol, 166
Diophantine approximation, 92
Diophantine equation, 33
discrete logarithm problem, 167
discriminant, 250
   cell, 251
   fundamental, 272
   triad, 250
divides, 14
divisor-sum function, 64

Eisenstein integer, 105
equivalence
   of binary quadratic forms, 249
   proper, 249
Euclidean algorithm, 26
Euclidean domain, 99

factor, 12
factorial, 72
factoring, 48
Fermat prime, 85
Fermat's Last Theorem, 95
Fermat's Little Theorem, 158
floor, 10
Ford circle, 87
fraction, 76

Gaussian integer, 104
golden ratio, 97
greatest common divisor, 31

Hadwiger-Finsler inequality, 264
Hasse diagram, 14
home triad, 239
homogeneous, 36
hyperbolic plane, 247

identity permutation, 204
inaccessible, 230
indefinite, 281
inert, 113, 116, 117
infinity-gon, 237
integer, 10
inversion
   of a permutation, 208
irreducible
   polynomials mod p, 145
isometry
   improper, 288
   of a binary quadratic form, 253

Jacobi symbol, 221

kissing fractions, 87

lake, 274, 281
lax superbasis, 237
least common multiple, 35
Legendre symbol, 202
lies above, 115
lies below, 114
lifting
   of congruences, 182
lonely, 194
loop
   simple, 263

Markov invariant, 296
Markov number, 297
Markov spectrum
   integer, 296
Markov triple, 297
max, 61
measures, 12
mediant, 86
Mersenne prime, 51
Miller-Rabin primality test, 162
min, 61
modulus, 130
monic, 146
multiple, 12

multiplicative, 64

natural number, 1
neighbor
    binary quadratic forms, 257

opposite
    quadratic form, 256
orientation
    of a superbasis, 239

pair, 8
part
    imaginary, 100
    real, 100
partner
    $a$, 196
path
    simple, 263
Pell's equation, 281
perfect, 67
permutation, 204
pigeonhole principle, 175
Pingala's algorithm, 161
polynomial
    constant, 140
    linear, 140
    mod p, 140
price, 92
primagon, 112
prime
    Gaussian or Eisenstein integer, 110
    integer, 110
prime decomposition, 56
prime gaps, 54
prime number, 48
prime number theorem, 71
primitive
    binary quadratic form, 270
primitive root
    mod p, 165
primitive vector, 231
principal
    binary quadratic form, 250

private key
    RSA, 186
proper divisors, 67
public key
    RSA, 186
Pythagorean triple, 81

quadratic reciprocity, 215
quadratic residue, 193

ramified, 113, 116, 117
rational numbers, 75
reduced, 76
    definite binary quadratic form, 278
    Gauss, 277
    indefinite binary quadratic form, 303
    Lagrange, 276
reduction operator
    for indefinite quadratic forms, 303
regular polygon, 85
relatively prime, 62
represent
    by a quadratic form, 244
representatives
    mod n, 128
Riemann hypothesis, 53
river, 281
riverbend number, 294
root
    mod p, 140
RSA cryptosystem, 186

semiotics, 76
Sieve of Eratosthenes, 47
sign
    of a cycle, 207
    of a permutation, 207
simplify
    mod n, 128
Sophie Germain prime, 171, 220
special orthogonal group, 253
split, 113, 116, 117
square numbers, 4
square root

mod n, 184
square-free, 261
square-scaling, 261
squares
   mod p, 193
subset, 8
superbasis
   lax, 238
   strict, 238

topograph
   domain, 232
   range, 244
totient, 156
transcendental, 83
transposition, 204
   adjacent, 209
tree, 263
triad, 237
triangular number, 5

unit, 48
   Eisenstein, 105
   Gaussian, 104
   polynomial mod p, 143

vector
   lax, 232

walking home, 234
water, 281
well, 262
   double, 262
   single, 262
witness
   perceptive, 162
   primality, 160

zero polynomial, 140

# Index of Names

This index lists the mathematicians mentioned in the text. Superscripts on dates refer to the following: uncertainty (?), lower bound (>), upper bound (<). For example, $1444-1544^{>?}$ means a birth-year of 1444 and a death-year that may be after 1544 (i.e., one source claims a death after 1544, but it is not well-attested).

The abbreviation *c.* for *circa* means approximately and refers to both birth and death. The abbreviation *fl.* for *floruit* means flourished, and refers to a time at which the person is known to have produced mathematical works.

Āryabhata (476–550CE), 18, 42, 188

Abu Mansur 'Abd al-Qahir ibn Tahir al-Baghdadi (?–1037), 188
Adleman, Leonard (1945–), 186
Agrawal, Manindra (1966–), 169
Alcuin of York (c.735–804CE), 18
Artin, Emil (1898–1962), 219

Bézout, Étienne (1730–1783), 43, 44
Bachet de Méziriac (1581–1638), 43, 94, 148, 254
Baker, Alan (1939–), 272, 307
Bhāskara I (c.600–680CE), 188
Bhāskara II (1114–1185), 300
Brahmagupta (c.598 – $665^{>}$), 300

Cebotarev, Nikolai Grigorievich (1894–1947), 141
Charles Levieux, Baron de la Vallée Poussin (1866–1962), 71
Chebyshev, Pafnuty Lvovich (1821–1894), 117, 121, 301
Chen Jingrun (1933–1996), 58
Cohen, Henri (1947–), 294
Conway, John Horton (1937–), 225, 238, 244, 263, 276

Descartes, René (1596 – 1650), 78, 94
Dickson, Leonard Eugene (1874–1954), 188
Diophantus (fl.c.$80^{?}$CE(Knorr), fl.c.$240^{?}$CE(Tannery) ), 33, 43, 94, 225, 254
Dirichlet, Johann Peter Gustav Lejeune (1805–1859), 95, 97, 219, 277

Eisenstein, Ferdinand Gotthold Max (1823–1852), 99, 120, 219
Eratosthenes of Cyrene, (c.276BCE– 195BCE), 71
Euclid of Alexandria (fl.c.300BCE), 19, 25, 42, 50, 67, 68, 70, 78, 82, 85, 94
Euler, Leonhard (1705–1783), 62, 69, 95, 148, 149, 158, 188, 189, 193, 201, 218, 225, 255, 301

Fermat, Pierre de (1601? or 1607? – 1665), 85, 94, 116, 148, 158, 200, 254, 278
Fibonacci, born Leonardo Bonacci (c.1170–1250), 188, 254
Ford Sr., Lester Randolph (1886–1967), 95
Frey, Gerhard (1944–), 95

Gauss, Carl Friedrich (1777–1855), 18, 53, 70, 71, 85, 99, 120, 148, 149, 188, 193, 201, 211, 219, 255, 272, 276, 301
Germain, Marie-Sophie (1776–1831), 95, 171
Goldbach, Christian (1690–1764), 201
Green, Ben (1977–), 51

Hadamard, Jacques Salomon (1865 – 1963), 71
Hecke, Erich (1887–1947), 118, 121
Heegner, Kurt (1893–1965), 272, 307
Helfgott, Harald (1977–), 58
Hensel, Kurt (1861–1941), 189
Hilbert, David (1862–1943), 211, 219
Hurwitz, Adolf (1859–1919), 95

Jacobi, Carl Gustav Jacob (1804–1851), 64, 149

Korkin, Aleksandr Nikolayevich (1837–1908), 301

Lagrange, Joseph-Louis (1736–1813), 148, 149, 195, 201, 255, 276, 301
Lamé, Gabriel Léon Jean Baptiste (1795–1870), 95
Lang, Serge (1927–2005), 95
Langlands, Robert Phelan (1936–), 211, 219
Lebesgue, Victor-Amédée (1791–1875) (not Henri, of Lebesgue measure fame), 95
Legendre, Adrien-Marie (1752–1833), 95, 150, 193, 202, 218, 219, 255, 277
Lemmermeyer, Franz (1962–), 218, 277
Lenstra, Hendrik (1949–), 169, 294
Liouville, Joseph (1809–1882), 95
Liu Hui (fl. 263CE), 42

Markov, Andrey Andreyevich (1856–1922), 296, 301
Maynard, James (1987–), 71
Mersenne, Marin (1588–1648), 68, 116, 148, 200
Miller, Gary Lee (c.1950 (PhD 1975)–), 169

Nīlakaṇṭha (1444–1544$^{>?}$CE), 19
Nichomachus of Gerasa (c. 60–140CE), 19, 67, 71
Nicolas Bourbaki, pseudonym for a group (1934–), 1

Pascal, Blaise (1623–1662), 117
Piṅgala (fl.c.300–100$^?$BCE), 161

Rabin, Michael O. (1931–), 169
Ribet, Kenneth (1948–), 95
Riemann, Bernhard (1826–1866), 71
Rivest, Ron (1947–), 186
Roth, Klaus Friedrich (1925–2015), 95

Sarnak, Peter Clive (1953–), 121
Serre, Jean-Pierre (1926–), 95
Shamir, Adi (1952–), 186
Shanks, Daniel (1917–1996), 298
Shimura, Goro (1930–), 95
Siegel, Carl Ludwig (1896–1981), 95
Stark, Harold (1939–), 272, 307
Sunzi (c.220–420CE), 174, 188
Sunzi (c. 220–420CE), 42
Sylvester, James Joseph (1814–1897), 189

Taniyama, Yutaka (1927–1958), 95
Tao, Terence (1975–), 51
Taylor, Richard (1962–), 95
Thue, Axel (1863–1922), 95

Waring, Edward (1736–1798), 195
Wiles, Andrew John (1953–), 95
Wilson, John (1741–1793), 195

Zhang, Yitang (1955–), 54, 71
Zolotarev, Yegor Ivanovich (1847–1878), 204, 210, 211, 219, 301

# Bibliography

Manindra Agrawal, Neeraj Kayal, and Nitin Saxena. PRIMES is in P. *Annals of mathematics*, 160(2):781–793, 2004.

Martin Aigner. *Markov's theorem and 100 years of the uniqueness conjecture*. Springer, 2015.

William R Alford, Andrew Granville, and Carl Pomerance. There are infinitely many Carmichael numbers. *Annals of Mathematics*, 139(3):703–722, 1994.

Claudi Alsina and Roger B Nelsen. Geometric proofs of the Weitzenböck and Hadwiger-Finsler inequalities. *Mathematics Magazine*, 81(3):216–219, 2008.

Nesmith C Ankeny. The least quadratic non residue. *Annals of mathematics*, 55(1):65–72, 1952.

A.A. Krishnaswami Ayyangar. New light on Bhaskara's chakravala or cyclic method of solving indeterminate equations of the second degree in two variables. *J. Indian Math. Soc*, 18:225–248, 1929.

C.G. Bachet. *Problèmes plaisants et délectables qui se font par les nombres*. P. Rigaud, 1624.

E. Bézout. *Cours de mathématiques a l'usage des gardes du pavillon et de la marine*. 1767.

Jean Bourgain, Alexander Gamburd, and Peter Sarnak. Markoff triples and strong approximation. *Comptes Rendus Mathematique*, 354(2):131–135, 2016.

Ezra Brown. Class numbers of real quadratic number fields. *Transactions of the American Mathematical Society*, 190:99–107, 1974.

R. Creighton Buck. Sherlock Holmes in Babylon. *The American Mathematical Monthly*, 87(5):335–345, 1980.

Maarten Bullynck. Modular arithmetic before C.F. Gauss: Systematizations and discussions on remainder problems in 18th-century Germany. *Historia Mathematica*, 36(1):48–72, 2009.

Oliver Byrne. *The first six books of the Elements of Euclid in which coloured diagrams and symbols are used instead of letters for the greater ease of learners*. Taschen GmbH, Cologne, 2010. Facsimile of the first edition of 1847, The accompanying commentary volume contains two essays by Werner Oechslin.

J.W.S. Cassels. The Markoff chain. *Annals of Mathematics*, 50(3):676–685, 1949.

J.W.S. Cassels. *An introduction to Diophantine approximation*. Number 45 in Cambridge Tracts in Mathematics and Mathematical Physics. Cambridge University Press, New York, 1957.

Pafnuty Lvovich Chebyshev. Lettre de M. le Professeur Tchébychev à M. Fuss sur un nouveaux théorème relatif aux nombres premiers contenus dans les formes $4n + 1$ et $4n + 3$. *Bulletin de la Class Physico-mathematique de l'Academie Imperiale des Sciences de Saint-Pétersbourg*, 11:208, 1853.

Karine Chemla. A Chinese canon in mathematics and its two layers of commentaries: reading a collection of texts as shaped by actors. In *Looking at it from Asia: the Processes that Shaped the Sources of History of Science*, pages 169–210. Springer, 2010.

Jing Run Chen. On the representation of a larger even integer as the sum of a prime and the product of at most two primes. *Sci. Sinica*, 16:157–176, 1973.

Jean Christianidis. The way of Diophantus: Some clarifications on Diophantus' method of solution. *Historia Mathematica*, 34(3):289–305, 2007.

Henri Cohen and Hendrik W Lenstra Jr. Heuristics on class groups of number fields. In *Number Theory Noordwijkerhout 1983*, pages 33–62. Springer, 1984.

John Horton Conway and Francis Y.C. Fung. *The sensual (quadratic) form*. Number 26 in Carus Mathematical Monographs. MAA, 1997.

David A Cox. *Primes of the form $x^2 + ny^2$: Fermat, class field theory, and complex multiplication, second edition*. Pure and applied mathematics. John Wiley & Sons, 2013.

Harald Cramér. On the order of magnitude of the difference between consecutive prime numbers. *Acta Arithmetica*, 2(1):23–46, 1936.

Christopher Cullen. The Suàn shù shū,"Writings on reckoning": Rewriting the history of early Chinese mathematics in the light of an excavated manuscript. *Historia mathematica*, 34(1):10–44, 2007.

Bibhutibhusan Datta and Avadhesh Narayan Singh. *History of Hindu mathematics*. Asia Publishing House; Bombay, 1935.

Boris Nikolaevich Delone. *The St. Petersburg school of number theory*, volume 26 of *History of Mathematics*. American Mathematical Society, 2005.

René Descartes. *The Geometry of René Descartes: with a Facsimile of the First Edition*. Dover Publications, 2012.

Leonard E. Dickson. *Introduction to the Theory of Numbers*. University of Chicago Press, 1929.

Leonard E. Dickson. *History of the Theory of Numbers*. Number 86 in AMS Chelsea Publishing. American Mathematical Society, 1999.

Whitfield Diffie and Martin Hellman. New directions in cryptography. *IEEE transactions on Information Theory*, 22(6):644–654, 1976.

P.G.L. Dirichlet. Beweis des Satzes, dass jede unbegrenzte arithmetische Progression, deren erstes Glied und Differenz ganze Zahlen ohne gemeinschaftlichen Factor sind, unendliche viele Primzahlen enthält. *Abh. Akad. d. Wiss. Berlin*, 48:45–71, 1837.

P.G.L. Dirichlet, L. Kronecker, L. Fuchs, and Deutsche Akademie der Wissenschaften zu Berlin. *G. Lejeune Dirichlet's Werke*, volume 225 of *AMS Chelsea Publishing*. American Mathematical Society, 1969.

M.L. D'Ooge. *Nichomachus of Gerasa, Introduction to Arithmetic: Translated into English.* University of Michigan studies: Humanistic series. Macmillan Company, 1926.

Gotthold Eisenstein. Beweis des Reciprocitätssatzes für die cubischen Reste in der Theorie der aus dritten Wurzeln der Einheit zusammengesetzten complexen Zahlen. *Journal für die reine und angewandte Mathematik*, 27:289–310, 1844.

Noam Elkies. The ABC's of number theory. *Harvard College Mathematics Review*, 1(1):57–76, 2007.

Leonhard Euler. E29: De solutione problematum diophanteorum per numeros integros. *Commentarii academiae scientiarum Petropolitanae*, 6:175–188, 1738.

Leonhard Euler. E54: Theorematum quorundam ad numeros primos spectantium demonstratio. *Commentarii academiae scientiarum Petropolitanae*, 8:141–146, 1741.

Leonhard Euler. E72: Variae observationes circa series infinitas. *Commentarii academiae scientiarum Petropolitanae*, 9:160–188, 1744.

Leonhard Euler. E164: Theoremata circa divisores numerorum in hac forma $paa \pm qbb$ contentorum. *Commentarii academiae scientiarum Petropolitanae*, 14:151–181, 1751.

Leonhard Euler. E228: De numeris, qui sunt aggregata duorum quadratorum. *Novi Commentarii academiae scientiarum Petropolitanae*, 4:3–40, 1758.

Leonhard Euler. E241: Demonstratio theorematis fermatiani omnem numerum primum formae $4n + 1$ esse summam duorum quadratorum. *Novi Commentarii academiae scientiarum Petropolitanae*, 5:3–13, 1760a.

Leonhard Euler. E242: Demonstratio theorematis fermatiani omnem numerum sive integrum sive fractum esse summam quatuor pauciorumve quadratorum. *Novi Commentarii academiae scientiarum Petropolitanae*, 5:13–58, 1760b.

Leonhard Euler. E262: Theoremata circa residua ex divisione potestatum relicta. *Novi Commentarii academiae scientiarum Petropolitanae*, 7:49–82, 1761.

Leonhard Euler. E271: Theoremata arithmetica nova methodo demonstrata. *Novi Commentarii academiae scientiarum Petropolitanae*, 8:74–104, 1763.

Leonhard Euler. E323: De usu novi algorithmi in problemate pelliano solvendo. *Novi Commentarii academiae scientiarum Petropolitanae*, 11:29–66, 1767.

Leonhard Euler. E552: Observationes circa divisionem quadratorum per numeros primos. *Opuscula varii argumenti*, 1:64–84, 1783.

Leonhard Euler. E598: De insigni promotione scientiae numerorum. *Opuscula Analytica*, 2:275–314, 1785.

Leonhard Euler. *E792: Tractatus de numerorum doctrina capita sedecim, quae supersunt.* Number 2 in Commentationes arithmeticae. 1849.

R. Fitzpatrick. *Euclid's Elements in Greek.* Lulu.com, 2006.

L. R. Ford. Fractions. *Amer. Math. Monthly*, 45(9):586–601, 1938.

David H Fowler. Ratio in early Greek mathematics. *Bulletin of the American Mathematical Society*, 1(6):807–846, 1979.

Günther Frei. The unpublished Section Eight: On the way to function fields over a finite field. In Catherine Goldstein, Norbert Schappacher, and Joachim Schwermer, editors, *The shaping of arithmetic after CF Gauss's Disquisitiones Arithmeticae*, pages 159–198. Springer, 2007.

G.A. Freiman. Diophantine approximations and the geometry of numbers (markov's problem). *Kalinin. Gosudarstv. Univ., Kalinin*, 1975.

C.F. Gauss. *Disquisitiones arithmeticae*. Translated into English by Arthur A. Clarke, S. J. Yale University Press, New Haven, Conn.-London, 1966.

C.F. Gauss, E. Schering, and Königliche Gesellschaft der Wissenschaften zu Göttingen. *Carl Friedrich Gauss Werke*, volume 2. Gedruckt in der Dieterichschen Universitäts-Buchdruckerei, 1876.

Jason Gower and Samuel Wagstaff Jr. Square form factorization. *Mathematics of computation*, 77(261):551–588, 2008.

John Hadley and David Singmaster. Problems to sharpen the young. *The Mathematical Gazette*, 76(475): 102–126, 1992.

Marshall Hall Jr. The Markoff spectrum. *Acta Arithmetica*, 18:387–399, 1971.

Brian Hayes. Gauss's day of reckoning: A famous story about the boy wonder of mathematics has taken on a life of its own. *American Scientist*, 94(3):200, 2006.

E. Hecke. Eine neue Art von Zetafunktionen und ihre Beziehungen zur Verteilung der Primzahlen. *Math. Z.*, 6(1-2):11–51, 1920.

Hans Heilbronn. On the class-number in imaginary quadratic fields. *The Quarterly Journal of Mathematics*, (1):150–160, 1934.

Harald Helfgott. The ternary Goldbach problem. *ArXiv e-prints*, (1501.05438), January 2015.

Kurt Hensel. Neue Grundlagen der Arithmetik. *Journal für die reine und angewandte Mathematik*, 127:51–84, 1904.

Adolf Hurwitz. Über die angenäherte Darstellung der Irrationalzahlen durch rationale Brüche. *Mathematische Annalen*, 39(2):279–284, 1891.

C.G.J. Jacobi. Note sur la decomposition d'un nombre donne en quatre carrés. *Journal für die reine und angewandte Mathematik*, 3:191, 1828.

Shen Kangsheng. Historical development of the chinese remainder theorem. *Archive for history of exact sciences*, 38(4):285–305, 1988.

Victor Katz, editor. *The mathematics of Egypt, Mesopotamia, China, India, and Islam: A sourcebook*. Princeton University Press, Princeton, NJ, 2007.

Erica Klarreich. Together and alone, closing the prime gap. *Quanta Magazine*, 19, 2013.

A. Korkinge and G. Zolotareff. Sur les formes quadratiques positives. *Math. Ann.*, 11(2):242–292, 1877.

J.L. Lagrange. Démonstration d'un théoreme nouveau concernant les nombres premiers. *Académie royale des sciences et belles lettres*, 1771.

J.L. Lagrange. *Oeuvres. Tome 1*. Georg Olms Verlag, Hildesheim-New York, 1973a. Publiées par les soins de J.-A. Serret, Avec une notice sur la vie et les ouvrages de J.-L. Lagrange par J.-B. J. Delambre, Nachdruck der Ausgabe Paris 1867.

J.L. Lagrange. *Oeuvres. Tome 2*. Georg Olms Verlag, Hildesheim-New York, 1973b. Publiées par les soins de J.-A. Serret, Nachdruck der Ausgabe Paris 1868.

G. Lakoff and R.E. Núñez. *Where Mathematics Comes from: How the Embodied Mind Brings Mathematics Into Being*. Basic Books, 2000.

Youness Lamzouri, Xiannan Li, and Kannan Soundararajan. Conditional bounds for the least quadratic non-residue and related problems. *Mathematics of Computation*, 84(295):2391–2412, 2015.

Adrien-Marie Legendre. *Essai sur la théorie des nombres*. 1808.

Franz Lemmermeyer. The development of the principal genus theorem. In Catherine Goldstein, Norbert Schappacher, and Joachim Schwermer, editors, *The shaping of arithmetic after CF Gauss's Disquisitiones Arithmeticae*, pages 529–561. Springer, 2007.

Franz Lemmermeyer. *Reciprocity laws: from Euler to Eisenstein*. Springer, 2013.

Andrei Andreevich Markov. On binary quadratic forms with positive determinant. *Uspekhi Matematicheskikh Nauk*, 3(5):7–51, 1948.

James Maynard. Small gaps between primes. *Annals of Mathematics*, 181(1):383–413, 2015.

Gary L Miller. Riemann's hypothesis and tests for primality. *Journal of computer and system sciences*, 13(3):300–317, 1976.

Hermann Minkowski. *Geometrie der Zahlen (2 vol.)*. Teubner, Leipzig, 1910. reprint, Chelsea, New York, 1953.

Roger B. Nelsen. Visual gems of number theory. *Math Horizons*, 15(3):7–31, 2008.

Otto Neugebauer. *The exact sciences in antiquity*. Dover Publications, second edition edition, 1969.

Andrew Odlyzko, Michael Rubinstein, and Marek Wolf. Jumping champions. *Experimental Mathematics*, 8(2):107–118, 1999.

David Pengelley and Fred Richman. Did Euclid need the Euclidean algorithm to prove unique factorization? *The American mathematical monthly*, 113(3):196–205, 2006.

Herbert Pieper. On Euler's contributions to the four-squares theorem. *Historia mathematica*, 20(1):12–18, 1993.

Kim Plofker. *Mathematics in India*. Princeton University Press, 2009.

Paul Pollack. Revisiting Gauss's analogue of the prime number theorem for polynomials over a finite field. *Finite Fields and Their Applications*, 16(4):290–299, 2010.

Bjorn Poonen. Undecidability in number theory. *Notices of the AMS*, 55(3), 2008.

Michael O. Rabin. Probabilistic algorithm for testing primality. *Journal of number theory*, 12(1):128–138, 1980.

K. Ramasubramanian and M. D. Srinivas. Development of calculus in India. In *Studies in the history of Indian mathematics*, volume 5 of *Cult. Hist. Math.*, pages 201–286. Hindustan Book Agency, New Delhi, 2010.

R. Rashed and A. Armstrong. *The Development of Arabic Mathematics: Between Arithmetic and Algebra*. Boston Studies in the Philosophy and History of Science. Springer Netherlands, 2013.

Roshi Rashed. Les travaux perdus de Diophante. I, II. *Rev. Hist. Sci.*, 27:97–122; ibid. 28 (1975), 3–30, 1974.

K. A. Ribet. On modular representations of $\mathrm{Gal}(\overline{\mathbf{Q}}/\mathbf{Q})$ arising from modular forms. *Invent. Math.*, 100(2): 431–476, 1990.

Ronald L Rivest, Adi Shamir, and Leonard Adleman. A method for obtaining digital signatures and public-key cryptosystems. *Communications of the ACM*, 21(2):120–126, 1978.

Eleanor Robson. Neither Sherlock Holmes nor Babylon: A reassessment of Plimpton 322. *Historia Mathematica*, 28(3):167–206, 2001.

K. F. Roth. Rational approximations to algebraic numbers. *Mathematika*, 2:1–20; corrigendum, 168, 1955.

Michael Rubinstein and Peter Sarnak. Chebyshev's bias. *Experimental Mathematics*, 3(3):173–197, 1994.

Norbert Schappacher. "Wer war Diophant?". *Math. Semesterber.*, 45(2):141–156, 1998. See also Diophantus of Alexandria: a Text and its History, published online 2005.

M. Schmitz. The life of Gotthold Ferdinand Eisenstein. *Res. Lett. Inf. Math. Sci.*, 6:1–13, 2004.

Lowell Schoenfeld. Sharper bounds for the Chebyshev functions $\theta(x)$ and $\psi(x)$, ii. *Math. Comp*, 30(134): 337–360, 1976.

Jacques Sesiano. *Books IV to VII of Diophantus' Arithmetica in the Arabic translation attributed to Qusṭā ibn Lūqā*, volume 3 of *Sources in the History of Mathematics and Physical Sciences*. Springer-Verlag, New York, 1982.

Daniel Shanks. Incredible identities. *Fibonacci Quarterly*, 12(3):271–280, 1974.

Simon Singh. *Fermat's enigma: The epic quest to solve the world's greatest mathematical problem, with a foreword by John Lynch*. Walker and Company, New York, 1997.

Man-Keung Siu and Yip-Cheung Chan. On Alexander Wylie's *jottings* on the science of the Chinese arithmetic. *Proceedings of HPM 2012*, 2012.

H.M. Stark. The Gauss class-number problems. In William Duke and Yuri Tschinkel, editors, *Analytic number theory: a tribute to Gauss and Dirichlet*, volume 7 of *Clay Mathematics Proceedings*, 2007.

Martin D. Stern. A Mediaeval derivation of the sum of an arithmetic progression. *The Mathematical Gazette*, 74(468):157–159, 1990.

James Joseph Sylvester. On certain ternary cubic-form equations. *American Journal of Mathematics*, 2(4): 357–393, 1879.

Christian Marinus Taisbak. Perfect numbers a mathematical pun? An analysis of the last theorem in the ninth book of Euclid's Elements. *Centaurus*, 20(4):269–275, 1976.

P. Tannery and C. Henry, editors. *Oeuvres de Fermat*. Gauthier-Villars et fils, 1891.

Richard Taylor and Andrew Wiles. Ring-theoretic properties of certain Hecke algebras. *Ann. of Math. (2)*, 141(3):553–572, 1995.

Yuri Tschinkel. About the cover: On the distribution of primes–Gauss' tables. *Bulletin of the American Mathematical Society*, 43(1):89–92, 2006.

E.R. Tufte. *Visual Explanations: Images and Quantities, Evidence and Narrative*. Graphics Press, 1997.

P. Von Finsler and H. Hadwiger. Einige Relationen im Dreieck. *Commentarii Mathematici Helvetici*, 10(1): 316–326, 1937.

André Weil. *Number theory: An approach through history from Hammurapi to Legendre, Reprint of the 1984 edition*. Modern Birkhäuser Classics. Birkhäuser Boston, Inc., Boston, MA, 2007.

Andrew Wiles. Modular elliptic curves and Fermat's last theorem. *Ann. of Math. (2)*, 141(3):443–551, 1995.

Alexander Wylie and Henri Cordier. *Chinese researches*. 1897.

Don Zagier. The first 50 million prime numbers. *The Mathematical Intelligencer*, 1:7–19, 1977.

Yitang Zhang. Bounded gaps between primes. *Annals of Mathematics*, 179(3):1121–1174, 2014.

G. Zolotarev. Nouvelle démonstration de la loi de réciprocité de Legendre. *Nouvelles Annales de Mathématiques*, 11(2), 1872.